Molecular and Cellular Genetics

PRINCIPLES OF MEDICAL BIOLOGY
A Multi-Volume Work, Volume 5

Editors: **E. EDWARD BITTAR,** *Department of Physiology,*
University of Wisconsin, Madison
NEVILLE BITTAR, *Department of Medicine,*
University of Wisconsin, Madison

Principles of Medical Biology
A Multi-Volume Work

Edited by **E. Edward Bittar,** *Department of Physiology, University of Wisconsin, Madison and* **Neville Bittar,** *Department of Medicine University of Wisconsin, Madison*

This work provides:

* A holistic treatment of the main medical disciplines. The basic sciences including most of the achievements in cell and molecular biology have been blended with pathology and clinical medicine. Thus, a special feature is that departmental barriers have been overcome.

* The subject matter covered in preclinical and clinical courses has been reduced by almost one-third without sacrificing any of the essentials of a sound medical education. This information base thus represents an integrated core curriculum.

* The movement towards reform in medical teaching calls for the adoption of an integrated core curriculum involving small-group teaching and the recognition of the student as an active learner.

* There are increasing indications that the traditional education system in which the teacher plays the role of expert and the student that of a passive learner is undergoing reform in many medical schools. The trend can only grow.

* Medical biology as the new profession has the power to simplify the problem of reductionism.

* Over 700 internationally acclaimed medical scientists, pathologists, clinical investigators, clinicians and bioethicists are participants in this undertaking.

Molecular and Cellular Genetics

Edited by **E. EDWARD BITTAR**
Department of Physiology
University of Wisconsin
Madison, Wisconsin

NEVILLE BITTAR
Department of Medicine
University of Wisconsin
Madison, Wisconsin

 JAI PRESS INC.

Greenwich, Connecticut   *London, England*

Library of Congress Cataloging-in-Publication Data

Molecular and cellular genetics / editors, E. Edward Bittar, Neville
Bittar.
    p. cm.—(Principles of medical biology ; v. 5)
    Includes bibliographical references and index.
    ISBN 1-55938-809-9
    1. Molecular genetics. 2. Biochemical genetics. I. Bittar, E.
Edward. II. Bittar, Neville. III. Series.
    [DNLM: 1. Genetics, Biochemical. QH 430 M7161 1996]
QH442.M635   1996
574.87'328—dc20
DNLM/DLC
for Library of Congress   96-20598
                                                    CIP

Copyright © 1996 by JAI PRESS INC.
55 Old Post Road, No. 2
Greenwich, Connecticut 06836

JAI PRESS LTD.
38 Tavistock Street
Covent Garden
London WC2E 7PB
England

ISBN: 1-55938-809-9
Library of Congress Catalog No.: 96-20598

Printed and bound in Great Britain by
CPI Antony Rowe, Chippenham and Eastbourne

CONTENTS

Chapter 17
The Genome Project—A Commentary

Chapter 18
Transgenic Regulation in Laboratory Animals

LIST OF CONTRIBUTORS

Roger L.P. Adams

Department of Biochemistry
University of Glasgow
Glasgow, Scotland

J. Julian Blow

DNA Replication Control Laboratory
ICRF Clare Hall Laboratories
Hertfordshire, England

Paul M. Brickell

Department of Molecular Pathology
University College London
Medical School
London, England

James P.J. Chong

DNA Replication Control Laboratory
ICRF Clare Hall Laboratories
Hertfordshire, England

Andrew R. Collins

Rowett Research Institute
Aberdeen, Scotland

Karin A. Eidne

MRC Reproductive Biology Unit
Centre for Reproductive Biology
Edinburgh, Scotland

Elena Faccenda

MRC Reproductive Biology Unit
Centre for Reproductive Biology
Edinburgh, Scotland

Paul S. Freemont

Protein Structure Laboratory
Imperial Cancer Research Fund
London, England

John D. Hawkins

Department of Biochemistry
Medical College of St. Bartholomew's
Hospital
University of London
London, England

Roger D. Kornberg Department of Structural Biology
Stanford University
School of Medicine
Stanford, California

David S. Latchman Division of Molecular Pathology
University College London
Medical School
London, England

Darryl R.J. Macer Institute for Biological Sciences
University of Tsukuba
Ibaraki, Japan

Erich A. Nigg Swiss Institute for Experimental Cancer
Research (ISREC)
Epalinges, Switzerland

Ian R. Phillips Department of Biochemistry
Queen Mary and Westfield College
University of London
London, England

Bryan M. Turner Anatomy Department
University of Birmingham
Medical School
Birmingham, England

Linda J. Van Eldik Departments of Pharmacology and
Cell Biology
Vanderbilt University
Nashville, Tennessee

Ram S. Verma Division of Genetics
The Long Island Hospital
SUNY Health Science Center
Brooklyn, New York

Alice Vrielink Department of Biochemistry
McGill University
Montreal, Canada

PREFACE

The tools of molecular biology have revolutionized our understanding of gene structure and function and changed the teaching of genetics in a fundamental way. The transition from classical genetics to molecular genetics was initiated by two discoveries. One was the discovery that DNA has a complementary double-helix structure and the other that a universal genetic code does exist. Both led to the acceptance of the central dogma that RNA molecules are made on DNA templates.

The last twenty years have seen remarkable growth in our knowledge of molecular genetics, most of which is the outcome of recombinant DNA technology. This technology which is not limited to cloning, sequencing, and expression has created a biotechnology industry of its own, the purpose of which is to develop new diagnostic and therapeutic approaches in medicine. Both industries in collaboration with the biomedical community are now engaged in laying down the foundation of molecular medicine.

The present volume seeks to provide a coherent account of the new science of molecular genetics. Its content however is by no means exhaustive, partly because of the publication explosion but more because of space restrictions. A rudimentary knowledge of genetics on the reader's part is assumed. Quite understandably, considerable emphasis is placed on major technical advances but not without expounding numerous new ideas and phenomena including alternative splicing, PCR, DNA methylation, genomic imprinting, and so on.

Our special thanks are due to the authors for their scholarly contributions and for their patience and helpful cooperation. Our thanks are also due to Ms. Lauren Manjoney and the staff members of JAI Press for their skill and courtesy.

E. EDWARD BITTAR
NEVILLE BITTAR

Chapter 1

The Nucleosome

ROGER D. KORNBERG

INTRODUCTION

The structure and function of eukaryote chromosomes is one of the great frontiers of modern biology. It encompasses a vast unexplored area, rich in fundamental interest and in clinical implications. The subject is nowhere better illustrated than in the behavior of chromosomes through the cell cycle. At metaphase, during cell division, chromosomes are highly condensed: the average human chromosome, whose DNA is on the order of 50 mm in length, is condensed to about 5 μ, for a packing ratio (length of DNA divided by length of chromosome) of approximately

Principles of Medical Biology, Volume 5
Molecular and Cellular Genetics, pages 1–10.
Copyright © 1996 by JAI Press Inc.
All rights of reproduction in any form reserved.
ISBN: 1-55938-809-9

10,000. During interphase, when gene transcription and other DNA events occur, chromosomes decondense; the packing ratio falls to about 10. This dramatic cycle of expansion and contraction is closely linked to chromosome function. For example, the control of chromosome condensation is believed to represent a key element in the regulation of gene expression, and the attainment of a highly condensed state is indispensable to chromosome segregation. What is the pattern of DNA folding or coiling in the condensed state? What chromosomal proteins bring about this condensation? How is condensation controlled in a regionally and temporally specific manner? These are central problems in chromosome biology today.

It has long been apparent from light microscopy that the condensation of chromosomal material is not achieved in a single step. Rather, there is a hierarchy of levels of condensation, with the overall packing ratio given by the product of the ratios at the individual stages. This multiplicative mechanism underlies the extraordinary degree of chromosome condensation in metaphase. It also confers important functional advantages: it allows the decondensation of one region of a chromosome without effect on others, enabling selective, or differential, gene expression. Our current understanding of chromosome structure and function is primarily at the first level of condensation, arising from DNA coiling in the *nucleosome*, the fundamental particle of the chromosome. This chapter describes the discovery of the nucleosome, its structure, and its involvement in gene regulation.

THE NUCLEOSOME: HYPOTHESIS

Studies of chromosomal material, or *chromatin*, began with chemical analyses about 100 years ago, which revealed roughly equal weights of DNA and a distinctive type of basic protein termed *histone*. This work was both seminal and, at the same time, held back further progress for decades, because the chromatin was derived from calf thymus. The choice of tissue was motivated by its high proportion of chromatin, reflecting the small size of thymus cells, whose volume is almost entirely occupied by the nucleus. Thymus cells are largely inactive and destined for destruction during development of the immune response. Perhaps because of this cell fate, the thymus is rich in degradative enzymes, such as proteases, whose action on the histones gave the appearance of remarkable heterogeneity. It seemed that there were dozens of histone types, which together with other evidence led to the notion of histones as specific gene repressors.

A breakthrough came in the 1950s when histones were extracted with strong denaturants, such as concentrated acid or guanidine hydrochloride, in preparation for the newly devised methods of protein sequencing. The denaturants prevented proteolysis, and sequencing revealed both a limited variety of histone types and their remarkable evolutionary conservation. There are four types of histone, termed H2A, H2B, H3, and H4 in all eukaryotes, and a fifth type, H1, in many eukaryotes as well. H3 and H4 are the most highly conserved, with as few as two amino acid changes in H4 across the vast evolutionary span from plants to animals. Despite these findings, the notion of heterogeneity in chromatin persisted. Although the five types of histone were roughly equimolar in most tissues and organisms, there appeared to be sufficient variation in the ratios to rule out a unique repeating order. Much effort was devoted to analyzing the interactions of histones with one another in order to elucidate their arrangements in chromatin. The basis of the previous breakthrough was at the same time the cause of the next impasse. The use of strong denaturants in the isolation of the histones resulted in their extensive aggregation, giving the appearance of innumerable patterns of histone-histone association.

The final breakthrough came in the early 1970s, when the extraction of histones from chromatin with concentrated salt solutions, in the presence of powerful protease inhibitors, enabled their isolation in the native state and the ensuing discovery of unique oligomeric forms. In particular, H3 and H4 were isolated as a double dimer, or $(H3)_2(H4)_2$ tetramer. The principles of chromatin structure were deduced from this finding as follows. First, the precise stoichiometry of H3 and H4 in the tetramer suggested that deviations from equimolar amounts in previous analyses were due to experimental error. Supposing the same to be true for H2A and H2B led to the expectation of a repeating arrangement in chromatin based on pairs of the four major histones, or a total of eight histone molecules in the repeating unit. Second, based on the approximately equal weights of histones and DNA found in most tissues, eight histones were expected to be associated with about 200 DNA base pairs. Third, the nature of the H3/H4 tetramer, reminiscent of other soluble proteins, suggested the arrangement of histones and DNA, as well as the overall shape of the complex. A familiar tetrameric protein of similar size and stoichiometry, hemoglobin, had been studied extensively by X-ray crystallography, revealing a compact, roughly spherical or "globular" structure. In particular, globular proteins are not penetrated by holes 25Å in diameter, the size of double helical DNA. A globular assembly of histones would therefore be expected to form a roughly spherical particle, with DNA wrapped on the outside. Several lines of evidence regarding H1, including an amount of this histone roughly half that of the others in most tissues, led to the expectation that a single molecule of H1 would be associated with the DNA on the outside of the particle. A chain of such particles, connected by DNA, would constitute a *chromatin fiber*.

The idea thus emerged of the *nucleosome*, as a fundamental particle of the eukaryote chromosome. Acceptance of this idea finally laid to rest the notion of heterogeneity in types or arrangements of histones, although variation, arising from posttranslational modifications of histones, sequence preferences in histone-DNA interaction, and binding of other proteins may nonetheless be of great biological significance. We turn first here to the proof of existence of the nucleosome and its three-dimensional structure before addressing functional implications.

THE NUCLEOSOME: EXPERIMENTAL CONFIRMATION

The existence of the nucleosome was most clearly demonstrated by a combination of biochemistry and electron microscopy. Biochemical studies employed nucleases, especially micrococcal or staphylococcal nuclease, to cleave the DNA between nucleosomes in a chromatin fiber. Digestion for a brief time or with a limited amount of nuclease, followed by extraction of the DNA and gel electrophoresis, gave a ladder of bands corresponding to multiples of a smallest or unit size. Comparison with DNA fragments of known length disclosed variation of the unit size among tissues and organisms from about 170–240 base pairs, with a typical value of about 200 base pairs. Alternatively, the nuclease-digested chromatin was fractionated as native protein-DNA complexes by sedimentation in a sucrose gradient. A series of peaks was obtained, and subsequent extraction and gel electrophoresis showed that the first peak contained only DNA of the 200 base pair unit size, the second peak contained only DNA of twice this size, and so forth. Finally, electron microscopy revealed individual, roughly spherical particles, about 100Å in diameter, in the first peak, pairs of particles in the second peak, etc. The one-to-one correspondence between multiples of the unit DNA length and numbers of particles seen in the electron microscope proved the relationship between them. The idea of a repeating particle as the basis of chromatin structure thus became a reality.

The identification of a histone octamer as the protein moiety of the nucleosome was established by chemical crosslinking. The histones in the isolated nucleosome particle could be connected by reaction with a small, doubly reactive chemical compound to form a molecule having eight times the mass of a single histone. The octamer could also be isolated in the native state by extraction and maintenance in concentrated salt solutions. The protein in this form was crystallized and its structure determined by X-ray diffraction at atomic resolution (see below).

STRUCTURE OF THE NUCLEOSOME

Upon extensive digestion with micrococcal nuclease, DNA of the nucleosome monomer is degraded from the ends, pausing at about 166 base pairs, and reaching a limit at about 146 base pairs. During the conversion from 166 to 146 base pairs, H1 is released. The particle with 166 base pairs that includes H1 is sometimes called the chromatosome. The limit digestion product, consisting of 146 base pairs of DNA complexed with a histone octamer, is termed the *core particle* of the nucleosome. The difference between the repeat length of 170–240 base pairs, revealed by brief nuclease digestion (see above), and the core length of 146 base pairs is termed the *linker*. The repeat length may be thought of as the center-to-center distance of core particles along a chromatin fiber, while the linker, which ranges from about 25–75 base pairs, represents the DNA connecting the core particles.

In contrast with the repeat length ranging from 170–240 base pairs, the core length of 146 base pairs is invariant. Crystallization of core particles from many organisms further attests to the uniformity of nucleosomes, and has enabled structure determination to 7Å resolution in work published to date. The results show DNA in the classical B form coiled in a left-handed superhelix around the histone octamer, forming a disk about 55Å in height along the superhelix axis and about 110Å in diameter. From the circumference and the value of 3.4Å per base pair for B-DNA, it may be calculated that 166 and 146 base pairs of DNA correspond to 2 and to 1 3/4 turns of the superhelix, respectively. Digestion from 166 to 146 base pairs removes 10 base pairs from each of the two ends of superhelix, which lie on the same side of the particle, where DNA enters and exits the nucleosome. H1, released in the course of digestion, may bind to the DNA at this unique location

The structure of the histone octamer conforms well to that deduced from the analysis of core particles (see Figure 1). The octamer is seen to consist of H3/H4 and H2A/H2B dimers, arrayed in a protein superhelix, whose dimensions match those of the DNA superhelix. Moreover, the pattern of positively charged amino acids on the surface of the octamer coincides with the paths of the DNA strands. The order of histone dimers in the protein superhelix is (H2A/H2B)-(H3/H4)-(H3/H4)-(H2A/H2B), so the octamer contains a central H3/H4 tetramer, which binds the central turn of DNA. Within the histone dimers, there is no clear division of protein density, but rather the two polypeptides interdigitate extensively. The

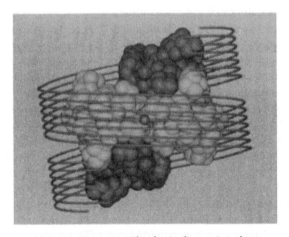

Figure 1. Structure of the nucleosome. The three-dimensional structure of the histone octamer (Arents et al., 1991) is shown in white (histones H3 and H4) and blue (histones H2A and H2B). Parallel lines in green define a ribbon having the width of duplex DNA (about 20Å) and indicate the left-handed superhelical path around the histone octamer in the nucleosome. (Courtesy of E. Moudrianakis).

two types of dimer are largely alpha-helical and exhibit a common tertiary fold. The amino-terminal 15–40 residues of the histones are not seen in the octamer electron density map but appear, at least in some cases, to be present in the core particle map. These regions may constitute separately folded domains, flexibly linked to the rest of the protein structure, constrained in location only upon binding DNA.

DNA TOPOLOGY AND BENDING IN THE NUCLEOSOME

Coiling of a coil, such as that of the DNA double helix in the nucleosome, is commonly referred to as supercoiling. Measurement of the degree of supercoiling in a nucleosome presents the problem that on extraction free of protein, DNA will tend to straighten out and its supercoiling will be lost. This problem may be overcome by constraining the ends of the DNA, for example, by joining them to form a circle. Naturally occurring circular chromosomes, such as viral "minichromosomes," which are packaged in chromatin virtually identical with that of the host cell, have therefore been employed to measure the supercoiling associated with nucleosomes. The results show about 1.1 left-handed supercoils per nucleosome, less than the value of about 1.8 expected from the structural evidence for nearly two turns of DNA superhelix in the nucleosome. This discrepancy is commonly referred to as the "linking number paradox" (the degree of supercoiling in a circular DNA can be expressed in terms of a mathematical quantity called the linking number). One possible resolution of the paradox is that both left-handed supercoiling and also some compensatory right-handed twisting about the double helix axis occur as DNA is incorporated in a nucleosome.

The compaction of DNA in nucleosomes may be near the maximum achievable without disruption of the B structure. The DNA is bent tightly, on a radius of curvature of about 30Å, compared with the 20Å diameter of the double helix, and the superhelical turns are closely spaced, less than 10Å apart on the surface of the nucleosome. An important consequence of the tight bending is a degree of DNA sequence preference in the formation of nucleosomes. While almost all DNA sequences can be accommodated, some are favored over others. In particular, A/T-rich sequences are preferred where the minor groove of the double helix faces inwards towards the surface of the nucleosome, while G/C-rich sequences tend to be found where the minor groove faces outwards. DNA conforming to these requirements, having A/T- and G/C-rich sequences interspersed with the periodicity of the double helix, may be bound several orders of magnitude more tightly by the histone octamer than random or unfavorable sequences. At the other extreme, homopolymeric sequences, such as poly(dA)-poly(dT), which may be rather rigid and resist bending, appear to be excluded from nucleosomes.

The question immediately arises whether sequence preference in nucleosome formation plays any role in chromosome function *in vivo*. Nucleosomes are found in nonrandom locations near promoters of transcribed genes and in the vicinity of other chromosomal elements, such as centromeres and origins of DNA replication. Such *nucleosome positioning* could be important for covering up certain DNA sequences and exposing others to the action of transcription factors, enzymes, and so forth. Several recent findings are consistent with the involvement of nucleosome positioning in chromosome function along these lines. On the other hand, it remains to be determined whether nucleosome positioning is a cause or a consequence of functional activity. Moreover, random locations of nucleosomes are frequently found and may prevail in most chromosomal regions.

NUCLEOSOMES AND TRANSCRIPTION

Formation of a nucleosome on a promoter inhibits the initiation of transcription *in vitro*. By contrast, nucleosomes downstream of a transcription start site fail to block RNA chain elongation. The histone octamer is displaced, at least transiently, by polymerase passage. An inhibitory effect of nucleosomes on transcription has also been demonstrated *in vivo* by the construction of yeast strains in which the expression of a histone gene is controlled by an inducing agent such as galactose. Shift to growth in the absence of inducer causes the gradual loss of nucleosomes and the onset of transcription of many previously inactive genes. The nucleosome may therefore be regarded as a general repressor of transcription.

Repression by nucleosomes is attributed to promoter occlusion: wrapping of promoter DNA around the histone octamer interferes with the binding of RNA polymerase and accessory factors. Repression is overcome *in vivo* by the exposure of promoters in nucleosome-free regions. Virtually every transcriptionally active promoter so far analyzed has been found in a *nuclease-hypersensitive site*, a region of typically a few hundred base pairs that is more susceptible to cleavage by DNases than is the surrounding chromatin. Several lines of evidence indicate that the DNA of a nuclease-hypersensitive site is essentially naked, devoid of histones and most other proteins. The transcribed region downstream of a promoter is usually more susceptible to DNase digestion as well, though less than a nuclease-hypersensitive site. Transcribed regions generally remain packaged with histones in nucleosomes, but the distribution of nucleosomes and even their structure appears altered, to an extent that varies from one gene to another. These changes in the chromatin of transcribed regions and their variation may be a consequence of polymerase passage: the changes could reflect the transient displacement of histone octamers, while the magnitude of the changes depends on the relative rates of displacement and nucleosome reassembly.

The mechanism by which nuclease-hypersensitive sites are formed is an important object of current research. Genetic studies in yeast have indicated the possible involvement of a multiprotein assembly, termed the SWI/SNF complex because it includes the products of the *SWI1*, *SWI2*, *SWI3*, *SNF5*, and *SNF6* genes. The SWI2 protein is a DNA-dependent ATPase, whose activity is required for interaction of the complex with nucleosomes *in vitro*. This interaction can expose sites for

DNA-binding proteins that are obscured by the histone octamer, but it does not appear to cause displacement of the octamer, so it cannot alone account for the appearance of nucleosome-free regions *in vivo*.

Genetic studies in yeast have also implicated the amino-terminal regions of the histones in the process of gene activation. Deletion of residues 4–30 of H3 increases transcription of inducible promoters, while deletion of residues 4–28 of H4 decreases transcription. The amino-terminal regions are of particular interest since they are the target of many posttranslational modifications of histones that accompany gene activation and repression. For example, lysine residues at positions 5, 8, 12, and 16 of H4 undergo acetylation, and changing these residues to arginine, which cannot be acetylated, greatly diminishes inducible transcription.

BEYOND THE NUCLEOSOME

The packing ratio of DNA in the nucleosome core particle is about 10, several orders of magnitude less than that in the most condensed states of eukaryote chromosomes (see below). There are, at present, few indications of how nucleosomes are packed in condensed states or of how decondensation occurs for activity. Current evidence points to coiling of the chain of nucleosomes, forming a fiber about 30 nm (300Å) in diameter, in which the packing ratio is about 50. No experimental indications or concrete proposals for the further folding of 30 nm fibers have yet emerged.

The condensation of chromatin in higher order structures may be directly involved in gene activation and repression. Two types of DNA sequences have been described that define large chromosomal domains containing sets of coordinately regulated genes. A single copy of a sequence termed a locus control region (LCR) is required for appropriate expression of all genes in the 80,000 base pair human globin locus. The LCR is also responsible for increased susceptibility to DNase I digestion of the entire locus in cells active in globin gene transcription. It may be imagined that the LCR drives decondensation of a large domain, making all genes within it available for transcription and more exposed to nuclease attack.

The second type of sequence appears to delimit the range of action of an LCR and thus to define the boundaries of a large chromosomal domain. Such a sequence was first found bracketing a 24,000 base pair region of enhanced susceptibility to DNase I digestion containing the transcriptionally active chicken lysozyme gene. Sequences with similar properties have since been found at the globin locus and elsewhere.

Finally, domains of specially condensed chromatin, termed heterochromatin, are responsible for the stable maintenance of some genes in a transcriptionally repressed state. Most is known about this mechanism of repression from genetic studies in yeast, where the DNA sequence directing the process is termed a silencer, and some of the proteins involved are denoted SIR for silent information regulator. The genetic studies have further demonstrated the central role of the nucleosome in repression by heterochromatin and the involvement of the amino-terminal regions of the histones in the process.

SUMMARY

Studies of the histones, a family of abundant, conserved nuclear proteins, revealed the existence of the nucleosome, as a fundamental particle of the chromosome. The nucleosome comprises a core, containing 146 base pairs of DNA, invariant among organisms, and a linker, which ranges from 25–75 base pairs depending on the source. Core DNA is wrapped in 1 3/4 turns of a left-handed superhelix around an octamer comprising two each of histones H2A, H2B, H3, and H4, forming a disk-shaped particle 55Å in height and 110Å in diameter. Linker DNA connects one core to the next in a chromatin fiber. Nucleosomes inhibit transcription by promoter occlusion, and are removed in the process of gene activation. Nucleosomes also enable transcriptional repression by chromatin condensation.

REFERENCES

Arents, G., Burlingame, R.W., Wang, B.-C., Love, W.E., & Moudrianakis, E.N. (1991). The nucleosomal core histone octamer at 3.1 Å resolution: A tripartite protein assembly and a left-handed superhelix. Proc. Natl. Acad. Sci. USA 88, 10148–10152.

Cairns, B.R., Kim, Y.J., Sayre, M.H., & Kornberg, R.D. (1994). A multisubunit complex containing the *SWI1/ADR6, SWI2/SNF2, SWI3, SNF5,* and *SNF6* gene products isolated from yeast. Proc. Natl. Acad. Sci. USA 91, 1950–1954.

Drew, H.R., & Travers, A.A. (1986). Sequence periodicities in chicken nucleosome core DNA. J. Mol. Biol. 191, 659–675.

Durrin, L.K., Mann, R.K., Kayne, P.S., & Grunstein, M. (1991). Yeast histone H4 N-terminal sequence is required for promoter activation *in vivo.* Cell 65, 1023–1031.

Felsenfeld, G. (1992). Chromatin as an essential part of the transcriptional mechanism. Nature 355, 219–224.

Finch, J.T., Noll, M., & Kornberg, R.D. (1975). Electron microscopy of defined lengths of chromatin. Proc. Natl. Acad. Sci. USA 72, 3320–3322.

Grosveld, F., van Assendelt, G.B., Greaves, D.R., & Kollias, G. (1987). Position-independent, high-level expression of the human β-globin gene in transgenic mice. Cell 51, 975–985.

Grunstein, M. (1990). Histone function in transcription. Ann. Rev. Cell Biol. 6, 643–678.

Han, M., & Grunstein, M. (1988). Nucleosome loss activates yeast downstream promoters *in vivo.* Cell 55, 1137–1145.

Kornberg, R.D. (1974). Chromatin structure: A repeating unit of histones and DNA. Science 184, 868–871.

Kornberg, R.D. (1977). Structure of chromatin. Ann. Rev. Biochem. 46, 931–954.

Kornberg, R.D., & Lorch, Y. (1991). Irresistible force meets immovable object: Transcription and the nucleosome. Cell 67, 833–836.

Kornberg, R.D., & Lorch, Y. (1992). Chromatin structure and transcription. Ann. Rev. Cell Biol. 8, 563–587.

Kornberg, R.D., & Thomas, J.O. (1974). Chromatin structure: Oligomers of the histones. Science 184, 865–868.

Lorch, Y., LaPointe, J.W., & Kornberg, R.D. (1987). Nucleosomes inhibit the initiation of transcription but allow chain elongation with the displacement of histones. Cell 49, 208–210.

Richmond, T.J., Finch, J.T., Rushton, B., Rhodes, D., & Klug, A. (1984). Structure of the nucleosome core particle at 7 Å resolution. Nature 311, 532–537.

Simpson, R.T. (1991). Nucleosome positioning: Occurrence, mechanisms, and functional consequences. Prog. Nucleic Acids Res. Mol. Biol. 40, 143–184.

Thomas, J.O., & Kornberg, R.D. (1975). An octamer of histones in chromatin and free in solution. Proc. Natl. Acad. Sci. USA 72, 2626–2630.

Travers, A.A., & Klug, A. (1987). The bending of DNA in nucleosomes and its wider implications. Philos. Trans. R. Soc. Lond. B. 317, 537–561.

van Holde, K.E. (1989). In: Chromatin. Springer-Verlag, New York.

Widom, J., & Klug, A. (1985). Structure of the 300Å chromatin filament: X-ray diffraction from oriented samples. Cell 43, 207–213.

Winston, F., & Carlson, M. (1992). Trends Genet. 8, 387–391.

Chapter 2

DNA Replication and its Control

J. JULIAN BLOW and JAMES P.J. CHONG

Principles of Medical Biology, Volume 5
Molecular and Cellular Genetics, pages 11–31.
Copyright © 1996 by JAI Press Inc.
All rights of reproduction in any form reserved.
ISBN: 1-55938-809-9

INTRODUCTION

The cell division cycle describes the events that are required for a single cell to divide and generate two copies of itself. Most cellular components, such as ribosomes or mitochondria, are present in each cell in multiple copies, and their numbers increase more or less continuously during cell growth. Chromosomes, however, are normally present in each cell only as a single copy. Their duplication is strictly controlled and normally occurs only as a prelude to cell division. In eukaryotic cells chromosomes are duplicated during a discrete period of the cell division cycle called S-phase, bounded by the presynthetic G1-phase, and the postsynthetic G2-phase.

The basic mechanism of DNA duplication has been highly conserved during evolution. This process, termed "semi-conservative replication," involves the sequential polymerization of nucleotides onto the 3′ end of a nascent DNA strand by a DNA polymerase enzyme. Since eukaryotic chromosomes consist of double-stranded DNA, both of the parental strands are replicated at the same time. This function is performed by a complex of proteins (the replication fork), which act together to progress along the chromosome, replicating both strands of DNA as they go. The various activities present at the replication fork have to work efficiently and coordinately, in order to minimize the number of errors introduced into the newly-replicated DNA.

There appears to be little external control over the progress of active replication forks, and therefore replication control must primarily be exerted over the initiation of active replication forks. Both the timing and location of initiation events must be controlled so that all of the chromosome is replicated precisely once during each S-phase, since any sections of DNA either left unreplicated, or replicated more than once, would be potentially lethal to the cell.

THE REPLICATION FORK

The replication fork has a complex task to perform in accurately replicating the two strands of duplex DNA (Figure 1A). First, a helicase must unwind the double-stranded DNA ahead of the fork to allow the new complementary strands to be synthesized. For this to occur, the normal chromatin structure of chromosomal DNA must be temporarily disassembled. Torsional stress produced by this unwinding must be relieved by topoisomerases (nicking-closing enzymes; see Table 1). Once unwound, DNA complementary to both template strands is synthesized. Since the two strands of duplex DNA run in opposite directions and since new deoxyribonucleotides can only be added to the 3′ end of a strand, only one of the nascent strands (the "leading strand") can be synthesized in the same direction as the replication fork is moving. The other strand (the "lagging strand") is synthesized in the opposite direction. DNA polymerases cannot start a new DNA chain without a free 3′-OH group to add nucleotides to. Therefore, another enzyme, a primase,

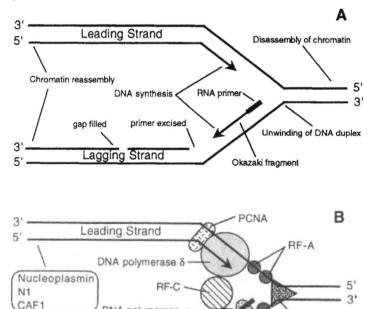

Figure 1. Activities and proteins involved in the eukaryotic replication fork. **A.** The events occurring on the leading and lagging strands of a replication fork progressing towards the right of the figure. Each single strand of DNA is shown by a line; single stranded RNA is shown by a heavy line. Arrows show the direction of chain growth. **B.** Proteins present at the SV40 replication fork. DNA polymerase δ and PCNA are shown on the leading strand with DNA polymerase α/primase on the lagging strand. Interactions between the polymerases and RF-C are hypothetical, but are consistent with results obtained from prokaryotic systems (Alberts, 1984). CAF1 is a chromatin assembly factor present in the crude SV40 system (Smith and Stillman, 1989); nucleoplasmin and N1 are proteins with similar functions that act together in chromatin assembly in *Xenopus* (Dilworth et al., 1987).

synthesizes a short RNA primer (usually 8–10 nucleotides long) which is then elongated by the DNA polymerase, until it reaches the 5' end of the previous RNA primer. In eukaryotes these lagging strand chains, called "Okazaki fragments," are generally 140–200 nucleotides long. To generate a continuous chain on the lagging strand, the Okazaki fragments must be processed and joined together. This involves excision of the RNA primer, filling in the gap by a DNA polymerase utilizing the next fragment as a primer, and then ligation of the contiguous fragments by a DNA ligase. Newly replicated DNA must then be complexed with histone proteins to restore the normal chromatin structure of eukaryotic DNA.

Table 1. Cellular Proteins Required for SV40 DNA Replication *In Vitro*

Protein	Function
DNA polymerase α -primase complex	Lagging strand DNA synthesis Initiation of leading strands
DNA polymerase δ	Leading strand DNA synthesis
PCNA	Polymerase δ auxiliary factor. Required for leading strand DNA synthesis and correct processing Okazaki fragments.
RF-A	Single-stranded DNA binding protein. Required for unwinding the origin of replication at initiation and needed for fork progression.
RF-C	Required for leading strand DNA synthesis and correct processing of Okazaki fragments
Topoisomerases	Required for fork progression. Type II topoisomerase also required for decantenation of daughter strands.
CAF1	Chromatin assembly

The best picture currently available for the composition of the eukaryotic replication fork comes from studies of the SV40 virus, a small double-stranded DNA virus that infects monkey and human cells. Complete replication of the entire viral genome requires only two viral components: a 90 kDa virus-encoded protein called the SV40 large tumor antigen (T antigen), and the SV40 minimal origin of replication, 65 base pairs of DNA, to which T antigen can bind. The majority of the proteins involved in replicating the SV40 genome are provided by the host cell, and so this system probably gives a fairly accurate picture of what normal eukaryotic replication forks look like. Replication of DNA molecules containing the SV40 origin sequence can be accomplished *in vitro* using extracts prepared from monkey or human cells grown in culture, and supplemented with T antigen (Li and Kelly, 1984; reviewed in Stillman, 1989). This system has been extensively fractionated to identify host proteins involved in SV40 DNA replication. The basic processes of initiation and replication fork progression have been achieved *in vitro* with purified enzymes (Tsurimoto et al., 1990; Weinberg et al., 1990). Table 1 lists host cell DNA polymerases and accessory factors with known functions in SV40 DNA replication.

DNA Polymerases

DNA polymerases synthesize new DNA chains from free nucleotides. Five eukaryotic polymerases have been identified: α, β, γ, δ, and ε. On the basis of sensitivity to inhibitors it can be concluded that neither β nor γ are required for chromosomal DNA replication. Polymerase γ is required for mitochondrial DNA replication, and polymerase β is probably involved in DNA repair.

Full replication of the SV40 DNA in the cell-free system requires both α and δ polymerases, which perform different functions at the replication fork (Figure 1B) (Weinberg and Kelly, 1989; Tsurimoto et al., 1990; Weinberg et al., 1990). DNA on the leading strand is synthesized by the continuous activity of polymerase δ.

Normally DNA polymerase δ is nonprocessive, that is it only synthesizes a short stretch of DNA on binding to a DNA template before dissociating and diffusing away. However, another protein required for replication in the cell-free system called proliferating cell nuclear antigen (PCNA) acts as a polymerase δ processivity factor, causing it to remain attached to the growing strand of DNA, so that each binding event results in the synthesis of thousands of nucleotides of nascent DNA (Tan et al., 1986; Bravo et al., 1987; Prelich et al., 1987). Polymerase δ is also associated with a 3'–5' exonuclease activity, which can proofread and remove mismatched nucleotides, increasing the fidelity of leading strand synthesis. Okazaki fragments on the lagging strand are synthesized by DNA polymerase α, which is also a relatively nonpro-cessive polymerase. Polymerase α is found tightly associated with a primase, which is capable of synthesizing short RNA primers on single-stranded DNA. The DNA polymerase then takes over from the primase, and synthesizes DNA on the RNA primer, until the gap between adjacent RNA primers is filled (Figure 1A).

Evidence for the involvement of both α and δ DNA polymerases in normal chromosomal DNA replication comes from genetic studies in yeast. Three genes encoding polymerases α, δ, and ε have been identified in yeast. Mutations in any of these three genes prevent the cell from completing DNA replication (Boulet et al., 1989; Sitney et al., 1989; Morrison et al., 1990; Araki et al., 1992). This suggests that in addition to α and δ, polymerase ε also has a necessary function at the chromosomal replication fork. Polymerase ε has a proofreading 3'–5' exonuclease activity, which increases the fidelity with which it replicates DNA. Since polymerase α apparently lacks an exonuclease, polymerase δ is a better candidate for replication of the bulk of the lagging strand DNA. This has led to the suggestion that the role of polymerase α on the lagging strand may in fact be very brief, extending the RNA primer with just enough DNA to then allow polymerase ε to take over and finish replication of the Okazaki fragment at high fidelity (Hübscher and Thömmes, 1992). Alternatively, polymerase ε may be involved in some other replicative function such as filling the gaps on the lagging strand left after the RNA primers have been excised (Figure 1A).

Other Replication Factors in the SV40 System

As mentioned above, the high processivity of polymerase δ on the leading strand is dependent on the accessory protein PCNA. Another replication factor, called RF-C, is also required for leading strand DNA synthesis in the SV40 cell-free system (Tsurimoto and Stillman, 1989). In the absence of either PCNA or RF-C only lagging strand DNA is synthesized, but the Okazaki fragments are not ligated together, and nascent DNA accumulates as fragments 140–200 nt long. This suggests that processing of Okazaki fragments into full-length DNA is dependent on the presence of a fully active replication fork which coordinates leading strand synthesis with Okazaki fragment processing. This is consistent with models of replication forks in prokaryotes (Alberts, 1984) (Figure 1B).

. PCNA and RF-C also play important roles in the setting up of the replication fork during the initiation process. Polymerase α and its associated primase lay down the first nucleotides at the site of replication initiation, which will subsequently become the leading strand. PCNA and RF-C form a complex on the 3' end of this nascent strand, preventing polymerase α from extending it any further. This permits polymerase δ to take over the task of leading strand synthesis, while polymerase α is recycled to perform lagging strand synthesis (Tsurimoto and Stillman, 1991a,b).

INITIATION OF REPLICATION FORKS: REPLICATION ORIGINS

In order to replicate their large genomes, eukaryotic cells initiate replication forks at 10^3–10^5 replication origins scattered throughout the genome. These replication forks normally move along the DNA until they encounter another fork coming from the opposite direction. At this point the forks terminate, and the replication proteins disassemble from the DNA. The involvement of two forks with opposite polarity in termination ensures that all the DNA between adjacent replication origins is fully replicated. Since there is little control over replication fork progression, the initiation of replication forks at origins of replication is the key controlling step. There is considerable variability in the timing and location of initiation events in different cell types. In multicellular organisms, the total length of S-phase (and the whole cell division cycle itself) often varies widely throughout embryonic development, with early embryonic cells tending to have shorter S-phases than later or more differentiated cells. The shorter S-phase of embryonic cells is achieved both by increasing the synchrony with which initiation occurs, and by reducing the distance between adjacent replication origins.

Timing of Initiation Events During S-Phase

One of the simplest replication patterns is seen in the embryo of the fruit fly, *Drosophila melanogaster* (Blumenthal et al., 1973). The early embryo has an S-phase duration of only 3.4 minutes. Electron microscopy of replication forks shows that replication origins are spaced an average of 7.9 kb along the DNA, and all replication forks initiate at the start of S-phase. In contrast, adult *Drosophila* tissue culture cells have a typical S-phase duration of 7 hours. This longer S-phase length is due to both an increase in the distance between adjacent replication origins (spaced on average more than 30 kb apart), and a lowered synchrony of initiation from different origins. This replication pattern is depicted schematically in Figure 2. Similar variation in S-phase organization during development is seen in a range of different organisms, where earlier embryonic cells have shorter S-phases

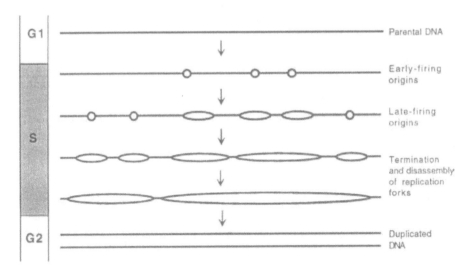

Figure 2. Replication scheme of a chromosome segment from a mammalian tissue culture cell. A single chromosome section is shown at different stages of replication, from G1 through S to G2. Three origins are shown firing early in S-phase, and three origins firing in mid S-phase.

more adult or more differentiated cells. The increased spacing between replication origins that occurs during later embryonic development means that some replication origins used in the early embryo are not used in more adult cells. Consistent with this, early embryos display a higher degree of promiscuity towards DNA sequences they can use for replication origins than do adult cells from the same organism (see below).

Autonomously Replicating Sequences in *Saccharomyces cerevisiae*

In the yeast *Saccharomyces cerevisiae*, specific DNA sequences have been shown to function as origins of DNA replication. These sequences were originally determined by their ability to support the extrachromosomal replication in yeast cells of small DNA molecules containing them. The replication of these plasmids was found to be dependent on short sections of yeast chromosomal DNA (<200 bp) called autonomously replicating sequences (ARSs) which were sufficient to support extrachromosomal replication. ARSs were subsequently shown to act as origins of replication both in plasmids and in their natural chromosomal environment (Brewer and Fangman, 1987; Huberman et al., 1987; Huberman et al., 1988). However, not all the sequences identified as ARSs actually act as replication origins when in their original chromosomal context, nor do all ARSs function in every cell cycle (Linskens and Huberman, 1988).

Analysis of ARSs showed them to be composed of a highly conserved core consensus (or A) element of 11 bp (5'-^T_ATTTA$^T_C$$^A_GTTT^T_A$-3'), embedded in flanking sequences which enhance the ability of the ARS to function as an origin. In the case of a particular yeast origin called ARS1, the flanking sequences have been analyzed and shown to contain a 3' B element (divided into three regions, B1, B2, and B3), and a 5' C element (Marahrens and Stillman, 1992). Other components of the flanking sequence may contribute to the ease with which the DNA can be unwound during initiation (Umek and Kowalski, 1988).

Recently a multiprotein complex has been purified from yeast which binds with great sequence specificity to the ARS1 yeast origin of replication (Bell and Stillman, 1992). This complex is called origin recognition complex (ORC) and there is a striking correlation between its ability to bind specific DNA sequences and the ability of these sequences to function as origins of replication. *In vivo* genomic footprinting studies show that the pattern of proteins bound over this region in the intact chromosome corresponds almost exactly to the footprint that ORC makes *in vitro* (Diffley and Cocker, 1992). Since ORC footprints most copies of ARS1 chromatin prepared from cells distributed throughout the cell cycle, this also suggests that ORC is bound to these origins for most of the cell division cycle.

Another factor called ARS binding factor 1 (ABF1) (Shore et al., 1987; Diffley and Stillman, 1988) can be seen binding to the B3 element of ARS1 by genomic footprinting (Diffley and Cocker, 1992). Further, the C element may facilitate the accurate positioning of a nucleosome close to the 5' end of the A element. This leads to the scenario depicted in Figure 3 where ORC is bound to DNA in the A and B1 regions of ARS1, close to a 5' nucleosome (possibly positioned by the C

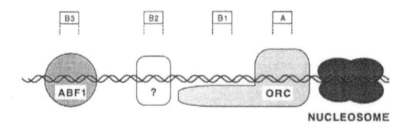

Figure 3. Proteins bound to the ARS1 replication origin in budding yeast. Four DNA sequence elements A, B1, B2, and B3 that affect origin function are indicated on a stretch of double-stranded chromosomal DNA. ORC is bound to the A element, and also protects one side of the DNA over the B1 element. ABF1 is bound to the B3 element. A positioned nucleosome is found to the right of the A element. Reprinted from Blow, 1992a.

element) with ABF1 bound at B3. Mutational studies of ARS1 show that although the B3 element improves the efficiency of origin usage, it is not absolutely essential, and can also be substituted by binding sites for other transcriptional activators (Marahrens and Stillman, 1992), or removed completely without abolishing ORC binding. This may suggest a role for transcription factors such as ABF1 in helping to guide ORC onto the A element in the presence of competition from other nonspecific or nucleosomal proteins.

The Effect of Origin Context on Timing of Initiation of DNA Replication

As described above, not all replication origins within a cell initiate DNA replication ("fire") at the same time. Specific groups of origins fire at characteristic times within S-phase. Early-firing origins are typically found in actively transcribed euchromatin, which is relatively loosely packed in the nucleus, and which is likely to be more accessible to proteins involved in both transcription and replication. Analysis of early-firing origins shows that the DNA encoding housekeeping genes (those genes expressed in all cell types) is replicated early (Goldman et al., 1984). Other early-firing origins can be identified as those near genes which are being expressed in a tissue-specific manner (Goldman et al., 1984; Calza et al., 1984). The effect of the degree of DNA condensation on timing of origin firing is best illustrated in the replication of the two X chromosomes of female cells. Both chromosomes contain essentially the same sequences, but the inactivated X chromosome is highly condensed, and consequently replicates much later in S-phase than the active copy (reviewed in Fangman and Brewer, 1992).

The initiation timing of specific origins in yeast appears to be independent of the DNA sequence of the ARS, but instead depends on the larger-scale chromosomal context. When telomeric origin sequences from yeast chromosomes are transplanted to a centromeric position on the chromosome they change from being late-firing to early-firing (Ferguson and Fangman, 1992). The opposite has a similar effect: moving early-firing origins from their centromeric position to a telomeric position causes them to initiate late. These results are consistent with a position-dependent effect of chromatin structure. Origins in mammalian cells appear to be organized into groups or clusters within the nucleus (Nakamura et al., 1986), further suggesting a role for high-order structure in the initiation of replication.

Physically Defined Origins in Higher Eukaryotes

Plasmids containing inserts of chromosomal DNA do not seem capable of autonomous replication when transfected into tissue culture cells of higher eukaryotes, making the identification of origin elements by the ARS assay impossible. This result may be due to the complexity of replication origins in these cell types, or to some mechanism that can detect the nonchromosomal nature of the transfected

DNA. Despite this, higher eukaryotic cells do appear to have reproducible replication origins. A number of techniques have now been developed for mapping initiation sites on the chromosomes of higher eukaryotes (reviewed in Vassilev and DePamphilis, 1992).

Chinese hamster ovary cells containing an amplified region of DNA incorporating the dihydrofolate reductase (DHFR) gene have been used to study the best-characterized mammalian origin of replication, which is found distal to this gene (Heintz et al., 1983; Leu and Hamlin, 1989). Even here, the key features required for origin function have not been identified. The exact mechanism of replication origin initiation is not understood, and different origin-mapping techniques have yielded conflicting results. Techniques that define an "origin of bidirectional replication," where replication products on one DNA strand switch from continuous leading strand products to discontinuous Okazaki fragments, have identified such a site, less than 450 bp long, downstream of the DHFR gene (Burhans et al., 1990). However, when 2-D gel techniques are used to identify small replication bubbles (presumably early products of the initiation process), initiation is seen taking place in this region over a broad 28 kb "initiation zone" (Vaughn et al., 1990). While these techniques select different populations of replication intermediates, this represents a significant discrepancy which needs to be resolved (see Vassilev and DePamphilis, 1992 for a discussion of the different techniques).

A model proposed by Linskens and Huberman (1990) incorporates aspects of both these results. This suggests that DNA replication is initiated at many points within the origin region by the frequent unwinding of the DNA to form microbubbles, small replication bubbles that are observed in 2-D gels (Vaughn et al., 1990), but the majority of which perform unidirectional replication moving away from a central origin of bidirectional replication. The microbubbles eventually fuse, to generate a single larger replication bubble that moves away from the initiation zone in the conventional manner described above for the SV40 fork. The details of how these early events would be controlled are currently unclear, though the model is consistent with certain observations made by electron microscopy of replicating DNA (Micheli et al., 1982).

Replication Origins in Embryos

In contrast to the results obtained in tissue culture cells, the early embryos of a number of organisms, including frogs and fruit flies, are capable of replicating a wide range of DNA molecules introduced into them (Harland and Laskey, 1980; Steller and Pirrotta, 1985; Crevel and Cotterill, 1991). In *Xenopus*, where the reaction can be studied in a cell-free system prepared from unfertilized eggs, replication bubbles can be seen at a large number of different sites on replicating DNA (Hyrien and Méchali, 1992; Mahbubani et al., 1992). This suggests that the embryo can potentially use a very large number of DNA sequences to initiate replication forks, and is reminiscent of the abundant replication bubbles observed within the initiation zone of the DHFR gene (Vaughn et al., 1990). However, despite the abundance of potential initiation sites in the *Xenopus* system, a typical 15 kb plasmid initiates replication at only a single randomly-positioned origin (Mahbubani et al., 1992). Interestingly, the density of replication origins observed in this study (one per 15 kb plasmid) is approximately equal to the interorigin spacing seen in chromosomes in the *Xenopus* embryo.

On the chromosomal DNA in these early embryos, the location of initiation sites is unlikely to be completely random, but is presumably influenced by other chromosomal features such as chromosomal coiling or looping (Buongiorno-Nardelli et al., 1982). The distribution of initiation sites in fruit fly embryos has been carefully studied by electron microscopy (Blumenthal et al., 1973). In this exceedingly short cell cycle (10 minutes), adjacent initiation sites seem to be spaced periodically, at integral multiples of 3.4 kb intervals. This would suggest that the physical organization of DNA may be more important than its sequence in determining initiation sites in these cell types.

REPLICATION CONTROL

In principle, replication might be controlled in the cell cycle by limiting the availability of some or all of the factors required at the replication fork. However, although there is some cell cycle control over their levels, most replication fork activities are present in the cell throughout the cell cycle. Part of the reason for this is that many are also likely to be involved in other cellular processes such as DNA repair. The continued presence of replication fork functions throughout the cell cycle means that DNA replication must be controlled by a relatively small number of S-phase inducers whose function is to regulate when and where initiation takes place.

Cell Fusion Studies

In the classic studies by Rao and Johnson (1970), HeLa cells at different stages of the cell cycle were fused and the replication timing of the different nuclei was measured. The results are summarized in Figure 4. In fusions of G1 and S-phase cells, the G1 nucleus was induced to enter S-phase earlier than it would normally have done while the S-phase nucleus continued replication as normal. The more S-phase cells fused to a single G1 cell, the faster the G1 nucleus entered S-phase, consistent with the activity of an S-phase inducer. These S-phase inducers can apparently act only on G1 nuclei, and cannot induce G2 nuclei to undergo another round of DNA replication, since in G1/G2 and S-phase/G2 HeLa cell fusions G2 nuclei were not induced to replicate a second time. Possible mechanisms preventing the re-replication of G2 nuclei are discussed in more detail below.

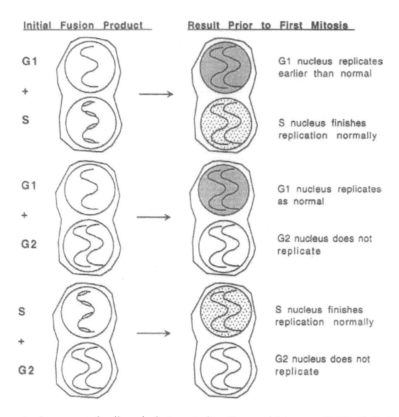

Figure 4. Summary of cell cycle fusion studies (Rao and Johnson, 1970). Cells from G1, S, and G2 stages of the cell cycle were fused, giving hybrids with two nuclei. G1 nuclei are indicated with a single chromosome, S nuclei with a chromosome containing replication bubbles, and G2 with a double chromosome. Prior to the first mitosis after fusion, all nuclei have fully replicated, and the extent of replication performed in the hybrid is indicated by the degree of shading in each nucleus. Reprinted from Blow, 1995.

These experiments suggest that replication control can be divided into two distinct components: the DNA template in the nucleus, and activities present in the cytoplasm which act on this nuclear substrate (Blow, 1995). The nucleus can be in either of two states: either capable or incapable of responding to S-phase inducers by undergoing DNA replication. Similarly, the cytoplasm provides two types of activity, one that induces a responsive (G1) nucleus to undergo S-phase (S-phase inducers), and another (related to induction of M-phase, as discussed below) that changes a refractory G2 nucleus into a responsive G1 nucleus. Once S-phase has been induced, the progress of the nucleus through the different stages of DNA replication occurs without further signals from the cytoplasm.

A Nuclear Envelope Requirement for DNA Replication

Cell-free extracts of *Xenopus* eggs can support the basic events of the cell cycle, including passage through S-phase and mitosis, and have been used to study the cell cycle control of DNA replication. When DNA is added to these cell-free extracts, it is replicated precisely once during each *in vitro* interphase (Blow and Laskey, 1986). The cell-free system also assembles exogenously added DNA into structures resembling normal interphase nuclei. These *in vitro*-assembled nuclei are surrounded by a double unit envelope studded with nuclear pores, and are capable of selectively importing soluble nuclear proteins. Possession of a complete nuclear envelope and the selective import of nuclear proteins is required before the initiation of DNA replication can occur in this system (Blow and Watson, 1987; Newport, 1987; Cox, 1992). A similar dependence of initiation on nuclear assembly is seen in a cell-free extract of *Drosophila* eggs (Crevel and Cotterill, 1991), and possibly represents a fundamental requirement for chromosome replication in eukaryotes.

The dependence of initiation on the presence and function of the nuclear envelope means that the *Xenopus* system can be used to study the way that cytoplasmic signals induce nuclei to initiate DNA replication. Using flow cytometry to monitor the timing and rate of replication (Blow and Watson, 1987), different nuclei in the same extract were seen to replicate at different times, with some nuclei finishing replication before other nuclei had started. However, once they had entered S-phase, each nucleus replicated relatively fast, each undergoing a burst of about 100,000 near-synchronous initiation events, so that all the DNA within a nucleus was replicated as a single coordinated unit. Further experiments indicate that the feature that defines this unit of replication is indeed the nuclear envelope (Leno and Laskey, 1991). These experiments add further weight to the idea that selective nuclear accumulation of replication proteins is an important mechanism by which initiation is controlled.

Prevention of Re-Replication in a Single S-Phase

The cell fusion experiments summarized in Figure 4 show that only G1 nuclei can initiate DNA replication in response to cytoplasmic signals; G2 nuclei must reach a G1 state by passing through mitosis before they can respond positively to such signals. In the *Xenopus* system, it was shown that G2 (replicated) nuclei were incapable of undergoing a further round of replication on readdition to fresh extract; if, however, they were allowed to progress into mitosis (consequently undergoing nuclear envelope breakdown and chromosome condensation), they then became replication-competent (Blow and Laskey, 1988). Some process during mitosis therefore permitted refractory G2 nuclei to revert to the responsive G1 state. On investigation, it was found that agents that caused nuclear envelope permeabilization, such as lysolecithin or phospholipase, left the G2 nuclei capable of re-replicating in fresh extract. Similar results have been obtained in cell-free extracts of *Drosophila* eggs (Crevel and Cotterill, 1991) and on introduction of mammalian tissue culture nuclei into *Xenopus* eggs (De Roeper et al., 1977) or egg extracts (Leno et al., 1992).

Figure 5 outlines a model proposed to explain these results (Blow and Laskey, 1988). An essential replication factor, called licensing factor, can bind to DNA during mitosis. Licensing factor cannot cross the nuclear envelope, so that once nuclear assembly is complete, licensing factor is present either free in the cytoplasm or bound to DNA in the nucleus. On entry into S-phase, each molecule of licensing factor bound to DNA supports a single initiation event, after which it is inactivated or destroyed. Thus in G2, no active licensing factor remains in the nucleus, and the nuclear envelope must become transiently permeable (as normally occurs during mitosis) to allow DNA to be licensed for a further round of replication.

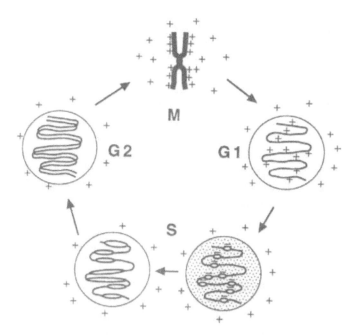

Figure 5. Model for the control of DNA replication in *Xenopus* egg extracts. The replication of a single chromosome during a single cell cycle is shown. During mitosis (M), licensing factor (+) binds to origins of replication on the condensed chromosome. On exit from mitosis, the chromosome decondenses and is completely surrounded by the nuclear envelope (G1). During S-phase, initiation events occur only at sites of bound licensing factor (S). Each licensing factor can only support a single initiation event, after which it is inactivated or destroyed (–). Reprinted from Blow, 1992b.

A number of observations suggest that the licensing factor model may apply to a wide range of eukaryotic organisms. The product of the yeast CDC46 gene, which is required for a very early event in DNA replication, behaves as though it were a licensing factor homologue (Hennessy et al., 1990). CDC46 protein is nuclear in G1, and apparently chromatin-associated, but disappears from the nucleus at the start of S-phase. It is seen in the cytoplasm throughout G2 and mitosis, but at the end of mitosis it translocates back to the nucleus (yeast do not undergo nuclear envelope breakdown during mitosis). Other genetic data suggest that CDC46 interacts directly with yeast replication origins (Chen et al., 1992). Members of the CDC46 gene family are found throughout the eukaryotic kingdom, with homologues in fission yeast, *Xenopus* (Coxon et al., 1992), mouse, and humans (Thömmes et al., 1992). Recent work has shown the highly conserved MCM/P1 family of proteins (of which CDC46 is a member), to have an essential function in the licensing of DNA for replication (Chong et al., 1995; Madine et al., 1995; Kobota et al., 1995; Chong et al., 1996).

One prediction of the licensing factor model is that changes are made to replication origins at the end of mitosis which persist throughout G1 and which are required for their subsequent activity in S-phase. Further observations in both yeast and *Xenopus* support this idea. ORC, the yeast origin recognition complex, is bound to the ARS1 origin throughout most of the cell cycle (Diffley and Cocker, 1992). Therefore the binding of ORC to ARS1 identifies it as a potential origin of replication, but does not as such trigger initiation. Initiation presumably occurs under the influence of cell cycle-dependent functions that would either bind to or modify the origin complex in such a way as to activate the initiation process. An attractive way of modulating ORC in this way would be by cell cycle regulated kinases, such as the cdc2-like kinases, as described below. Similarly, in the *Xenopus* cell-free system the replication protein RP-A (see above) binds to chromatin at a small number of discrete foci prior to nuclear assembly (Adachi and Laemmli, 1992). It is at these foci that initiation appears to occur. Thus it seems likely that the presence of RP-A at these sites is an indication that the DNA is licensed for subsequent initiation.

Cyclin-Dependent Kinases in Chromosome Replication

Among the activities generated on commitment to cell division are the S-phase inducers that cause G1 nuclei to undergo DNA replication. Some of the basic activities that occur during this process have been identified in yeast, with close parallels in higher eukaryotes. The event called START results in a yeast cell becoming committed to DNA replication and cell division. START itself corresponds to a requirement for the protein product of the cdc2 gene (in fission yeast) or the CDC28 gene (in budding yeast), which are close homologues of one another and members of the widespread family of cyclin dependent protein kinases (cdks).

The activity of these protein kinases is dependent on their physical association with a protein from the cyclin family. Yeast strains deficient in either cdc2/CDC28 or its activating cyclins will arrest at START (reviewed in Nasmyth, 1990; Forsburg and Nurse, 1991). Cdk and cyclin homologues that can functionally substitute for the yeast gene have been identified in a range of eukaryotic organisms, including humans, *Xenopus*, and *Drosophila* (Norbury and Nurse, 1992). The ability of these genes to rescue the yeast defect at START suggests that they may have a similar role in the commitment to replication in the organisms they come from.

A role for cdc2-like proteins in the S-phase of higher eukaryotes is supported by a number of experiments, though it is currently not clear which members of the cdk family are involved and exactly what their cyclin partners are. Depletion of cdk proteins from the *Xenopus* cell-free system leave it unable to replicate DNA, but capable of assembling nuclei and supporting replication fork elongation (Blow and Nurse, 1990; Fang and Newport, 1991). Replication can be restored by readdition of partially purified cdks. Similar results have been obtained by microinjection of anti-cdk antibodies into tissue culture cells (Pagano et al., 1993).

If cyclin-dependent kinases are involved in controlling S-phase in higher eukaryotes, they are likely to require activation by one or more cyclins. In mammalian cells cyclin A protein can be observed within the nucleus at about the time cells start S-phase (Pines and Hunter, 1991). Microinjection of antibodies directed against cyclin A (immunodepletion) left cells incapable of replicating DNA (Girard et al., 1991; Pagano et al., 1992). During S-phase, cyclin A appears to form part of a kinase with a specific cdk (cdk2), while associating with a different cdk (cdc2) later in G2 (Pagano et al., 1992).

The exact identity of the cyclin proteins required for S-phase induction is presently unclear. Although implicated in S-phase in mammalian cells, cyclin A does not appear to be necessary for S-phase in either *Drosophila* or *Xenopus* early embryos. *Drosophila* embryos lacking a cyclin A gene show no detectable S-phase defect (Knoblich and Lehner, 1993). Similarly, when protein synthesis is blocked in *Xenopus* embryos during metaphase, they are still capable of progressing through the subsequent S-phase and arrest in G2, although most of the cyclin A is destroyed as the embryo passes out of mitosis (Harland and Laskey, 1980; Minshull et al., 1990). The differing results from different experimental systems perhaps suggests that different cyclins in conjunction with different cdks may be employed at commitment to DNA replication in different cell types.

SUMMARY

In order for cells to divide, they must replicate their chromosomes precisely once. This replication is performed by a number of proteins that act together to form a replication fork. The major control over DNA replication occurs at replication origins, the sites where active replication forks are initiated. In the yeast *S. cerevisiae*, origins have been identified as ARSs which possess a defined DNA sequence. These ARSs are recognized by specific proteins including ORC and ABF1. The elements which define origins in higher eukaryotes have not yet been identified. The temporal sequence in which origins fire is dependent on both the chromosomal location of the origin and the transcriptional activity of the dividing cell, with euchromatin and actively-expressed genes tending to replicate early in S-phase.

DNA replication is an all-or-nothing event and occurs only after commitment to cell division. Cell fusion studies imply that S-phase inducers must be supplied to a nucleus as a signal to start replication. In addition, cells which have replicated their DNA but not yet undergone mitosis, will not perform further replication, even in the presence of these S-phase inducers. This is currently best explained by the action of a licensing factor, an origin-binding protein which cannot cross the nuclear envelope and so requires a transient nuclear permeabilization in order to make the DNA replication competent. Cyclin-dependent kinases provide good candidates for the S-phase inducers, as cell-cycle regulated activities which can potentially act upon these origin complexes and actually induce initiation to take place. Precise identification of the proteins involved is still to be made.

ACKNOWLEDGMENTS

J.J.B. is a Lister Institute Centenary Research Fellow.

ABBREVIATIONS

ABF1, ARS Binding Factor 1; ARS, Autonomously Replicating Sequence; bp, base pairs; cdk, cyclin-dependent kinase; DHFR, Dihydrofolate reductase; kb, kilobases (1 kb = 1000 bp); nt, nucleotides; ORC, Origin Recognition Complex.

REFERENCES

Adachi, Y., & Laemmli, U.K. (1992). Identification of nuclear pre-replication centers poised for DNA synthesis in *Xenopus* egg extracts: Immunolocalization study of replication protein A. J. Cell Biol. 119, 1–15.

Alberts, B.M. (1984). The DNA enzymology of protein machines. CSH Symp. Quant. Biol. 49, 1–12.

Araki, H., Ropp, P.A., Johnson, A.L., Johnston, L.H., Morrison, A., & Sugino, A. (1992). DNA polymerase II, the probable homolog of mammalian DNA polymerase ε, replicates chromosomal DNA in the yeast *Saccharomyces cerevisiae*. EMBO J. 11, 733–740.

Bell, S.P., & Stillman, B. (1992). ATP-dependent recognition of eukaryotic origins of DNA replication by a multiprotein complex. Nature 357, 128–134.

Blow, J.J. (1992a). A protein complex present at origins of replication in yeast cells. Bioessays 14, 561–563.

Blow, J.J. (1992b). The regulation of chromosome replication. J. Pathol. 167, 175–179.

Blow, J.J. (1995). S phase and its regulation. In: Cell Cycle Control: Frontiers in Molecular Biology (Glover, D.M., Hutchison C., eds.), pp. 177–205. Oxford University Press, Oxford.

Blow, J.J., & Laskey, R.A. (1986). Initiation of DNA replication in nuclei and purified DNA by a cell-free extract of *Xenopus* eggs. Cell 47, 577–587.

Blow, J.J., & Laskey, R.A. (1988). A role for the nuclear envelope in controlling DNA replication within the cell cycle. Nature 332, 546–548.

Blow, J.J., & Nurse, P. (1990). A cdc2-like protein is involved in the initiation of DNA replication in *Xenopus* egg extracts. Cell 62, 855–862.

Blow, J.J., & Watson, J.V. (1987). Nuclei act as independent and integrated units of replication in a *Xenopus* cell-free DNA replication system. EMBO J. 6, 1997–2002.

Blumenthal, A.B., Kriegstein, H.J., & Hogness, D.S. (1973). The units of DNA replication in *Drosophila melanogaster* chromosomes. CSH Symp. Quant. Biol. 38, 205–223.

Boulet, A., Simon, M., Faye, G., Bauer, G.A., & Burgers, P.M. (1989). Structure and function of the *Saccharomyces cerevisiae* CDC2 gene encoding the large subunit of DNA polymerase III. EMBO J. 8, 1849–1854.

Bravo, R., Frank, R., Blundell, P.A., & Macdonald-Bravo, H. (1987). Cyclin/PCNA is the auxiliary protein of DNA polymerase α. Nature 326, 515–517.

Brewer, B.J., & Fangman, W.L. (1987). The localization of replication origins on ARS plasmids in *S. cerevisiae*. Cell 51, 463–471.

Buongiorno-Nardelli, M., Micheli, G., Carri, M.T., & Marilley, M. (1982). A relationship between replicon size and supercoiled loop domains in the eukaryotic genome. Nature 100–102.

Burhans, W.C., Vassilev, L.T., Caddie, M.S., Heintz, N.H., & DePamphilis, M.L. (1990). Identification of an origin of bidirectional DNA replication in mammalian chromosomes. Cell 62, 955–965.

Calza, R.E., Eckhardt, L.A., DeGiudice, T., & Schildkraut, C.L. (1984). Changes in gene position are accompanied by a change in time of replication. Cell 36, 689–696.

Chen, Y., Hennesy, K.M., Botstein, D., & Tye, B. (1992). CDC46/MCM5, a yeast protein whose subcellular localization is cell cycle-regulated, is involved in DNA replication at autonomously replicating sequences. Proc. Natl. Acad. Sci. USA 89, 10459–10463.

Chong, J.P.J., Mahbubani, H.M., Khoo, C-Y., & Blow, J.J. (1995). Purification of an MCM-containing complex as a component of the DNA replication licensing system. Nature 375, 418–421.

Chong, J.P.J., Thömmes, P., & Blow, J.J. (1996). The role of MCM/P1 proteins in the licensing of DNA replication. Trends Biochem. Sci. 21, 102–106.

Cox, L.S. (1992). DNA replication in cell-free extracts from *Xenopus* eggs is prevented by disrupting nuclear envelope function. J. Cell Sci. 101, 43–53.

Coxon, A., Maundrell, K., & Kearsey, S.E. (1992). Fission yeast cdc21⁺ belongs to a family of proteins involved in an early step of chromosome replication. Nucl. Acids Res. 20, 5571–5577.

Crevel, G., & Cotterill, S. (1991). DNA replication in cell-free extracts from *Drosophila melanogaster*. EMBO J. 10, 4361–4369.

De Roeper, A., Smith, J.A., Watt, R.A., & Barry, J.M. (1977). Chromatin dispersal and DNA synthesis in G1 and G2 HeLa cell nuclei injected into *Xenopus* eggs. Nature 265, 469–470.

Diffley, J.F.X., & Cocker, J.H. (1992). Protein-DNA interactions at a yeast replication origin. Nature 357, 169–172.

Diffley, J.F.X., & Stillman, B. (1988). Purification of a yeast protein that binds to origins of DNA replication and a transcriptional silencer. Proc. Natl. Acad. Sci. USA 85, 2120–2124.

Dilworth, S.M., Black, S.J., & Laskey, R.A. (1987). Two complexes that contain histones are required for nucleosome assembly in vitro. Cell 51, 1009–1018.

Fang, F., & Newport, J.W. (1991). Evidence that the G1-S and G2-M transitions are controlled by different cdc2 proteins in higher eukaryotes. Cell 66, 731–742.

Fangman, W.L., & Brewer, B.J. (1992). A question of time: Replication origins of eukaryotic chromosomes. Cell 71, 363–366.

Ferguson, B.M., & Fangman, W.L. (1992). A position effect on the time of replication origin activation in yeast. Cell 68, 333–339.

Forsburg, S.L., & Nurse, P. (1991). Cell cycle regulation in the yeasts *Saccharomyces cerevisiae* and *Schizosaccharomyces pombe*. Ann. Rev. Cell Biol. 7, 227–256.

Girard, F., Strausfeld, U., Fernandez, A., & Lamb, N.J. (1991). Cyclin A is required for the onset of DNA replication in mammalian fibroblasts. Cell 67, 1169–1179.

Goldman, M.A., Holmquist, G.P., Gray, M.C., Caston, L.A., & Nag, A. (1984). Replication timing of genes and middle repetitive sequences. Science 224, 686–692.

Harland, R.M., & Laskey, R.A. (1980). Regulated DNA replication of DNA microinjected into eggs of *Xenopus laevis*. Cell 21, 761–771.

Heintz, N.H., Millbrandt, J.D., Greisen, K.S., & Hamlin, J.L. (1983). Cloning of the initiation region of a mammalian chromosomal replicon. Nature 302, 439–441.

Hennessy, K.M., Clark, C.D., & Botstein, D. (1990). Subcellular localization of yeast CDC46 varies with the cell cycle. Genes Dev. 4, 2252–2263.

Huberman, J.A., Spotila, L.D., Nawotka, K.A., El-Assouli, S., & Davis, L.R. (1987). The in vivo replication origin of the yeast 2μm plasmid. Cell 51, 473–481.

Huberman, J.A., Zhu, J., Davis, L.R., & Newlon, C.S. (1988). Close association of a DNA replication origin and an ARS element on chromosome III of the yeast, *Saccharomyces cerevisiae*. Nucl. Acids Res. 16, 6373–6384.

Hübscher, U., & Thömmes, P. (1992). DNA polymerase ε: In search of a function. Trends Biochem. Sci. 17, 55–58.

Hyrien, O., & Méchali, M. (1992). Plasmid replication in *Xenopus* eggs and egg extracts: A 2D gel electrophoretic analysis. Nucl. Acids Res. 20, 1463–1469.

Knoblich, J.A., & Lehner, C.F. (1993). Synergistic action of *Drosophila* cyclins A and B during the G2-M transition. EMBO J. 12, 65–74.

Kubota, Y., Mimura, S., Nishimoto, S-I., Takisawa, H., & Nojima, H. (1995). Identification of the yeast MCM3-related protein as a component of *Xenopus* DNA replication licensing factor. Cell 81, 601–609.

Leno, G.H., Downes, C.S., & Laskey, R.A. (1992). The nuclear membrane prevents replication of human G2 nuclei but not G1 nuclei in *Xenopus* egg extract. Cell 69, 151–158.

Leno, G.H., & Laskey, R.A. (1991). The nuclear membrane determines the timing of DNA replication in *Xenopus* egg extracts. J. Cell Biol. 112, 557–566.

Leu, T.-H., & Hamlin, J.L. (1989). High-resolution mapping of replication fork movement through the amplified dihydrofolate reductase domain in CHO cells by in-gel renaturation analysis. Mol. Cell. Biol. 9, 523–531.

Li, J.J., & Kelly, T.J. (1984). Simian virus 40 DNA replication in vitro. Proc. Natl. Acad. Sci. USA 81, 6973–6977.

Linskens, M.H., & Huberman, J.A. (1988). Organization of replication of ribosomal DNA in *Saccharomyces cerevisiae*. Mol. Cell. Biol. 8, 4927–4935.

Linskens, M.H.K., & Huberman, J.A. (1990). The two faces of higher eukaryotic DNA replication origins. Cell 62, 845–847.

Madine, M.A., Khoo, C-Y., Mills, A.D., & Laskey, R.A. (1995). MCM3 complex required for cell cycle regulation of DNA replication in vertebrate cells. Nature 375, 421–424.

Mahbubani, H.M., Paull, T., Elder, J.K., & Blow, J.J. (1992). DNA replication initiates at multiple sites on plasmid DNA in *Xenopus* egg extracts. Nucl. Acids Res. 20, 1457–1462.

Marahrens, Y., & Stillman, B. (1992). A yeast chromosomal origin of DNA replication defined by multiple functional elements. Science 255, 817–823.

Micheli, G., Baldari, C.T., Carri, M.T., Di-Cello, G., & Buongiorno-Nardelli, M. (1982). An electron microscope study of chromosomal DNA replication in different eukaryotic systems. Exp. Cell Res. 137, 127–140.

Minshull, J., Golsteyn, R., Hill, C.S., & Hunt, T. (1990). The A- and B-type cyclin associated cdc2 kinases in *Xenopus* turn on and off at different times in the cell cycle. EMBO J. 9, 2865–2875.

Morrison, A., Araki, H., Clark, A.B., Hamatake, R.K., & Sugino, A. (1990). A third essential DNA polymerase in *S. cerevisiae*. Cell 62, 1143–1151.

Nakamura, H., Morita, T., & Sato, C. (1986). Structural organizations of replicon domains during DNA synthetic phase in mammalian cells. Exp. Cell Res. 165, 291–297.

Nasmyth, K. (1990). FAR-reaching discoveries about the regulation of START. Cell 63, 1117–1120.

Newport, J. (1987). Nuclear reconstitution in vitro: Stages of assembly around protein-free DNA. Cell 48, 205–217.

Norbury, C., & Nurse, P. (1992). Animal cell cycles and their control. Ann. Rev. Biochem. 61, 441–470.

Pagano, M., Pepperkok, R., Lukas, J., Baldin, V., Ansorge, W., Bartek, J., & Draetta, G. (1993). Regulation of the cell cycle by the cdk2 protein kinase in cultured human fibroblasts. J. Cell Biol. 121, 101–111.

Pagano, M., Pepperkok, R., Verde, F., Ansorge, W., & Draetta, G. (1992). Cyclin A is required at two points in the human cell cycle. EMBO J. 11, 961–971.

Pines, J., & Hunter, T. (1991). Human cyclins A and B1 are differentially located in the cell and undergo cell cycle-dependent nuclear transport. J. Cell Biol. 115, 1–17.

Prelich, G., Tan, C.K., Kostura, M., Mathews, M.B., So, A.G., Downey, K.M., & Stillman, B. (1987). Functional identity of proliferating cell nuclear antigen and a DNA polymerase-delta auxiliary protein. Nature 326, 517–520.

Rao, P.N., & Johnson, R.T. (1970). Mammalian cell fusion: Studies on the regulation of DNA synthesis and mitosis. Nature 225, 159–164.

Shore, D., Stillman, D.J., Brand, A.H., & Nasmyth, K.A. (1987). Identification of silencer binding proteins from yeast: Possible roles in SIR control and DNA replication. EMBO J. 6, 461–467.

Sitney, K.C., Budd, M.E., & Campbell, J.L. (1989). DNA polymerase III, a second essential DNA polymerase, is encoded by the S. cerevisiae CDC2 gene. Cell 56, 599–605.

Smith, S., & Stillman, B. (1989). Purification and characterization of CAF-1, a human cell factor required for chromatin assembly during DNA replication in vitro. Cell 58, 15–25.

Steller, H., & Pirrotta, V. (1985). Fate of DNA injected into early Drosophila embryos. Dev. Biol. 109, 54–62.

Stillman, B. (1989). Initiation of eukaryotic DNA replication in vitro. Ann. Rev. Cell Biol. 5, 197–245.

Tan, C.-K., Castillo, C., So, A.G., & Downey, K.M. (1986). An auxiliary protein for DNA polymerase-δ from fetal calf thymus. J. Biol. Chem. 261, 12310–12316.

Thömmes, P., Fett, R., Schray, B., Burkhart, R., Barnes, M., Kennedy, C., Brown, N.C., & Knippers, R. (1992). Properties of the nuclear P1 protein, a mammalian homologue of the yeast Mcm3 replication protein. Nucl. Acids Res. 20, 1069–1074.

Tsurimoto, T., Melendy, T., & Stillman, B. (1990). Sequential initiation of lagging and leading strand synthesis by two different polymerase complexes at the SV40 DNA replication origin. Nature 346, 534–539.

Tsurimoto, T., & Stillman, B. (1989). Purification of a cellular replication factor, RF-C, that is required for coordinated synthesis of leading and lagging strands during Simian Virus 40 DNA replication in vitro. Mol. Cell. Biol. 9, 609–619.

Tsurimoto, T., & Stillman, B. (1991a). Replication factors required for SV40 DNA replication in vitro. I. DNA structure-specific recognition of a primer-template junction by eukaryotic DNA polymerases and their accessory proteins. J. Biol. Chem. 266, 1950–1960.

Tsurimoto, T., & Stillman, B. (1991b). Replication factors required for SV40 DNA replication in vitro. II. Switching of DNA polymerase alpha and delta during initiation of leading and lagging strand synthesis. J. Biol. Chem. 266, 1961–1968.

Umek, R.M., & Kowalski, D. (1988). The ease of DNA unwinding as a determinant of initiation at yeast replication origins. Cell 52, 559–567.

Vassilev, L.T., & DePamphilis, M.L. (1992). Guide to identification of origins of DNA replication in eukaryotic cell chromosomes. Crit. Rev. Biochem. Mol. Biol. 27, 445–472.

Vaughn, J.P., Dijkwel, P.A., & Hamlin, J.L. (1990). Replication initiates in a broad zone in the amplified CHO Dihydrofolate Reductase domain. Cell 61, 1075–1087.

Weinberg, D.H., Collins, K.L., Simancek, P., Russo, A., Wold, M.S., Virshup, D.M., & Kelly, T.J. (1990). Reconstitution of simian virus 40 DNA replication with purified proteins. Proc. Natl. Acad. Sci. USA 87, 8692–8696.

Weinberg, D.H., & Kelly, T.J. (1989). Requirement for two DNA polymerases in the replication of simian virus 40 DNA in vitro. Proc. Natl. Acad. Sci. USA 86, 9742–9746.

RECOMMENDED READINGS

Chong, J.P.J., Thömmes, P., & Blow, J.J. (1996). The role of MCM/P1 proteins in the licensing of DNA replication. Trends Biochem. Sci. 21, 102–106.

Fangman, W.L., & Brewer, B.J. (1992). A question of time: Replication origins of eukaryotic chromosomes. Cell 71, 363–366.

Hübscher, U., & Thömmes, P. (1992). DNA polymerase ε: In search of a function. Trends Biochem. Sci. 17, 55–58.

Linskens, M.H.K., & Huberman, J.A. (1990). The two faces of higher eukaryotic DNA replication origins. Cell 62, 845–847.

Norbury, C., & Nurse, P. (1992). Animal cell cycles and their control. Ann. Rev. Biochem. 61, 441–470.

Stillman, B. (1989). Initiation of eukaryotic DNA replication in vitro. Ann. Rev. Cell Biol. 5, 197–245.

Tsurimoto, T., Melendy, T., & Stillman, B. (1990). Sequential initiation of lagging and leading strand synthesis by two different polymerase complexes at the SV40 DNA replication origin. Nature 346, 534–539.

Vassilev, L.T., & DePamphilis, M.L. (1992). Guide to identification of origins of DNA replication in eukaryotic cell chromosomes. Crit. Rev. Biochem. Mol. Biol. 27, 445–472.

Chapter 3

DNA Methylation

ROGER L.P. ADAMS

Principles of Medical Biology, Volume 5
Molecular and Cellular Genetics, pages 33–66.
Copyright © 1996 by JAI Press Inc.
All rights of reproduction in any form reserved.
ISBN: 1-55938-809-9

INTRODUCTION

Distribution of Minor Bases

As well as the four principal bases that contribute to the familiar structure of DNA, there are a number of other bases (minor bases) present in lesser amounts. Of these, the most important are 5-methylcytosine, 4N-methylcytosine and 6N-methyladenine (Figure 1). Although only the first of these is of significance in higher eukaryotes, we can, and have, learned much about the function of these minor methylated bases from a study of prokaryotes where methyladenine, as well as methylcytosine, has an important function. Table 1 gives some idea of the distribution and abundance of minor methylated bases in both prokaryotes and eukaryotes. There appears to be little pattern to these observations. Higher plants contain up to 6% 5-methylcytosine in their DNA, while some insects and lower eukaryotes have none detectable. Several species of bacteria lack 5-methylcytosine, a few contain N4-methylcytosine, but most contain methyladenine which is absent from higher eukaryotes.

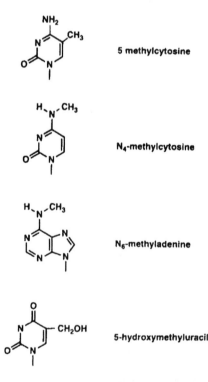

5 methylcytosine

N_4-methylcytosine

N_6-methyladenine

5-hydroxymethyluracil

Figure 1. Minor bases present in DNA.

Table 1. Distribution of Modified Bases

Species	Methylcytosine (% of Cytosine)	Methyladenine (% of Adenine)
E. coli strain C	1.0	2.1
Micrococcus luteus	0	0
Micoplasma arginini	0	2.0
Chlamydomonas	0.7	0.9
yeast	0	0
fruit fly	0	0
locust	0.95	—
salmon	9.1	—
frog	7.2	0
chick	4.5	—
mouse	3.8	—
wheat	31	0

Synthesis of Minor Bases

Thymine is a methylated base with a structure very similar to 5-methylcytosine but thymine is a major base making up 20–30% of the bases present in most DNA. As the deoxyribonucleoside triphosphate, dTTP, it is one of the substrates (along with dATP, dGTP, and dCTP) of DNA polymerase and is incorporated into newly synthesized DNA during replication. In certain organisms one of the normal, major bases is completely replaced by an unusual base. For instance, when bacteriophage T4 infects the bacterium *Escherichia coli,* it subverts the host cell's metabolism in such a way that newly synthesized DNA contains hydroxymethylcytosine instead of cytosine. This unusual base is incorporated as the triphosphate, hmdCTP, in place of dCTP during DNA replication. Such wholesale replacements will not concern us further in this chapter.

The minor methylated bases are not introduced into DNA along with the four major bases during DNA replication but, rather, they are added afterwards following the intervention of DNA methyltransferases (or methylases). Methyltransferases, as their name implies, transfer a methyl group from the methyl donor, S-adenosyl methionine (AdoMet), onto certain adenines or cytosines in polymeric DNA. The adenines and cytosines that form the target for these enzymes usually occur in certain specific, short nucleotide sequences in the DNA and methyltransferases from different species show different target site specificities. Generally the target sites are short, palindromic sequences; that is sequences that are identical on both strands when read in the 5' to 3' direction. Humans have a single DNA methyltransferase that transfers methyl groups to the 5' position on cytosines in the sequence 5'-CG-3'; whereas bacteria often have several DNA methyltransferases, the most ubiquitous being the Dam methylase that transfers methyl groups to the 6-amino position on adenines in the sequence 5'-GATC-3'. Higher plants contain

Self complementary methylase target sequences

$$
\begin{array}{cc}
\text{m} & \text{m} \\
\text{5'-CNG-3'} & \text{5'-GAATTC-3'} \\
\text{3'-GNC-5'} & \text{3'-CTTAAG-5'} \\
\text{m} & \text{m}
\end{array}
$$

Plant methylase   M.*Eco*R I

$$
\begin{array}{cc}
\text{m} & \text{m} \\
\text{5'-CCGG-3'} & \text{5'-CCGG-3'} \\
\text{3'-GGCC-5'} & \text{3'-GGCC-5'} \\
\text{m} & \text{m}
\end{array}
$$

M.*Hpa*II   M.*Msp*I

Figure 2. Target sites for DNA methyltransferases.

two activities: one that methylates the cytosine in CG dinucleotide target sequences and a second that acts on CNG target sequences (where N represents any base). The abundance of the CNG target sequence in plant DNA may explain the presence of the high levels of 5-methylcytosine in higher plants. Because of the self complementary nature of the target sites, methylated bases occur in pairs along the DNA (Figure 2).

Methyltransferases

The genes for DNA 5-cytosine methyltransferases have been cloned and sequenced from many organisms and the enzymes show several regions of sequence conservation (Figure 3). This demonstrates that these enzymes have probably

Region I $I^{D}/_{S}LF^{S}/_{A}GIG^{A}/_{G}F/_{I}..^{A}/_{G}....\mathbf{G}$

Region IV $DLL.^{G}/_{A}\mathbf{G}^{F}/_{P}\mathbf{PC}^{Q}/_{P}.\mathbf{FS}..G...GE$

Region VI $P.....\mathbf{ENVK}^{G}/_{N}L.....G$

Region VIII $^{N}/_{I}{}^{S}/_{D}..^{F}/_{Y}{}^{N}/_{G}{}^{V}/_{I}{}^{P}/_{A}\mathbf{Q.R}^{E}/_{K}\mathbf{R}...^{I}/_{V}{}^{G}/_{A}$

Region X $S...\mathbf{Y}^{K}/_{R}Q.\mathbf{GN}^{S}/_{A}{}^{I}/_{V}.V.V...^{F}/_{A}$

Figure 3. Conserved sequences found in cytosine-C5 DNA methyltransferases. The amino acid sequence is given for five of the 10 conserved regions. Those amino acids in bold are completely conserved among all the enzymes that have been sequenced while those in normal type are largely conserved. The data are taken from Klimasauskas et al. (1991).

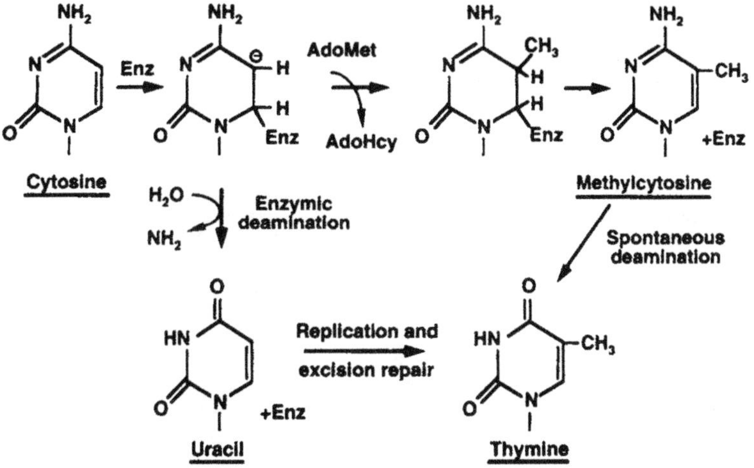

Figure 4. The reaction mechanism originally proposed by Santi for cytosine-C5 methylases is shown across the top of the figure (Chen et al., 1991).

evolved from a single methyltransferase present in some primordial organism. That this enzyme has been conserved highlights the importance of DNA methylation to the efficient survival of organisms at all stages of evolution. In addition to the conserved regions, enzymes from different species have unique specificity regions that determine the target site specificity.

At the active site, all these DNA 5-cytosine methylases have the amino acid proline followed by cysteine and this cysteine interacts covalently with carbon atom 6 of the target cytosine on the DNA. This activates carbon atom 5 so that it can accept a methyl group from AdoMet. On methyl group transfer the AdoMet is converted to S-adenosyl homocysteine (AdoHcy) and the covalent link between the enzyme and the DNA is broken (Figure 4, upper pathway). When these enzymes interact with DNA, they cause the cytosine in the target site to flip out of the double helix into a pocket in the enzyme where the methyltransfer takes place (Klimasauskas et al., 1994).

Methyltransferases act on DNA shortly after replication but, as we shall see, there is an important time window between the synthesis of DNA and its methylation. Within this time window, the parental strands of DNA carry the preexisting, methylated bases but the newly-synthesized daughter strands are, as yet, unmethylated. Such DNA is said to be *hemimethylated* as it carries methyl groups on one strand only. In prokaryotes, all target sites are efficiently methylated and any DNA which becomes introduced into the cell from outside (e.g., in the form of a virus DNA molecule) can become methylated *de novo*. Thus the DNA methyltransferase can act on DNA that contains no methylated bases in the target sites or on newly synthesized DNA that is hemimethylated.

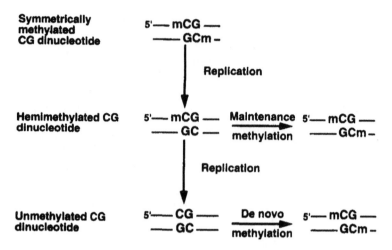

Figure 5. Maintenance and *de novo* methylation. Only if two rounds of replication occur in the absence of methylation will the DNA lose its methylation.

In eukaryotes not all target sites are methylated and the pattern of methylation is different in different tissues. This means that a particular target site may be methylated in the DNA of kidney cells but not methylated in the DNA of brain cells. The consequence of this less-than-complete methylation is that, following replication, the newly synthesized DNA will contain some unmethylated target sites and some hemimethylated target sites. The eukaryotic DNA methyltransferases show a greater ability to transfer methyl groups to hemimethylated sequences than to unmethylated target sequences and, because of this preference to act on hemimethylated DNA, the original pattern of methylated and unmethylated targets is maintained from cell generation to generation (Figure 5). This conservation of information in the form of the pattern of methylation is known as epigenetic inheritance since no permanent change in the genetic material is involved.

Methylation and DNA/Protein Interactions

The presence of methylated bases in DNA does not alter the coding potential of that DNA. Thus methylcytosine in the template strand of a DNA molecule will still lead to the incorporation of a guanine on transcription, and the presence of a methyladenine will still lead to the incorporation of a uracil into the RNA. The significance of the methylation lies in its potential to alter DNA/protein interactions. The 5 position of cytosine is exposed in the major groove of the DNA double helix adjacent to the 7 position of guanine that is the most sensitive site for modification by exogenous alkylating agents. In the situation where all target sites in the genome are fully methylated, there is no increase in the sequence information in that DNA. However, where only some of the target sites are methylated, DNA binding proteins may be able to discriminate between methylated and unmethylated sequences. This is particularly pertinent in eukaryotes where the sites methylated differ in the DNA from different tissues and at different times in development, but it also allows exogenous, unmethylated DNA (e.g., viral DNA) to be distinguished from the host cell's genomic DNA when this is methylated.

As the pattern of methylation differs between the DNA of different tissues, this pattern must be established during the development of the organism and we shall discuss later how this might be brought about. Once established, one can ask the question, "can the pattern be changed, either in a programmed manner or accidentally?" *Addition* of methyl groups to previously unmethylated regions (*de novo* methylation) is catalyzed inefficiently by the mammalian DNA methyltransferase but is a mechanism of inactivating infecting viral DNA (see below). It is also an important step in early embryonic development. *Removal* of methyl groups from 5-methylcytosine in DNA is unlikely to occur by a reversal of the synthetic mechanism as the nature of the C–C bond involved makes it resistant to cleavage. However, the methylated base could be excised from DNA and replaced with cytosine by a DNA repair mechanism. More commonly, *passive* demethylation can occur if two or more rounds of replication occur in the absence of DNA methylation (Figure 5).

Methylation and DNA Structure

Altering the nucleotide sequence of DNA by the introduction of methyl groups can lead to an alteration in the shape and structure of the DNA molecule (Zacharias, 1993). Although everyone is familiar with the normal B-form of DNA whose structure was elucidated by Watson and Crick, this basic structure varies depending on the base sequence. The bases can move with respect to one another by rolling, twisting, and sliding in any one of three dimensions and they do so to achieve a structure of maximum stability that depends on the base sequence. These stable structures may involve the DNA molecule taking up a curved conformation and the extent of distortion possible is dependent on the presence of methylcytosine or methyladenine. Furthermore, some regions of DNA may be able to take on a cruciform or even a triple stranded structure. Cruciform extrusion is dependent on the stability of the DNA duplex at the center of a palindromic sequence and this stability is also dependent on the presence of methylated bases. Thus methylcytosine increases duplex stability whereas methyladenine reduces stability with the consequence that the former reduces the chances of cruciform extrusion while the latter encourages cruciform extrusion. Methylcytosine-containing triple helices are more stable than the corresponding cytosine-containing triple helices.

Clearly, by altering the base sequence, methylation of DNA has the potential to alter the structure of the DNA molecule. One of the most extreme alterations that is known to occur entails the complete breakdown of the right-handed double helical structure typified by B-form DNA and the formation of a left-handed double helical structure known as Z-DNA. Z-DNA formation is favored by regions of DNA containing a run of alternating purine and pyrimidine nucleotides, particularly alternating mCG dinucleotides. It is possible that the methylation of a run of alternating CG nucleotides could be sufficient to swing the equilibrium in favor of the Z-form of DNA and such a major conformational change would have a dramatic effect on the ability of the DNA molecule to interact with proteins. Evidence for the formation of Z-DNA *in vivo* is not very strong but even minor changes in DNA conformation can affect DNA/protein interactions.

DNA RESTRICTION/MODIFICATION

Prokaryotic Systems

One of the major functions that methylation of DNA is known to perform is in distinguishing DNA that is required for expression from DNA that can be degraded. The clearest example of this is in the restriction of the host range of particular bacterial viruses (phage) and plasmids. When a phage infects a bacterial cell that is not its normal host, the phage DNA is degraded by a (restriction) endonuclease which thereby prevents the multiplication of the phage. The endonuclease is one of a family of enzymes that recognizes a short palindromic sequence in the DNA (e.g., CCGG or GAATTC) and cleaves this target sequence if it is not methylated. Host cell DNA is resistant to cleavage because it has been modified at all target

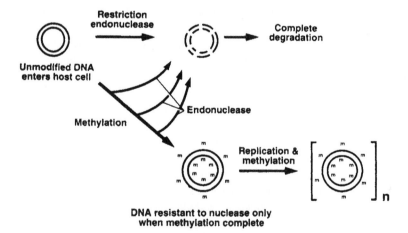

Figure 6. Restriction modification. On entering a host cell, unmodified DNA is subjected both to restriction endonucleases and to methylases. Only if the latter modify all the target sites on the incoming DNA will it become resistant to nuclease action. Otherwise, it will be degraded thereby restricting the phage growth.

Table 2. Classes of Methyltransferases

Class	Base Modified	Target	Maintenance/De Novo
Bacterial, type I (two subunits—large)	A	asymmetric	maintenance
Bacterial, type II (one subunit—small)	C or A	symmetric	*de novo*
Bacterial, type IIS (one or two subunits)	C or A	asymmetric	*de novo*
Bacterial, type III (one subunit—large)	A	asymmetric	one strand only
Mammalian (one subunit—large)	C	symmetric	maintenance
Plant	C	symmetric	maintenance

sequences by a cognate DNA methyltransferase. Restriction endonucleases have become the stock-in-trade of molecular biologists as they enable DNA to be cleaved at precise sequences prior to rejoining to form recombinant DNA molecules.

This growth restriction is very effective and fewer than one phage DNA molecule out of 10^5 escapes nuclease action. However, when foreign DNA enters a bacterial cell there is a race between the methylase that tries to modify all target sequences in the viral DNA and the endonuclease that needs cut only one target sequence to inactivate the virus. Occasionally a phage DNA molecule will become modified at all target sequences before the endonuclease has had time to act (Figure 6). When this happens, the phage DNA will proceed to replicate and produce new virus particles each of which will contain modified DNA. On subsequent infection of another, similar bacterial cell, this modified DNA will be resistant to nuclease action and the phage will reproduce rapidly. A different strain of bacteria will, however, contain a different restriction/modification system that will prevent infecting phage from surviving.

There are very many different endonucleases and cognate methylases and each bacterial strain has its characteristic complement of enzymes. In this way the host range of phage and plasmids is severely restricted. Restriction/modification enzymes fall into a number of different classes (Table 2) of which the type II enzymes are the simplest (Wilson, 1988). They consist of a small endonuclease and the cognate methylase, the genes for which are normally carried on a plasmid within the host cell. The methylase acts independently on each strand of the DNA substrate and so there is a significant chance of a *de novo* methylation event occurring. The Dam methylase resembles the type II methylases but there is no cognate endonuclease. The Dam methylase has other functions in that it is involved in DNA replication and mismatch repair.

The genes for the type I enzymes are present on the bacterial chromosome. Type I enzymes are complex molecules containing three subunits; a methylase, an endonuclease and a specificity subunit and these enzymes normally carry out only maintenance methylation. Thus methylation normally occurs immediately following replication and so the chances of a foreign DNA molecule becoming methylated *de novo* is much reduced. The target sites for type I enzymes are not palindromic but they do contain a site for methylase on each strand. The endonuclease present does not act at its recognition sequence but, rather, it cuts the DNA at some distant sequence. As well as having systems in which endonucleases cleave unmodified DNA, many bacteria have a second type of system (the *mcr* and *mrr* systems) that leads to cleavage of DNA having a particular distribution of methyl groups different from the host DNA. This leads to the degradation of most DNA that might enter the host bacterium as a result of infection, transformation, or conjugation unless that DNA has come from a bacterium of common restriction background (Raleigh, 1992).

Restriction/modification is an example of a primitive immunity system that is able to distinguish between self and non-self and to destroy the latter. To overcome this immunity was one of the problems encountered by those intending recombinant DNA technology as, for example, human DNA is modified at sequences that rendered it susceptible to degradation by the *mcr* A and B systems. Most cloning of human sequences is now done in strains lacking these restriction systems.

Many drug resistance genes are carried on plasmids and it is the promiscuity of plasmids that has led to the rapid spread of drug resistance throughout the strains of bacteria that commonly infect man. This promiscuity is limited to some small extent by the restriction phenomena just discussed.

Defense Systems in Eukaryotes

It has been proposed that DNA methylation also provides a defence mechanism against the expression of foreign DNA in cells of higher eukaryotes (Doerfler, 1992) although the mechanism proposed is almost the opposite to that applying in prokaryotes. In this case the foreign DNA would be targeted by the DNA methylase and once methylated the foreign DNA would be subjected to gene silencing rather than to degradation as is typical in the bacterial restriction systems. Even before much was understood about the bacterial restriction/modification systems, it had been noticed that some repeated sequences of DNA destined for degradation (or inactivation) became modified in eukaryotes (Sagar and Kitchin, 1975). The systems considered most significant at this time involved lower eukaryotes such as *Chlamydomonas* and some insects, but it was realized that a similar situation may apply in some mammalian systems. Thus, X-chromosome inactivation may be a mammalian example in which one copy of a duplicated chromosome is maintained in an inactive state as a result of DNA methylation.

One of the earliest observations on DNA methylation concerned the extent of methylation of ribosomal DNA (rDNA), i.e., the DNA that codes for ribosomal RNA. In the oocytes of *Xenopus laevis*, in addition to the normal chromosomal copies there are multiple copies of rDNA that are not associated with the chromosomes. These extra copies arise by a gene amplification mechanism and, in contrast to the chromosomal copies, they are not methylated. These extrachromosomal copies are transcribed to give the high levels of rRNA present in the *Xenopus* oocytes. Later this DNA disappears from the developing embryo, probably as a result of dilution. This failure to methylate extrachromosomal DNA has intrigued investigators for many years and results in the synthesis of unmethylated viral DNA following infection of mammalian cells with many viruses (e.g., SV40, HSV). It is only when viral DNA becomes integrated into the host cell chromosome that it becomes subject to a slow process of *de novo* methylation (see above).

More recently, Selker (1990) has elaborated this viewpoint by pointing out that methylation is generally associated with redundant DNA that is genetically inactive. This applies to repeated sequences in small genomes (e.g., that of *N. crassa*) as well as to satellite sequences in mammalian chromosomes. In *Neurospora*, duplicated sections of the haploid genome can arise at fertilization and these duplicated regions are subject to introduction of multiple point mutations which may arise largely as a result of methylation of cytosine followed by its deamination to thymine. This repeat induced point (RIP) mutation is clearly associated with the *de novo* methylation of most of the cytosines in the duplicated region which is transcriptionally inactive. Radman (Kricker et al., 1992) has broadened this argument by proposing that the association of duplicated regions leads to their extensive methylation and deamination. The consequent variation in nucleotide sequence is sufficient to reduce homologous recombination that might otherwise lead to gross chromosomal rearrangements that could be very deleterious to the cell. (Figure 7). A similar phenomenon may be the cosuppression of transgenes and their endogenous counterparts seen in higher plants and in *Neurospora*.

Of course, deamination of methylcytosine can occur not only at repeated regions but at any region that is methylated. It is probably the major cause of mutation and 30–40% of all human germline point mutations are thought to be methylation induced (Jones et al., 1992). Very efficient mechanisms exist to correct TG mismatches. One of these mechanisms (the major one) specifically excises the thymine leading to its replacement by cytosine thereby regenerating the correct CG base pair. However, less frequently the mismatch is corrected at random and this leads to the production of TA base pairs. It is believed that this random correction is the major source of the mutations arising as a consequence of methylcytosine deamination. An alternative source of mCG to TA mutations arises when the deamination product, TG, is replicated before repair can be effected (Figure 19).

A more direct mechanism of deamination has recently been proposed (Shen et al., 1992). It has been shown that a DNA methyltransferase can, in the presence of limiting AdoMet, enhance the rate of cytosine deamination thus leading to elevated levels of C to T mutations at target sequences (Figure 4, lower pathway). The *in vivo* significance of this reaction is not known, nor is it known whether such

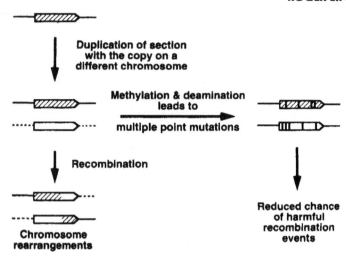

Figure 7. Methylation and deamination is often a consequence of gene duplication. These multiple point mutations will reduce the chance of recombination between the two copies of the gene with its potentially hazardous consequences.

enzyme-catalyzed deamination can occur following methyltransfer to convert a mCG into a TG base pair.

GENE EXPRESSION

Although the examples used in this section will be exclusively from the eukaryotes, there are situations in prokaryotes in which DNA methylation affects gene expression. However, as Holliday (1993) has lucidly pointed out, complex organisms require a second type of inheritance to be superimposed on the classic genetical system where the sequence of bases in DNA is clonally inherited at cell division. In higher eukaryotic organisms, the division of the zygote leads, during development, to clones of cells committed to particular patterns of gene expression. Once established, this differentiation is extremely stable, yet it is not the result of a genetic mutation. Rather, it is the result of *epigenetic* inheritance that can, at least in some instances, be affected by agents that interfere with DNA methylation.

The evidence that DNA methylation plays a role in gene inactivation falls into two categories. First, when the extent of methylation of genes in different tissues is investigated, an inverse correlation between expression and methylation is generally found. Thus, in a tissue in which it is expressed, a gene is undermethylated relative to the same gene in a tissue where it is not expressed. Although this is only a correlation, the evidence is supported by the second category of evidence that shows that when a gene is methylated *in vitro* and then introduced into cells it is less active than the control, unmethylated gene. These statements are, however, a gross oversimplification of the evidence which has been limited by the techniques available.

In Vivo Correlations

To measure the extent of *in vivo* methylation of a gene, the DNA isolated from various tissues is digested with restriction enzymes sensitive to methylation of CpG dinucleotides in their recognition sequences (Figure 8). The availability of such enzymes is limited but *Hpa*II has been widely used. *Hpa*II cuts the sequence CCGG but only when the inner cytosine is unmethylated. The isoschizomer, *Msp*I cuts the same sequence whether or not the inner cytosine is methylated and so provides an excellent control. The restricted DNA is fractionated by agarose gel electrophoresis

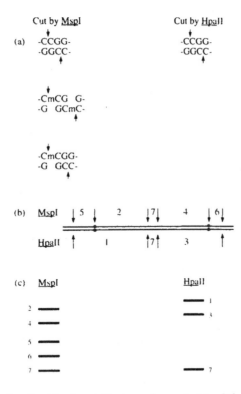

Figure 8. Target sites for *Hpa*II and *Msp*I are shown in (a) while (b) shows a typical gene with five CCGG sequences of which three are methylated. Part (c) shows the results of gel electrophoresis of the DNA after digestion with one of the enzymes where it is clear that methylated DNA gives rise to larger bands when digested with *Hpa*II. (The actual results would be obtained by treating a Southern blot with a radiolabelled probe corresponding to the gene.)

**Predicted frequency of -CG-dinucleotides in
vertebrate DNA** (G+C content = 40%)

$$= 0.2 \times 0.2 \qquad = 0.04 \qquad (1 \text{ in } 25)$$

Observed frequency $= 0.008$ $(1 \text{ in } 125)$

**Frequency of occurrence of
-CCGG- tetranucleotide**

$$= 0.008 \times 0.2 \times 0.2 = 0.00032 \qquad (1 \text{ in } 3125)$$

Figure 9.   Calculation of the frequency of occurrence of CG and CCGG sequences.

and transferred to a membrane where the gene under investigation can be high-lighted by hybridization to an appropriate, radioactive probe. This process is known as Southern blotting. If the restriction map of the gene is known, then the position of the bands generated by *Msp*I digestion can be predicted. Larger bands are produced by *Hpa*II if the gene is methylated. Comparison of the bands generated by digestion with *Hpa*II and *Msp*I will indicate the extent and location of methyl-lation of these target sequences within the gene.

There are several problems with this approach, the most serious being that the frequency of occurrence of *Hpa*II sites is very low. In vertebrate DNA the chance

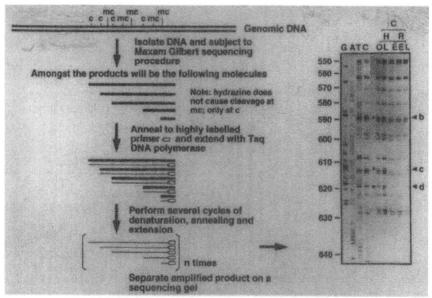

Figure 10.   The mechanism of genomic sequencing (Saluz and Jost, 1989). I would like to thank Hans Peter Saluz for the photograph of the sequencing gel which is part of the evidence used in the construction of Figure 12. This approach has now been superseded by the use of bisulphite genomic sequencing (Feil et al., 1994).

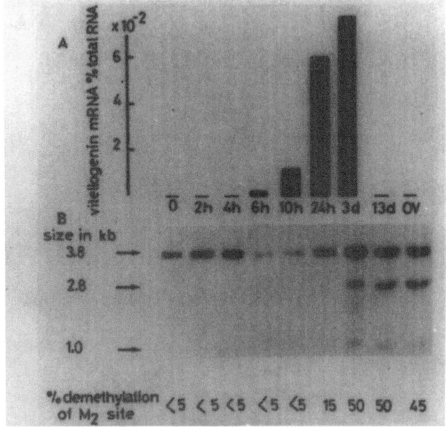

Figure 11. Time course of vitellogenin mRNA induction (A) and demethylation of the M2 site (-614) measured using sensitivity to *Hpa*II (From Wilks et al., 1984 with the permission of the authors and publishers). ·

of a *Hpa*II site occurring is once every 3,125 bases (Figure 9) and, moreover, *Hpa*II sites represent only 4% of the CpG dinucleotides in the genome. As some CpGs may be more important than others, a consideration of the methylation status of a chance *Hpa*II site may be irrelevant. Of course, several other enzymes that are sensitive to CpG methylation can be tried but the results reported are seldom absolutely clear. This difficulty can be overcome by genomic sequencing (Figure 10) but the technique is very difficult and has been applied in only a few situations. Nevertheless, the results that have been reported are intriguing.

The group of Jean-Pierre Jost in Basel have been studying the methylation of the avian vitellogenin gene as it is activated in the liver following injection of estrogen into immature roosters. In 1984, they reported that demethylation of an

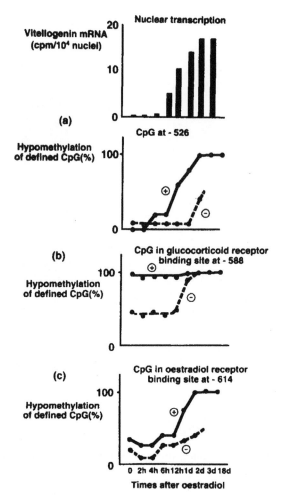

Figure 12. The same as Figure 11 but in this case the demethylation of three sites is demonstrated by genomic sequencing. The data is taken from Saluz et al. (1986).

*Hpa*II site near the estrogen receptor binding site occurred following enhanced transcription rather than simultaneously with enhanced transcription (Figure 11) (Wilks et al., 1984). However, in 1986, using genomic sequencing to look at this site and two others in the same region, they found the situation to be more complex (Figure 12) (Saluz et al., 1986). The proximal site (–526) and the estrogen receptor binding site (– 614) become demethylated concurrently with transcriptional activation but only on one strand. The second strand becomes demethylated only 36 hours later. As *Hpa*II will not cut either fully methylated or hemimethylated DNA its use indicates only when both strands have become demethylated, thus explaining

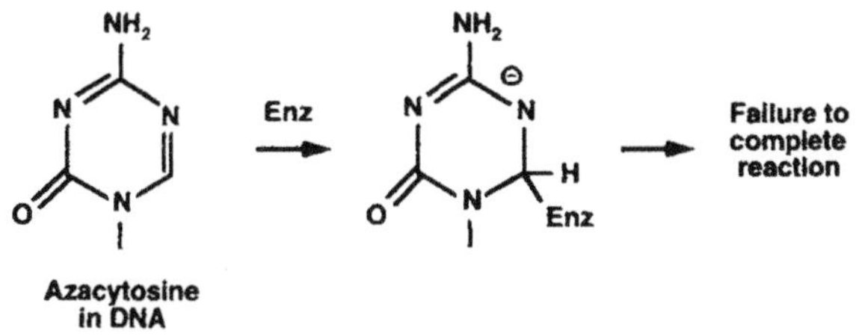

Figure 13. When the methyltransferase encounters an azacytosine in its target site in DNA, it becomes covalently bound to the DNA. The result is that the amount of free, active enzyme in the cell falls leading to a reduction in the subsequent methylation of the DNA.

the earlier result. Clearly, we have here a situation where hemimethylated DNA exists in the cell for more than the instant following replication. This is even more apparent at a site for binding the glucocorticoid receptor (−588) that is hemimethylated initially and only becomes fully demethylated 24 hours after estrogen injection.

The *in vivo* correlations described above are just that: correlations. The experiments provide no evidence as to whether DNA methylation prevents gene expression or whether nonexpressed genes become methylated. Some progress towards resolving this dilemma has come from the use of the nucleoside analogues azacytidine and azadeoxycytidine. These analogues are incorporated into nucleic acids and the latter is more specific in that it is incorporated only into DNA, not RNA. Incorporation of azacytosine into RNA causes the inhibition of protein synthesis and the deamination of azacytosine to azauridine leads to inhibition of nucleotide metabolism. When DNA methyltransferase attempts to methylate an azacytosine in a CpG dinucleotide in DNA it forms the covalent intermediate complex described in Figure 4 but is unable to complete the methyltransfer (Figure 13). This leads to a depletion of active enzyme and the consequent inability of the cell to methylate further, newly-synthesized DNA. In many cases this has been shown to be associated with major changes in gene expression. For example, when 10T1/2 cells (derived from mouse embryos) are treated with azacytidine considerable differentiation occurs, some cells forming adipocytes and others myotubules (Figure 14).

Transfection Studies

More direct results have been obtained from a comparison of the expression of methylated and control genes introduced into cells by transfection or injection. Expression can be measured shortly (24–48 hours) after transfection or transformed clones can be isolated in which the introduced gene has become integrated into the host chromosome and has replicated many times. In some cases the genes to be

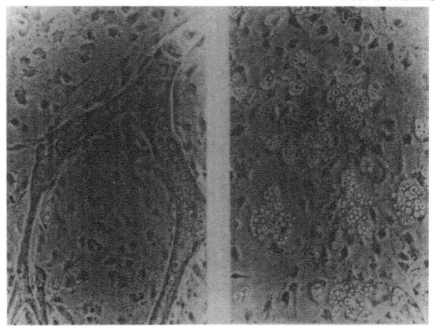

Figure 14. Differentiation of cultured mouse embryo cells on exposure to azacytidine for two weeks (myotubules) or four weeks (adipocytes). Reproduced from Taylor and Jones (1979) with permission of the authors and publishers.

studied can be obtained in methylated and unmethylated form by isolation from different tissues. For instance the *hprt* gene can be isolated from the inactive X-chromosome (when it is highly methylated) or the active X-chromosome (when it is largely unmethylated; see below). When introduced into cells lacking an active *hprt* gene only the gene from the active X-chromosome is expressed and capable of rescuing the cells from death when grown in HAT medium.

A different way in which methylated and control DNA can be prepared is by using methyl dCTP in a primer extension reaction with DNA polymerase. This leads to replacement of all cytosines by methylcytosine which will introduce artefacts though the extra methylcytosines are lost when such DNA replicates in the eukaryotic cells. This method has the advantage that specific regions of the construct can be methylated and so regional methylation can be studied (Figure 15). In general, it appears that methylation of the promoter regions of genes is more effective in inhibiting expression than is methylation of their 3' regions.

More often the methylated gene is obtained by the action of a bacterial methylase on a plasmid carrying the gene of interest—or the promoter of interest linked to a reporter gene such as that for chloramphenicol acetyl transferase (CAT). Many such experiments suffer from the same constraint as do the correlation experiments in

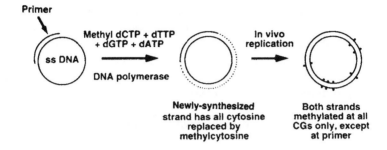

Figure 15. Synthesis of patch methylated DNA using methyl dCTP.

that, for example, M.*Hpa*II (the methylase cognate to the restriction enzyme *Hpa*II) will act on only a very limited fraction of the possible CpGs in the construct. Nevertheless, methylation is associated with inactivation in many instances.

The problem caused by the limited occurrence of a particular target sequence can be overcome by using a eukaryotic DNA methylase, but these enzymes have relatively low *de novo* activity and conditions to obtain 100% methylation have not been reliably determined. Fortunately, a prokaryotic methylase, M.*Sss*I, has recently been isolated that will act on the target site CpG and this has allowed more meaningful studies to be done (Bryans et al., 1992). In particular, we have been able to achieve regional methylation using M.*Sss*I which will act selectively on the duplex region present in a patched duplex molecule (Figure 16). Surprisingly, in a transient expression assay, it is clear that methylation of any region of the construct has an inhibitory effect and the effect is greater the greater the length of the methylated patch (Kass et al., 1993).

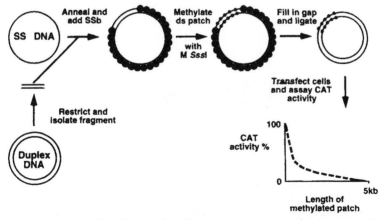

Figure 16. Synthesis of patch methylated DNA using M.*Sss*I and a demonstration that the inhibition of gene expression is dependent on the length of the methylated patch.

Perhaps to be expected is the finding that the effect of methylation is dependent on the nature of the recipient cells. For example, when a methylated actin gene is introduced into fibroblasts it is inactive, but the same methylated gene when introduced into myoblasts expresses actin as well as does the unmethylated control (Yisraeli et al., 1986). There must be factors present in the myoblasts (where the gene is normally expressed) that can overcome the normal inhibitory effect of methylation.

Possible Mechanisms of Gene Inactivation

A number of ways in which methylation of DNA might interfere with gene expression have been considered. The most obvious mechanism is when the presence of a methyl group at the recognition site of a positive transcription factor interferes with factor binding. Binding of transcription factors can be studied by gel retardation analysis and by footprinting, either *in vivo* or *in vitro*, and this

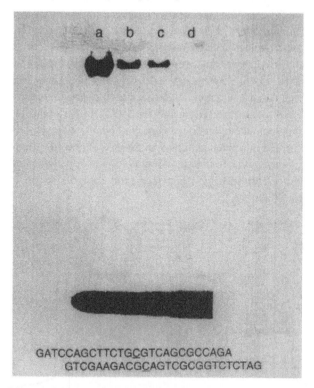

Figure 17. Gel retardation analysis of the binding of CREB to its target sequence. (a) Shows binding to the unmethylated target; (b) and (c) show binding to the hemimethylated target where either the upper strand (b) or the lower strand (c) is methylated. In (d) both strands were methylated. From Weih et al. (1991) with permission of the authors and publishers.

mechanism of action has been shown to be effective in some cases though it is seldom the only mechanism operative. Thus, although binding of the transcription factor CREB to its responsive element is prevented by methylation (Figure 17), it is also prevented by other factors that act even in the absence of CpG methylation (Weih et al., 1991). In addition, there are major transcription factors (e.g., Sp1) that will bind to their target site independently of whether or not it is methylated.

Different transcription factors may compete for the same or overlapping target sequences and the outcome may be quite different depending on which succeeds in binding. For example, at +10 in the chick vitellogenin gene there occurs a methylcytosine in the element ATTCACCTTCGCTATGAGGGGG which in the rooster is occupied by a protein, MDBP-2. On treatment of the rooster with estrogen, MDBP-2 declines in activity and is replaced by a different protein, NHP-1, that initiates the removal of the methylcytosine and its replacement with cytosine (Jost et al., 1991). This site-specfic demethylation occurs by removal of 5methyl dCMP and its replacement with dCMP in an excision repair reaction (Jost, 1993).

Methylated DNA may also be recognized by proteins that bring about deamination of the methylcytosine to thymine and this is the basis of the "ripping" reaction described earlier.

A more general mechanism of gene inactivation involves an alteration of chromatin structure to a form in which gene expression is prevented. Such changes may be initiated by the action of a protein (e.g., topoisomerase, histone H1 subtype, methylcytosine-binding protein) that acts preferentially on methylated DNA to seed the formation of the inactive chromatin.

Chromatin Structure

As well as the correlation between gene transcription and undermethylation, there is a correlation between transcription and sensitivity of isolated chromatin to nucleases. To be strictly accurate, the correlation is not with transcription but rather with the potential for transcription. Thus genes that are programmed to be active, or have recently been active, as well as those that are in the act of transcription are present in what is known as *active* chromatin. The characteristics of active chromatin are that it is more sensitive than is *inactive* chromatin to digestion by nucleases such as DNase I, micrococcal nuclease, or some restriction endonucleases. This increased sensitivity is believed to be a result of its being present in a more open conformation perhaps as a result of its containing less histone H1 and more acetylated histones H3 and H4 together with ubiquitinated histone H2B and a higher amount of the nonhistone proteins HMG 14 and HMG 17. The question arises, do these changes in chromatin structure result from a fundamental programming of the genome and, in turn, give rise to the undermethylation of the constituent DNA or is the reverse true; i.e., the programming is mediated by an undermethylation of DNA that brings about the formation of active chromatin. A number of experiments could be quoted to support either point of view and it is probable that both scenarios are correct in different situations.

If we consider the transfection studies referred to above, it is clear that methylated DNA is only inactive in transcription following its incorporation into inactive chromatin. At early times following transfection, methylated and unmethylated DNA are transcribed equally well (Figure 18). Later, unmethylated DNA is complexed into inactive chromatin whereas unmethylated DNA forms active chromatin. Moreover, where a plasmid has a methylated patch, the region of inactive chromatin is able to spread from this region all the way around the plasmid (Figure 16). Thus chromatin consists of domains of active chromatin interspersed with domains of inactive chromatin with each domain possibly covering several contiguous genes. Other experiments, such as those cited below under X-chromosome inactivation have, however, been interpreted as showing that methylation follows the formation of inactive chromatin rather than causing it.

Another characteristic of inactive chromatin is that its replication occurs late in S-phase while active genes are replicated early in S-phase when the levels of DNA methyltransferase might be limiting. In some situations an inactive, late-replicating

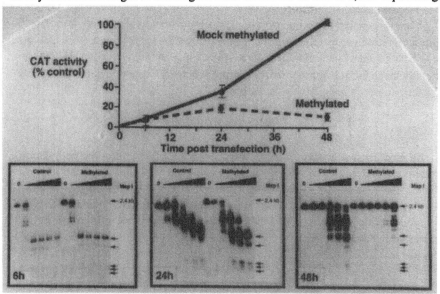

Figure 18. Time course of chromatin formation. On transfection of a CAT plasmid into mouse cells inactive chromatin is formed more quickly on methylated DNA thereby inhibiting gene expression. The graph shows the expression of CAT activity and the pictures show the sensitivity of the plasmid to digestion with *MspI* that is a measure of the amount of histone H1 associated with the minichromosome (from Bryans et al., 1992 and Kass et al., 1993).

gene can be activated and under these conditions it is now replicated early in S-phase. One way in which this reactivation can be brought about is by treatment with the demethylating agent azacytidine, showing that the presence of methylcytosines is essential for the maintenance of the inactive state of the gene.

Establishment of Patterns of DNA Methylation

Early Development and Genetic Imprinting

During embryonic development, a series of programmed events leads to the activation of particular genes and the deactivation of others at specific times in specific cell lineages ultimately leading to the patterns of expression typical of the tissues of the adult organism (Table 3). That the final transcriptional competence of a specific cell is related to its methylation status has been documented above. What is not clear is how the patterns of methylation are established during development. One must assume that at the same time that methylation is directing which proteins will be made, these very proteins are directing which cytosines will be methylated *de novo* and which demethylated during the process of differentiation. Only once the pattern has been established can the epigenetic stability be conserved via efficient maintenance methylation.

Genes in the gametes have a particular pattern of methylation that has been imposed during gametogenesis and this pattern may differ depending on whether the gene is present in the sperm or the oocyte. This means that two alleles in a given zygote may contain different patterns of methylation. Sometimes this leads to the inactivation of one allele—a situation referred to as allelic exclusion or functional hemizygosity. When this happens a single mutation in the active allele can lead to complete loss of function in the cell despite the continued presence of the remaining, normal (though inactive) allele. The existence of allelic exclusion makes it essential that both maternal and paternal gametes fuse to allow normal development as it is only in the normal zygote that a complete set of accessible information is available. The imprinting of the maternal and paternal genomes that occurs during gameto-

Table 3. Methylation and Development

Gametes	Highly methylated—except at CpG islands. Oocyte and sperm patterns different.
Zygote	Functional only with maternal and paternal alleles (functionally hemizygous if one allele inactive)
Blastula	Steady reduction of methylation, reaches minimum at blastula. Both X-chromosomes have unmethylated islands.
Implantation	Considerable *de novo* methylation. DNA methylase mutants defective. *One* X-chromosome inactivated at random.
Somatic cells	Patterns of methylation established by *de novo* methylation and demethylation. Thereafter, maintenance methylation operative.
Germ line	Demethylation is a very early event followed by gamete-specific methylation.

genesis may or may not be altered during subsequent embryonic development but once cell lineages are established the final pattern is maintained in the somatic cells of the offspring. The imprinting on a gene may change as that gene passes alternately through the male and female germ line (Surani, 1994).

Most of the DNA of the gametes is heavily methylated relative to somatic cell DNA but this average picture masks the situation in which some regions are highly methylated and others undermethylated. In sperm, while the satellite DNA sequences are unmethylated, the other repetitive sequences are fully methylated as are all sequences that are expressed in a tissue specific pattern. In contrast, the housekeeping genes have an unmethylated 5' CpG island (Razin and Cedar, 1993). DNA in the mature oocyte is somewhat less methylated than sperm DNA particularly in some oocyte and tissue specific genes even though the latter are not expressed in the oocyte. Considerable demethylation occurs during the first few divisions in the embryo despite the fact that the fertilized egg and early embryo (Monk et al., 1987, 1991) contain very high levels of DNA methyltransferase (this situation is similar to that in many tumors which have high levels of enzyme activity but low levels of methylcytosine). When genes containing methylated CpG islands are injected into oocytes, they rapidly lose their island methylation (Frank et al., 1991). DNA methyltransferase activity is considerably reduced in the blastula when the amount of methylcytosine in the DNA reaches a minimum. At the time of implantation, the normal embryo is able to carry out an efficient *de novo* methylation of most of the genome with the exception of the 5' CpG islands. Embryos lacking the DNA methyltransferase gene fail to develop correctly at about this time (Li et al., 1992). How the islands are protected is not known and the protection does not extend to the islands on the inactive X-chromosome as these do become methylated at this time. This demethylation and *de novo* methylation continues at a slower pace as the embryo develops until the somatic cells have established their characteristic patterns of methylation.

The assumption is that *de novo* methylation may well occur at all regions where the methylase can gain access to the DNA. That is, at all regions where access is not blocked by the presence of other DNA binding proteins, particularly the transcription factors that are associated with genes that are either active or that have the potential for activity. In such inaccessible regions, the DNA methyltransferase will not even be able to gain access in order to maintain the preexisting pattern of DNA methylation and this will lead to demethylation occurring in the same cells, side-by-side with *de novo* methylation. In other words, it appears that the pattern of methylation may be imposed by the pattern of DNA binding proteins present in the nucleus. One clear example of this is the methylation and inactivation of viral genes that become integrated into the host chromosome (see above) and a second example by the demethylation of actin genes introduced into myoblasts. What is far from clear, however, is how genes on the active X-chromosome are protected from methylation at the time when genes of identical sequence, in the same cell (and hence exposed to the same transcription factors) become methylated on the inactive X-chromosome.

X-Chromosome Inactivation

Female mammals have two X-chromosomes whereas males have an X-chromosome and a Y-chromosome. To avoid producing twice the amount of transcript from the genes on the two X-chromosomes, most of one of these is inactivated in female mammals. This is known as dosage compensation and the inactive X-chromosome is late replicating and readily visible in nuclear preparations as a highly condensed, densely staining region known as the Barr body. Inactivation occurs at about the time of implantation (day 5.5 in the mouse) and is initiated from the X-inactivation center where the *Xist* gene occurs (Rastan, 1994). In a random manner, the *Xist* gene becomes methylated on one X-chromosome but not on the other. The unmethylated *Xist* generates an RNA transcript that acts in cis to cause inactivation of the X-chromosome (Xi) from which it is transcribed. Methylation of this Xi chromosome then follows inactivation. The *Xist* gene on the other X-chromosome (Xa) is methylated and inactive and so inactivation of the rest of that Xa chromosome does not occur. Reactivation has to occur in the germ line and this is also a very early event in development. In the 13.5 day mouse embryo (the earliest in which primordial germ cells have been analyzed) all the DNA appears to be largely unmethylated and there then follows a series of methylation and demethylation events before the mature pattern is established.

There is one disease, fragile X mental retardation, that appears to be caused by a failure to reactivate a small region of the inactive X-chromosome during gametogenesis. This region contains the CpG island at the 5' end of the fragile mental retardation (FMR) gene and males that inherit this incompletely reactivated gene from their mothers show the affected phenotype (see Holliday, 1993). The CpG island in affected males is not only methylated but shows amplification of a CGG repeat and it is this amplification that is the initial silent *premutation* in the grandfather that prevents the reactivation of this region in the mother. One could speculate that the demethylation event involves a repair mechanism that encourages misalignment of repeats and their consequent amplification. Coupled with an inability of the demethylation mechanism to cope with a solid cluster of methylcytosines, this could possibly explain how the defective gene might arise.

X-inactivation may be just an extreme example of the changes that can occur to single genes during gametogenesis and early development. The difference being that the epigenetic changes occur to a whole chromosome and not to just an isolated gene. But even this is not completely true as there are certain genes that remain active on the otherwise inactive X-chromosome. In fully differentiated, somatic cells, the inactivated genes are never normally reactivated. However, they can be partly reactivated by treatment of cells with azacytidine which brings about demethylation of the DNA. Clearly in this case, one of the major factors in maintaining the inactive state is the methylation of the DNA.

Abnormal Methylation as a Cause of Cancer and Aging

Abnormal patterns of gene expression can lead to abnormal metabolism which may be manifest as cell death or cell transformation to the cancerous state. As DNA methylation plays some part in the control of gene expression, changes in the normal pattern of DNA methylation could, by affecting gene expression, lead to either an increased rate of cell aging or to cancer. These changes in DNA methylation may, themselves, be brought about by agents that damage the DNA without causing mutation (e.g., 5-azacytidine by preventing methylation can act as a potent carcinogen) (Carr et al., 1984) and it has been suggested that a demethylation event observed on virus infection might result in transformation of surviving cells (Macnab et al., 1988). Feeding rats a methyl-deficient diet brings about an increased incidence of liver tumors in which the DNA is hypomethylated (Wainfan and Poirer, 1992). On the other hand, many tumors, despite their low 5–methylcytosine content, have dramatically elevated levels of DNA methyltransferase that might be a cause of de novo methylation of an anti-oncogene (El-Deiry et al., 1991; Laird et al., 1995). This could lead to reduced expression of the tumor suppressor gene or to its complete inactivation by mutation (see below). In either case the end result would be uncontrolled growth that would further encourage the accumulation of chromosomal abnormalities.

It is instructive to compare the situation in the long-lived human with the short-lived mouse (Holliday, 1989, 1993). If the same controls were in force in the two species, then there would seem to be little reason why a mouse should ever develop a tumor in its short life or, on the other hand, not live to threescore year and ten! As, clearly, neither of these situations occurs, we must assume that the maintenance of methylation patterns (so-called epigenetic control) is much less effective in the mouse than in the human. Certainly, in primary cell culture, mouse cells rapidly lose methylcytosine from their DNA until a stable clone arises that is immortal and this, thereafter, maintains a constant methylcytosine content (Wilson and Jones, 1983; Spruck et al., 1993). There is, however, no evidence to suggest that the immortalization event is directly dependent on an increased ability to maintain epigenetic control. Yet one might imagine that the DNA methylase in the immortal mouse cell and in human cells is just that little bit more effective at maintaining a stable pattern of DNA methylation than is the normal mouse enzyme.

Despite what has been written above, the major effect of DNA methylation on cancer is mediated by the deamination of methylcytosine, a reaction that occurs spontaneously and which, if not corrected, leads to the conversion of a CG:CG dinucleotide pair to a TG:CA dinucleotide pair in half the daughter cells (Figure 19). Even one of the repair mechanisms that can correct a T:G mismatch leads to

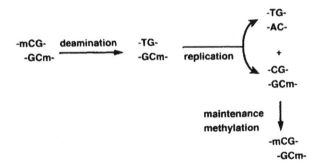

Figure 19. The consequence of replicating a deaminated DNA sequence prior to repair is that one of the daughter cells will contain a mutation.

mutation. That these events are of significance is attested to by the deficiency of CG dinucleotides in vertebrate DNA and the elevated levels of TG and CA dinucleotides relative to that expected from a consideration of the overall base composition of the DNA. Furthermore, deamination of methylcytosine is probably the major cause of mutation in all organisms. The site of action of the Dcm methylase in *E. coli* (CmC(A/T)GG is a hotspot for mutation and 30–40% of all

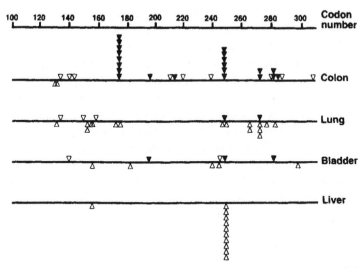

Figure 20. Sites of mutations in the p53 gene found in human tumors of the colon, lung, bladder and liver are shown by triangles. Solid triangles show those mutations likely to have arisen as a result of methylcytosine deamination. Transitions arising by other mechanisms are shown by triangles above the line and transversions are shown by triangles below the line. Reproduced from Jones et al. (1992) with permission of the authors and publisher.

human germline point mutations are thought to be methylation induced (Jones et al., 1992). This is particularly pertinent as a study of 52 colonic cancers has shown that nearly 50% of them were associated with a CG to TG transition in the p53 anti-oncogene (Figure 20). p53 is a protein that in the phosphorylated form regulates passage of cells into S-phase. Most mutations prevent the phosphorylation of p53, or disable it in some other way, with the result that cells are no longer arrested at the transition point in late G1-phase but proceed to divide without control. That a similar study of lung cancers showed only 7% to be associated with CG deamination in the p53 gene is attributed to the additional influence of benz(a)pyrene, a constituent of tobacco smoke, in causing G to T transversions.

REPAIR AND REPLICATION of DNA

Repair of Mismatches Arising at Replication

When an incorrect base is introduced into DNA at replication, it is efficiently removed by the proofreading activity associated with DNA polymerases. Occasionally, however, proofreading fails and the incorrect base is incorporated into the replicated DNA. Where the base is abnormal (i.e., not one of the bases normally present in DNA), the excision repair system will remove it. It is when a normal (though incorrect) base is introduced that the cell has a problem in distinguishing which is the incorrect base at the mismatch. Methylation helps solve this problem. Before the DNA methyltransferase has had time to act, newly replicated DNA is present in a hemimethylated form. During this time between synthesis and methylation of DNA, the repair mechanism is able to distinguish between the parental and daughter DNA strands on the basis of their methylation status: the daughter strand is transiently unmethylated. Most is known about this mismatch repair mechanism in bacteria but it is assumed that a similar mechanism operates in eukaryotes. In *E. coli* the mismatch is recognized by a protein that activates helicases to travel along the DNA, in both directions (Figure 21). When the helicases encounter a hemimethylated target site they, in turn, activate a nuclease to cut the unmethylated strand. The unmethylated fragment is then removed and the gap filled in by the action of DNA polymerase.

Mismatches Arising by Methylcytosine Deamination

Mismatches can also arise in mature DNA by deamination particularly of methylcytosine which produces thymine, a normal base (Figure 4). Mismatches arising by this route are corrected by a different mechanism to that described above (Weibauer et al., 1993). In both routes, however, DNA methylation plays a part in recognizing which is the incorrect base to be replaced.

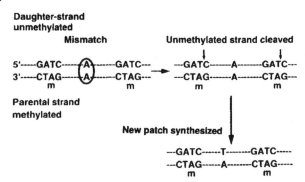

Figure 21. Mismatch repair in *E. coli*. The mismatch is recognized by MutS and the unmethylated strand cleaved by MutH at GATC target sites. The intervening DNA is replaced by synthesis of a large patch by DNA polymerase.

In vertebrates most methylcytosine deaminations occur in the sequence mCG resulting in a T:G mismatch. Such mismatches are recognized by a specific enzyme that first removes the thymine and replaces it with a cytosine. The mechanism involves cleavage of the DNA chain with removal of the sugar phosphate and the insertion of a single nucleotide by DNA polymerase (Figure 22). An alternative, and fortunately less active, repair system recognizes the T:G mismatch but repairs it at random such that 50% of the product contains a T:A mutation rather than the correct C:G base pair. This, rather than the replication of unrepaired deamination products, may be the major source of the mutation hotspots. Overall, it has been estimated that in eukaryotes 4% of T:G mismatches are converted to T:A mutations and, as deamination of methylcytosine is a common event (perhaps occurring as often as 200 times a day per cell), this accounts for its importance in tumorogenesis. Of course, T:G mismatches could arise by some mechanism other than mC deamination but they would largely be corrected as though the T were the incorrect base even though, statistically, this is true only 50% of the time (except when mC deaminations are involved).

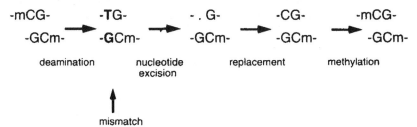

Figure 22. In very short patch repair a single nucleotide is removed from the DNA and replaced by one complementary to the template strand. The major very short patch repair system acting on TG mismatches specifically removes the thymine produced by methylcytosine deamination and replaces it with a cytosine.

Table 4. CG Frequency in Vertebrate DNA

DNA	Base Composition (%G+C)	CG Frequency (%)		O/E	TG(CA) Frequency (%)	
		Expected	Observed		Expected	Observed
a whole genome	40	4	0.8	0.2	6.0	7.3
b exons	57.7	8.3	3.4	0.41	4.5	8.3
c introns	48.8	6.0	2.0	0.33	6.5	7.1
d CpG island	70	12.3	11.6	0.95	5.3	7.0

Note: The table shows the observed frequency of dinucleotides compared with that expected if they occurred at random. Figures are derived from various sources.

CG Deficiency and CpG Islands

In eukaryotes, deamination of methylcytosine leads initially to T:G mismatches that may lead, as we have seen, to some progeny cells having a TG:CA dinucleotide pair where the parental cells had an mCG:mCG pair. From a study of dinucleotide frequencies, it is clear that this process has occurred to a significant extent over evolutionary time. This has resulted in the vertebrate genome in particular being deficient in the CG dinucleotide and containing an excess of TG and CA dinucleotides (Table 4). This characteristic applies particularly to the long stretches of noncoding spacer DNA but there are other regions of the genome that fail to show this deficiency. Not only are these regions not deficient in CpG dinucleotides but they are also enriched in C+G bases relative to A+T bases. These CpG islands occur towards the 5' ends of many, particularly housekeeping genes (Bird, 1986). They are also typical of the genes coding for the stable RNAs (ribosomal and transfer RNA). It has been proposed that the reason these CpG islands are not deficient in CG is a consequence of their being unmethylated in the germ line. This means there could be no conversion of mCG to TG in a manner that would be passed on to future generations. This in itself does not explain the G+C richness of the islands although deamination reactions are much less likely to occur in the more stable G+C-rich environment (Adams et al., 1987). The CpG islands usually stretch from the 5' flanking region of the gene into the first and often the second exon of the gene. Failure to methylate these regions in housekeeping genes is in keeping with the activity of these genes in all cells including germ line cells.

Initiation of Replication

Mismatch repair is not the only system that takes advantage of the time window following replication when the DNA is hemimethylated (Messer and Noyer-Weidner, 1988). In *E. coli* there is a cluster of GATC sites around the origin of replication. These are the targets for the action of the Dam methylase that methylates the adenine base. The sites around the origin of replication are particularly slow to become methylated as they are protected from methyltransferase activity by being complexed with a membrane-associated protein. Only the hemimethylated origin is sequestered in this way with the result that the newly-replicated origin is protected from reinitiation of replication while, at the same time, the two new origins are segregated to opposite ends of the bacterial cell. When the hemimethylated origins are eventually released, conditions are not appropriate for initiation of replication and the Dam methylase acts to methylate the newly synthesized strand (Messer and Noyer-Weidner, 1988). Although Dam⁻ cells are viable they show defects in replication initiation and chromosome segregation. However, methylated phage do not survive in these cells as the hemimethylated origin that is produced in the first round of replication is never released from its membrane attachment site.

There is no evidence for involvement of DNA methylation in replication in higher eukaryotes but in *Tetrahymena* the minichromosomes containing tandem repeats of ribosomal DNA have two origins of replication and the active one is chosen on the basis of its methylation status.

SUMMARY

Methylation of DNA is a postreplicative modification and the study of DNA methylation impinges on all areas of DNA metabolism. This is not surprising as a modified base that influences DNA/protein interactions not only provides a fifth base but a fifth base that can be generated under controlled conditions. In this way the modified base does not affect the genetic potential of the cell as this is still encoded in the sequence of the four major bases.

The extra base is put to use in many different situations and we are constantly learning new functions for this almost ubiquitous modification. One of the functions first recognized was in the restriction:modification phenomenon by which bacteria can regulate the host range of viruses. Other functions of methylation in bacteria are in DNA replication and chromosome segregation and in postreplicative mismatch repair. It is possible that these functions are confined to prokaryotes as the eukaryotes appear to achieve similar ends by other means though the mismatch repair systems that eliminate the products of methylcytosine deamination are universal. Novel uses for DNA methylation have arisen in eukaryotes where a role in the organization of chromatin structure and the control of gene expression are now clearly established. Furthermore, methylation is of central importance in the normal development of the mammalian embryo. Defects in the control of methylation or the incorrect repair of methylcytosine deamination products have been implicated in tumorogenesis and cell aging. Very little is known of the function of the considerable amounts of methylcytosine present in the DNA of plants which are in marked contrast to the negligible amounts found in some insects and lower eukaryotes.

Although methylation of DNA does not affect the base sequence that contains the genetic code, it does allow epigenetic changes to be registered and these can be inherited from generation to generation. This may not be far removed from a direct effect of the environment on the genetic potential of the organism and it is possible that we are only now beginning to understand the importance of DNA methylation.

ACKNOWLEDGMENTS

I would like to thank the Department of Biochemistry and Glasgow University for supporting my work over many years and the various grant-giving bodies that have provided funds for my research. In particular, I would like to thank the students and technicians who have not only done most of the laboratory work but have also helped to make the laboratory an enjoyable place in which to work.

REFERENCES

Adams, R.L.P., Davis, T., Rinaldi, A., & Eason, R. (1987). CpG deficiency, dinucleotide distributions and nucleosome positioning. Eur. J. Biochem. 165, 107–115.

Bird, A.P. (1986). CpG-rich islands and the function of DNA methylation. Nature 321, 209–213.

Bryans, M., Kass, S., Seivwright, C., & Adams, R.L.P. (1992). Vector methylation inhibits transcription from the SV40 early promoter. FEBS Lett. 309, 97–102.

Carr, B.I., Garrett Reilly, J., Smith, S.S., Winberg, C., & Riggs, A.D. (1984). The tumorogenicity of 5-azacytidine in the male Fischer rat. Carcinogenesis 5, 1583–1590.

Chen, L., MacMillan, A.M., Chang, W., Ezaz-Nikpay, K., Lane, W.S., & Verdine, G.L. (1991). Direct identification of the active-site nucleophile in a DNA (cytosine-5)-methyltransferase. Biochemistry 30, 11018–11025.

Doerfler, W. (1992). DNA methylation: Eukaryotic defence against the transcription of foreign genes? Microbial Pathogenesis 12, 1–8.

El-Deiry, W.S., Nelkin, B.D., Celano, P., Yen, R-W.C., Falco, J.P., Hamilton, S.R., & Baylin, S.B. (1991). High expression of the DNA methyltransferase gene characterizes human neoplastic cells and progression stages of colon cancer. Proc. Natl. Acad. Sci. USA 88, 3470–3474.

Feil, R., Charlton, J., Bird, A.P., Walter, J., & Reik, W. (1994). Methylation analysis on individual chromosomes: Improved protocol for bisulphite genomic sequencing. Nucl. Acids Res. 22, 695–696.

Frank, D., Keshet, I., Shani, M., Levine, A., Razin, A., & Cedar, H. (1991). Demethylation of CpG islands in embryonic cells. Nature 351, 239–241.

Holliday, R. (1989). Chromosome error propagation and cancer. Trends Genet. 5, 42–45.

Holliday, R. (1993). Epigenetic inheritance based on DNA methylation. In: DNA Methylation: Molecular Biology and Biological Significance (Jost, J.-P., & Saluz, H.P., eds.), pp. 452–468. Birkhäuser Verlag, Basel.

Jones P.A., Rideout III, W.M., Shen, J.-C., Spruck, C.H., & Tsai, Y.C. (1992). Methylation, mutation and cancer. BioEssays 14, 33–36

Jost, J-P. (1993). Nuclear extracts of chicken embryos promote an active demethylation of DNA by excision repair of 5-methyldeoxycytidine. Proc. Natl. Acad. Sci. USA 90, 4684–4688.

Jost, J.-P., Saluz, H.P., & Pawlak, A. (1991). Estradiol down regulates the binding activity of an avian vitellogenin gene repressor (MDBP-2) and triggers a gradual demethylation of the mCpG of its binding site. Nucl. Acids Res. 19, 5771–5775.

Kass, S., Goddard, J.P., & Adams, R.L.P. (1993). Inactive chromatin spreads from a focus of methylation. Mol. Cell. Biol. 13, 7372–7379.

Klimasauskas, S., Nelson, J.L., & Roberts, R.J. (1991). The methylase specificity domain of cytosine C5 methylases. Nucl. Acids Res. 19, 6183–6190.

Klimasauskas, S., Kumar, S., Roberts, R.R., & Cheng, X. (1994). *HhaI* methyltransferase flips its target base out of the DNA helix. Cell 76, 357–369.

Kricker, M.C., Drake, J.W., & Radman, M. (1992). Duplicated-target DNA methylation and mutagenesis in the evolution of eukaryotic chromosomes Proc. Natl. Acad. Sci. USA 89, 1075–1079.

Laird, P.W., Jackson-Grusby, L., Fazeli, A., Dickinson, S.L., Jung, W.E., Li, E., Weinberg, R., & Jaenisch, R. (1995). Suppression of intestinal neoplasia by DNA hypomethylation. Cell 81, 197–205.

Li, E., Bestor, T.H., & Jaenisch, R. (1992). Targeted mutation of the DNA methyltransferase gene results in embryonic lethality. Cell 69, 915–926.

Macnab, J.C.M., Adams, R.L.P., Rinaldi, A., Orr, A., & Clark, L. (1988). Hypomethylation of host cell DNA synthesised after infection or transformation of cells by herpes simplex virus. Mol. Cell. Biol. 8, 1443–1448.

Messer, W., & Noyer-Weidner, M. (1988). Timing and targetting: The biological functions of Dam methylation in *E. coli.* Cell 54, 735–737.

Monk, M., Boubelik, M., & Lehnert, S. (1987). Temporal and regional changes in DNA methylation in the embryonic, extraembryonic and germ cell lineages during mouse embryo development. Development 99, 371–382.

Monk, M., Adams, R.L.P., & Rinaldi, A. (1991). Decrease in DNA methylase activity during preimplantation development in the mouse. Development 112, 189–192.

Raleigh E.A. (1992). Organization and function of the *mcr*BC genes of *E. coli* K-12. Mol. Microbiol. 6, 1079–1086.

Rastan, S. (1994). X-chromosome inactivation and the *Xist* gene. Curr. Opin. Genet. Develop. 4, 292–297.

Razin, A., & Cedar, H. (1993). DNA methylation and embryogenesis. In: DNA Methylation: Molecular Biology and Biological Significance (Jost, J.-P., & Saluz, H.P., eds.), pp. 343–357. Birkhäuser Verlag, Basel.

Sagar, R., & Kitchin, R. (1975). Selective silencing of eukaryotic DNA. Science 189, 426–433.

Saluz, H.P., Jiricny, J., & Jost, J.P. (1986). Genomic sequencing reveals a positive correlation between the kinetics of strand-specific demethylation of the overlapping estradiol/glucocorticoid receptor binding sites and the rate of avian vitellogenin mRNA synthesis. Proc. Natl. Acad. Sci. USA 83, 7167–7171.

Saluz, H.P., & Jost, J.-P. (1989). Genomic sequencing and *in vivo* footprinting. Anal. Biochem. 176, 201–208.

Selker, E.U. (1990). DNA methylation and chromatin structure: A view from below. Trends Biochem. Sci. 15, 103–107.

Shen, J.C., Rideout III, W.M., & Jones, P.A. (1993). High frequency mutagenesis by a DNA methyltransferase, Cell 71, 1073–1080.

Spruck, C.H., Rideout, W.M., & Jones, P.A. (1992). DNA methylation and cancer. In: DNA Methylation: Molecular Biology and Biological Significance (Jost, J.-P., & Saluz, H.P., eds.), pp. 487–509. Birkhäuser Verlag, Basel.

Surani, M.A. (1994). Genomic imprinting: Control of gene expression by epigenetic inheritance. Curr. Opin. Cell Biol. 6, 390–395.

Taylor, S.M., & Jones, P.A. (1979). Multiple new phenotypes induced in 10T$\frac{1}{2}$ and 3T3 cells treated with azacytidine. Cell 17, 771–779.

Wainfan, E., & Poirier, L.A. (1992). Methyl groups in carcinogenesis: Effects on DNA methylation and gene expression. Cancer Res. 52, 2071s–2077s.

Weibauer, K., Neddermann, P., Hughes, M., & Jiricny, J. (1993). The repair of 5-methylcytosine deamination damage. In: DNA Methylation: Molecular Biology and Biological Significance (Jost, J.-P., & Saluz, H.P., eds.), pp. 510–522. Birkhäuser Verlag, Basel.

Weih, F. Nitsch, D., Reik, A., Schütz, G., & Becker, P.B. (1991). Analysis of CpG methylation and genomic footprinting at the tyrosine aminotransferase gene: DNA methylation alone is not sufficient to prevent protein binding *in vivo.* EMBO J. 10, 2559–2567.

Wilks, A., Seldran, M., & Jost, J.-P. (1984). An estrogen-dependent demethylation at the 5' end of the chicken vitellogenin gene is independent of DNA synthesis. Nucl. Acids Res. 12, 1163–1177.

Wilson, G.G. (1988). Type II restriction-modification systems. Trends Genet. 4, 314–318.

Wilson, V.L., & Jones, P.A. (1983). DNA methylation decreases in aging but not in immortal cells. Science 220, 1055–1057.

Yisraeli, J., Adelstein, R.S., Melloul, D., Nudel, U., Yaffe, D., & Cedar, H. (1986). Muscle-specific activation of a methylated chimeric actin gene. Cell 46, 409–416.

Zacharias, W. (1993). Methylation of cytosine influences the DNA structure. In: DNA Methylation: Molecular Biology and Biological Significance (Jost, J.-P., & Saluz, H.P., eds.), pp. 27–38. Birkhäuser Verlag, Basel.

RECOMMENDED READINGS

Adams, R.L.P., & Burdon, R.H. (1985). Molecular Biology of DNA Methylation. Springer Verlag, New York.

Adams, R.L.P. (1990). DNA methylation: The effect of minor bases on DNA–protein interactions. Biochem. J. 265, 309–320.

DNA Methylation and Gene Regulation (1990). Phil. Trans. R. Soc. Lond. B 326, 179–338. Issue devoted to topic.

Holliday, R. (1989). A different kind of inheritance. Scientific American 260 (6), 40–48.

Jost, J.-P., & Saluz, H.P. (1992). In: DNA Methylation: Molecular Biology and Biological Significance. Birkhäuser Verlag, Basel.

Wilson, G.G., & Murray, N.E. (1991). Restriction and modification systems. Ann. Rev. Genet. 25, 585–627.

Chapter 4

Role of Histones and their Acetylation in Control of Gene Expression

BRYAN M. TURNER

Principles of Medical Biology, Volume 5
Molecular and Cellular Genetics, pages 67–83.
Copyright © 1996 by JAI Press Inc.
All rights of reproduction in any form reserved.
ISBN: 1-55938-809-9

INTRODUCTION

According to one authoritative source, there are just over 200 different cell types in an adult human (Alberts et al., 1994). With only rare exceptions, all the cells in a given individual contain the same complement of DNA and exactly the same set of genes. What type a cell becomes, whether it is a fibroblast, a lymphocyte, or a neuron, is determined by the genes it expresses, i.e., which are switched on and which are switched off. Some genes, the so-called "housekeeping" genes, are expressed by all cells. These include genes encoding vital structural proteins such as histones or actin and enzymes involved in basic metabolic pathways. Other genes are expressed only in certain cell types or only at certain stages of development. For example, the genes for the α and β chains of globin (the protein component of hemoglobin) are expressed only in erythrocyte precursors, while the gene for the ε type of globin chain is expressed only during embryonic life. The pattern of cell type-specific or stage-specific genes that a cell expresses determines what type of cell it is or will become. Understanding how these patterns of gene expression are set and changed in an orderly way is fundamental to understanding both the normal processes of differentiation and development and how these processes may be subverted in diseases such as cancer and in various, poorly understood, developmental abnormalities.

GENES ARE SUBJECT TO LONG-TERM AND SHORT-TERM CONTROL

Genes can be switched on and off very rapidly For example, when cells are subjected to adverse environmental conditions, such as elevated temperature, they respond within seconds by shutting down most RNA and protein synthesis and switching on transcription of a small family of genes, the heat shock or stress response genes. A similarly rapid induction of growth control genes is seen when lymphocytes are stimulated by mitogens or the appropriate antigen. On the other hand, changes in the pattern of gene expression as cells differentiate occur more slowly and may require one or more rounds of cell division before they are complete.

It is important that cells retain their characteristic pattern of gene expression unless they are specifically instructed to change it. When skin cells divide and multiply during wound healing one hopes that they will remain as skin cells and not differentiate into hepatocytes or lens cells. In other words, cells must have a memory mechanism that, as a default setting, retains the pattern of gene expression appropriate to that cell from one cell generation to the next. This is not a trivial problem. During the cell cycle proteins are stripped from DNA by the passage of DNA polymerase through the entire genome during S-phase and chromatin undergoes major structural changes through mitosis. How then does the cell mark those genes that are on and those that are off?

CHROMATIN IS A MAJOR DETERMINANT OF GENE EXPRESSION IN EUKARYOTES

In bacteria, genes are switched on and off in response to environmental signals, such as the availability, or otherwise, of nutrients. Bacteria such as *Escherichia coli* have only a single RNA polymerase but this enzyme is directed towards the required genes by association with ancillary proteins, called sigma factors, which confer upon it the ability to recognize DNA sequences in the promotors of these genes. By changing the availability of these factors and of other, sometimes negative, regulators in response to environmental cues, bacteria are able to maintain an appropriate pattern of gene expression. In recent years it has become clear that eukaryotic organisms also use ancillary, DNA-binding proteins, generally called transcription factors, to regulate transcription of specific genes. There is now a large and growing number of eukaryotic transcription factors. Some of these proteins seem to be required for transcription of one, or just a few, genes while others, such as TFIID and NF1, bind to DNA sequence elements (the TATA box and CAAT box, respectively) found in the promotors of many genes. The number of different proteins required to initiate transcription of eukaryotic genes increases monthly but what is becoming clear is that the initiation complex always involves several proteins in addition to RNA polymerase, and that protein-protein as well as protein-DNA interactions are of crucial importance in assembling this complex.

Regulating gene expression in eukaryotes must accommodate one further complicating factor not present in bacteria, namely that eukaryotes package their DNA by complexing it with small basic proteins, the histones, to form chromatin. The basic unit of chromatin structure, the nucleosome core particle, consists of eight histones (two each of H2A, H2B, H3, and H4) around which are wrapped 146 base pairs of DNA. It is shown diagrammatically in Figure 1 and is described in Chapter 1 by Kornberg. Virtually all DNA, including coding DNA, in a typical eukaryotic cell is packaged as nucleosomes and it is very likely that the close association of histones with DNA will affect the ability of transcription factors and other DNA binding proteins to recognize and bind to specific DNA sequences. So chromatin, and the nucleosome in particular, must be major determinants of transcriptional initiation in eukaryotes. In addition, the necessity for the transcribing polymerase to somehow circumvent the nucleosomes that are known to be retained on actively transcribed genes gives a further possible stage at which control might be exercised through chromatin.

It would be easy to assume that the role of chromatin in regulating gene expression, though important, is essentially passive. That is, nucleosomes influence transcription by blocking access of transcription factors or polymerases to DNA or impeding polymerase transit. If that were so, then understanding the role of chromatin in transcription would be a matter of defining the mechanisms by which nucleosomes were positioned or displaced in order to obscure or reveal different parts of the genome. These processes are undoubtedly important. But recent results,

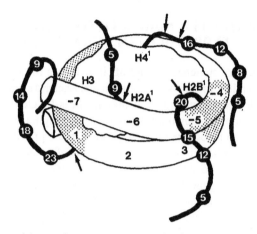

Figure 1. A diagram of the nucleosome core particle showing the sizes and probable positions of the amino-terminal domains of the core histones. Lysine residues which can be acetylated *in vivo* are shown as numbered circles. Trypsin cutting sites listed in Table 1 are indicated by arrows (the site at residue 23 of H2B is not shown). For the sake of clarity the drawing shows only one amino-terminal domain for each pair of histones. Each domain is shown fully extended (i.e., 3.6Å per amino acid residue), a situation which is unlikely to be realized *in vivo* for at least some of them. Numbers along the core DNA indicate each complete helical turn either side of the dyed axis. Regions of core DNA which become more susceptible to DNAase I cutting after removal of the histone amino-terminal tails are shaded. (Reprinted with permission from Journal of Cell Science).

particularly from the study of posttranslational modification of the core histones, suggest that the nucleosome may be capable of more subtle roles in regulating gene expression.

HISTONES ARE HIGHLY CONSERVED BUT EXTENSIVELY MODIFIED

The histones of the nucleosome core particle have been highly conserved through evolution, as has the core particle itself. One would be hard pressed, using physical or microscopical criteria, to distinguish nucleosomes from humans and plants. In view of this, it is initially puzzling to find that the core histones are subject to a wide array of enzyme-catalyzed, posttranslational modifications. These include acetylation, phosphorylation, methylation, ADP-ribosylation, and attachment of the small peptide ubiquitin. Not surprisingly these modifications have attracted considerable interest. The intimate association of histones with cellular DNA and the extreme conservation through evolution of both histones and the nucleosome core particle, make it likely that any change, particularly one which involves a change in net charge, will have an effect on chromatin function. Thus, posttranslational changes are potentially a major influence on events such as transcription, replication, DNA packaging through the cell cycle, and DNA repair. Of these changes, acetylation has been the most intensively studied, partly because it is one of the most frequent (e.g., over 25% of H4 molecules are acetylated in some human cell types), partly because it offers some experimental advantages over other modifications, such as chemical stability and relative ease of detection, but most of all because it has been consistently correlated with transcriptional activity.

THE HISTONE ACETYLATION-DEACETYLATION CYCLE

Acetylation of the histones which organize the nucleosome core particle is an ubiquitous posttranslational modification found in all animal and plant species so far examined. Acetylation occurs at specific lysine residues, all of which occur in the amino-terminal domains of the core histones (Figure 1).

The transfer of acetate groups from acetyl-CoA to histones and their subsequent removal is catalyzed by specific enzymes, the histone acetyl-transferases and deacetylases (Figure 2). The net level of acetylation of a histone at a particular genomic site is the result of an equilibrium between these two types of enzyme. Chalkley and colleagues (1980) showed that the steady-state level of acetylation and the rates at which acetate groups are turned over vary between different cell types, with half-lives ranging from a few minutes to several hours. The dynamic nature of the histone acetylation/deacetylation cycle is readily seen by treating cells with inhibitors of histone deacetylating enzymes, such as the salts of butyric or propionic acid. This results in the progressive accumulation of the more acetylated isoforms, often detectable within a matter of minutes (Figure 3).

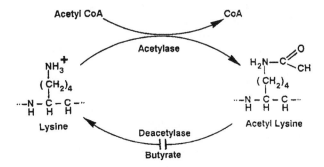

Figure 2. The enzyme-catalyzed acetylation and deacetylation of lysine residues. Short chain fatty acids (i.e., butyric acid) and their salts are inhibitors of the histone deacetylases. Growth of cultured cells in medium containing these inhibitors leads to accumulation of acetylated histones in chromatin.

INDIVIDUAL LYSINE RESIDUES ARE SELECTIVELY ACETYLATED

Because each acetate group added to a histone reduces its net positive charge by one (Figure 2), the acetylated isoforms differ in net charge and so can be separated by electrophoresis (Figure 3). However, each electrophoretic isoform is not necessarily homogeneous but may itself be a mixture of molecules in which the total number of acetate groups is the same, but the sites acetylated are different. For example, in the mono-acetylated isoform of H4 in Figure 3, the single acetate could

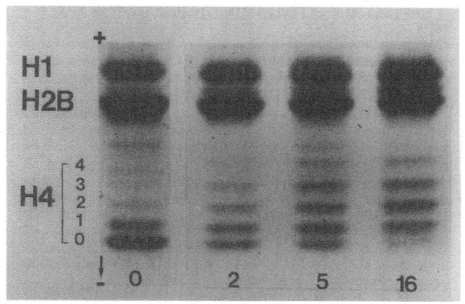

Figure 3. Separation of the acetylated isoforms of histones by electrophoresis. Histones isolated from a human cultured cell line have been electrophoresed under acidic conditions and have migrated towards the cathode (bottom of the figure). The gel has been stained with the dye Coomassie Blue to reveal proteins. Only the lower part of the gel containing histones H1, H2B, and H4 is shown. Histones migrate towards the cathode at a rate which depends largely upon their net positive charge. Reduction of this charge by acetylation of lysine residues slows the rate of migration. H4 isoforms with increasing numbers of acetate groups (indicated by the numbers 1 to 4 on the left) migrate progressively more slowly and are clearly separated. Cultured cells were grown in medium containing sodium butyrate for 0, 2, 5, and 16 hours, as indicated, prior to harvesting and isolation of histones. Butyrate inhibits histone deacetylases and can be seen to lead to progressive accumulation of the more acetylated H4 isoforms. Acetylated isoforms of H2B also accumulate in butyrate treated cells but are less clearly resolved than those of H4. H1 is not subject to modification by acetylation and butyrate has no effect on the electrophoretic behavior of this histone.

Table 1. Acetylation Sites in Mammalian Histones

Histone	Acetylation Sites (in order of use)	Trypsin Cleavage Sites (in chromatin)
H2A	5, 9 (minor)	11
H2B	12/15*, 20, 5	20, 23
H3	14, 23, 18, 9	26
H4	16, 8/12**, 5	17, 19

Notes: *In H2B, either lysine 12 or lysine 15 can be acetylated in the mono-ace-tylated isoform.
**In H4, either lysine 8 or lysine 12 can be acetylated after lysine 16. References to the original work on which the table is based can be found in Turner (1991).

be attached to lysine 5, 8, 12, or 16. The electrophoretic mobility would be the same in each case. It is important to recognize that the sites which are acetylated may be just as significant, in a functional sense, as the change in net charge. There are now many examples in the field of protein phosphorylation where attachment of phosphates to different amino acids on the same protein has quite different func-tional consequences. There is no *a priori* reason to suppose that acetylation will be any different. Such considerations have encouraged attempts to establish whether the different acetylation sites are used in a controlled or a random fashion. This has been tested with histone H4 by isolating the mono-, di-, tri, and tetra-acetylated isoforms and sequencing the amino-terminal region. In each of the several species tested so far, site usage has been shown to be nonrandom, though the frequency with which sites are used differs between species. Thus, in mono-acetylated H4 (H4Ac$_1$) from human or bovine cells, lysine 16 is the predominant acetylated site, while in cuttlefish testis lysine 12 is used exclusively. In human or bovine cells there is some flexibility in site usage, particularly in the more acetylated isoforms, while in cuttlefish lysines are acetylated in a fixed and invariant sequence.

The sequencing approach has been extended to include histones H3 and H2B from human and bovine cells and here too site usage is nonrandom but with a degree of flexibility reminiscent of that seen with H4. These results are summarized in Table 1. It should be emphasized that this type of analysis shows only the steady-state level of acetylation at each site for each of the acetylated isoforms. For each isoform this level may be the net result of several different acetylation-deace-tylation cycles, possibly occurring at different rates, in different parts of the genome, and catalyzed by different enzymes.

The fact that the pattern of H4 acetylation differs between species, while the sequence of the amino terminal region of H4 and, presumably, the structure of the nucleosome, are identical, suggests that differences in acetylation reflect the specificities and relative activities of the acetylating and deacetylating enzymes.

As yet, information on these important proteins is limited, though, at the time of writing, it seems they may soon succumb to the power of molecular genetics.

HISTONE AMINO-TERMINAL DOMAINS ARE NOT ESSENTIAL FOR THE ASSEMBLY OR STRUCTURAL INTEGRITY OF THE NUCLEOSOME CORE PARTICLE

Consideration of the role of histone acetylation in chromatin structure and function would ideally be based on a clear understanding of the role(s) of the core histone amino-terminal domains. Unfortunately, the biological function of these regions remains enigmatic, though some properties are well established, initially as a result of the work of Whitlock and Simpson (1977).

1. The amino-terminal domains of histones in nucleosome core particles are sensitive to proteolytic enzymes and are therefore likely to be exposed on the surface of the nucleosome. Trypsin cleavage sites on histones in the nucleosome core particle are indicated in Figure 1 and listed in Table 1.
2. Histone amino-terminal tails are not necessary for either the assembly of the core particle *in vitro* or its structural integrity. Core particles that are structurally normal (by sedimentation and nuclease digestion) can be assembled *in vitro* using histones prepared from trypsin-treated chromatin and therefore lacking their amino terminal domains (see Figure 1). There is evidence that the amino-terminal domains are required for the assembly *in vitro* of higher-order chromatin structures.
3. Removal of the amino-terminal tails exposes the core particle DNA at 20–35 and 60–80 bases from the 5′ end to attack by DNAase I, suggesting that tails interact with, and thereby protect, DNA in these regions. But only a few specific interactions between core DNA and histone amino-terminal regions have been detected and some of these may be transient.

The amino terminal domains in Figure 1 have been positioned in a way that is consistent with the presently available data. However, it is important to remember that in isolated core particles the amino-terminal tails are deprived of potential sites of interaction present in larger chromatin fragments, such as linker DNA, non-histone proteins, H1, etc. Such binding to core DNA as does occur may happen only through the absence of these other ligands and may not be a good guide to what happens *in vivo*.

HYPERACETYLATION OF CORE HISTONES CAUSES SUBTLE CHANGES IN CHROMATIN STRUCTURE

Extensive analysis by a variety of biophysical techniques has shown that control and hyperacetylated nucleosome core particles (usually isolated from butyrate-treated cells or assembled with histones prepared from such cells) exhibit only subtle differences, suggesting, at most, a partial unwinding or loosening of the core DNA. The induction of histone hyperacetylation with the deacetylase inhibitor sodium butyrate has long been known to increase the sensitivity of chromatin to DNAase I, but not to micrococcal nuclease. This presumably reflects changes in higher order chromatin structure, though it is not clear at what level of structure the effect is initiated. More recently, work from the laboratory of Bradbury has shown an effect of histone acetylation on the disposition of DNA around and between core octamers by assembling nucleosomes *in vitro* on circular plasmid DNA using histones with differing levels of acetylation.

THE RELATIONSHIP BETWEEN HISTONE ACETYLATION AND TRANSCRIPTION IS COMPLEX

The potential importance of histone acetylation was first championed by Allfrey and co-workers at the Rockefeller University in New York and from their very earliest experiments on stimulated lymphocytes, it was proposed that increased levels of histone acetylation are associated with increased transcriptional activity. Since then, various experimental approaches have shown that chromatin preparations enriched in transcribed genes are also enriched in acetylated histones and, conversely, that acetylated chromatin is enriched in transcribed genes. A particularly powerful approach has been used by Hebbes and co-workers at the University of Portsmouth. They have immunoprecipitated nucleosomes from chick embryo erythrocytes with an antibody to acetyl-lysine. The antibody-bound nucleosome fraction is enriched in acetylated histones. The DNA component of this fraction was shown to be enriched in sequences from the β-globin gene but depleted in sequences from the ovalbumin gene, which is not expressed in cells of the erythroid lineage. However, the β-globin gene was found to be enriched in the acetylated fraction even when nucleosomes were prepared from early embryos, in which this gene is not expressed. The important conclusion is that (in this system at least) histone acetylation marks genes which are *transcribable*, but not necessarily undergoing transcription.

A lack of any direct and general correlation between transcription and histone acetylation in mammalian cells was noted by Perry and Chalkley in 1982 and has been confirmed by recent immunoprecipitation experiments using antibodies to acetylated H4 (see below). O'Neill and Turner (University of Birmingham, UK) have shown that various genes in human cells are packaged with H4 having a moderate level of acetylation which does not vary with actual or potential transcriptional activity. In contrast, transcriptionally inactive (silent) heterochromatin is packaged with H4 having a very low level of acetylation. There is an interesting correspondence between these findings and results of experiments in the yeast *S. cerevisiae*. The immunoprecipitation approach has been used to show that the mating-type genes (see below) are packaged with underacetylated histones when silent and with acetylated histones when active. However, it is important to note that the level of acetylation is the same whether the genes are actually active or just potentially so. Collectively, results so far suggest that, in some systems at least, H4 acetylation can define regions of the genome that, over a relatively long term, are either potentially transcriptionally active or silent. Further examples are cited in the following sections.

EXPERIMENTAL APPROACHES TO STUDY HISTONE ACETYLATION AND TRANSCRIPTION

The extreme evolutionary conservation of histones and nucleosomes brings great benefits to the experimental scientist in that fundamental processes can be studied in a wide variety of experimental systems, each offering its own particular advantages. But there are dangers too, specifically of forgetting that yeast and humans (for example) might achieve the same end by different means and that conclusions drawn from experiments in one species may not be valid in another. With this caveat in mind, I will briefly describe three very different experimental approaches that are beginning to provide some clues about the ways in which histone acetylation might operate.

HISTONE ACETYLATION CAN FACILITATE TRANSCRIPTION FACTOR BINDING *IN VITRO*

Recent experiments in the laboratory of Wolffe at the National Institutes of Health, Bethesda, have attempted to measure directly the effect of histone acetylation on transcription factor binding. These used, as a model system, the 5S RNA gene from the toad *Xenopus*. This small gene was assembled into nucleosomes by mixing with core histones in buffers high in salt and urea and then dialyzing to progressively lower the salt, urea concentration, and permit nucleosome assembly. The 5S RNA gene is transcribed by Pol III and requires the transcription factor TFIIIA for activity. TFIIIA binds to a site near the center of the gene (not unusual for PolIII transcription factors). Binding of TFIIIA to its cognate sequence was much reduced when the gene was packaged into nucleosomes. However, this inhibition was relieved if the nucleosomes were assembled using either histones from which the amino terminal tails had been removed by proteolysis or histones in which the tails were highly acetylated (these having been isolated from cells grown in the presence of sodium butyrate). In the absence of evidence for unwrapping of core DNA, changes in helical twist or altered nucleosome positioning, all of which can influence protein binding, the authors conclude that acetylation alleviates inhibition by relieving steric hindrance to TFIIIA binding by the amino terminal tails of the core histones. At first sight this is not surprising. The histone tails are disposed on the nucleosome surface where they will almost inevitably encounter proteins scanning core DNA for a specific base sequence, and while acetylation may not leave the amino terminal tails waving in the solvent, the reduction in net charge it causes will certainly reduce the avidity with which they bind core DNA and facilitate their displacement. However, it is important to note that the abolition of transcription factor binding *in vitro* by positioned nucleosomes is by no means universal and more subtle mechanisms than simple steric hindrance may be involved. Some transcription factors, such as glucocorticoid receptor and the yeast factor GAL4, show (under some conditions at least) only modest reductions in binding to nucleosomal DNA while others, such as NF1 and heat shock factor, show almost complete abolition of binding. Histone acetylation may prove to facilitate binding transcription factors *in vitro* but further experiments will be required to establish the functional significance of this *in vivo*, where adjacent, contiguous DNA, auxiliary transcription factors, structural proteins, and enzymes all contribute to the final result.

THE POWER OF YEAST GENETICS HAS BEEN USED TO SHOW THAT ACETYLATABLE LYSINES IN H3 AND H4 HAVE CRUCIAL ROLES IN GENE REGULATION

It was initially surprising, even disappointing for some of us, to learn from Grunstein and colleagues at UCLA that mutants of the yeast *Saccharomyces cerevisiae*, with large deletions of the amino-terminal domains of either H2A, H2B, or H4, are not only viable, but grow in culture just as well as wild type cells. The inescapable conclusion is that the amino terminal domains of the core histones are not, individually, essential for the basic cellular functions necessary for growth and division of yeast cells, at least in the laboratory. However, two observations prevent the conclusion that the amino terminal domains of these histones are of no importance. The first is that double mutants in which the amino terminal domains of both H2A and H2B are deleted are not viable. In view of this, the viability of the single mutants may be just another example of the redundancy that evolution has built into living cells. Perhaps the amino terminal domain of H2A can substitute for H2B and vice versa. Secondly, the H4 deletion mutants showed reduced mating efficiency, the effect being particularly dramatic in those in which all acetylatable amino-terminal lysines were absent, i.e., deletions of residues 4–19.

To appreciate the significance of this last observation, it is necessary to understand that for yeast to mate successfully, two crucial genes called HMLα, and HMRa must be switched off. The cell chooses to become either mating type a or α (the nearest a yeast cell gets to being either male or female) by inserting either of these two genes into a new chromosomal locus (the MAT locus), where it is expressed. In those H4 mutants in which mating efficiency was severely depressed, the normally silent HMLα and HMRa genes were both constitutively transcribed (i.e., permanently switched on). The inappropriate expression of genes at these normally silent loci is presumably the cause of the severely reduced mating efficiency. It is important to note that this effect was not part of a general disruption of gene regulation; for example, the PHO gene in mutants was regulated just as in wild type cells by changes in phosphate concentration.

These observations have been extended by Grunstein and others by substituting various amino acids for some or all of the amino terminal lysine residues of H4. Mutants in which all four amino-terminal lysines were substituted by the nonacetylatable amino acid arginine (i.e., permanently positively charged) were not viable whereas substitution by glutamine (i.e., permanently neutral) gave a phenotype very similar to the 4–19 deletion mutant with severely reduced mating efficiency. Crucially, substitution of any or all of lysines 5, 8, or 12 with glycine was without measurable effect, whereas substitution of lysine 16 (or immediately adjacent residues) resulted in de-repression of the HMLα and HMRa genes and severely reduced mating efficiency. These results are consistent with the hypothesis that silencing of the HMLα and HMRa genes requires most of the amino terminal domain of H4 and that lysine 16 should not be acetylated. Acetylation of lysine 16 (mimicked in the mutants by substitution with a neutral amino acid) leads to activation of these genes and a consequent mating defect.

Further genetic and biochemical experiments have shown that the mechanism of silencing of the yeast mating type genes involves the interaction of defined nonhistone proteins not only with specific sites on DNA (the so-called Silencing Elements) but, directly or indirectly, with the amino-terminal region of H4. It is reasonable to propose that this interaction may be regulated by the acetylation status of H4, though this hypothesis remains to be proved by direct, biochemical analysis.

A similar experimental approach has been used to prepare and analyze mutants of the amino terminal region of histone H3. Deletions in this region (residues 4–15) or substitution of acetylatable lysines (residues 9, 14, and 18) did not effect silencing of the mating type genes to the same extent as H4 but they did allow hyperactivation of genes such as GAL1 that are required for metabolism of galactose and are regulated by the transcription factor GAL4. *In vivo*, control of GAL 1 involves not only GAL4 but positioned nucleosomes, the repressor protein MIG1, and a set of general activators of transcription. Several results suggest that the protein products of these genes form part of a multi-subunit complex that may act to antagonize the generally repressive effects of chromatin. Once again the results suggest, but do not yet prove, that the assembly or stability of a functionally important protein complex may be regulated by histone acetylation.

The yeast experiments clearly demonstrate the importance of considering individual histones when devising possible roles for histone acetylation. H3 and H4 clearly have different functions. They also suggest that individual sites of acetylation are important. However, the yeast genome is less than 1/200th the size of a typical mammalian genome and a large proportion is transcriptionally active. There is a high steady-state level of histone acetylation. Thus, despite the fact that the amino-terminal domains of yeast and human H4 have identical amino acid sequences, the question arises whether conclusions drawn from these experiments in yeast can be extended to higher eukaryotes. Recently quite a different experimental approach has shown that in higher eukaryotes, including humans, acetylation of histone H4 may play a role in the long-term control of gene expression similar to that identified in yeast.

IMMUNOMICROSCOPY SHOWS THAT ACETYLATED H4 ISOFORMS MARK FUNCTIONALLY DIFFERENT CHROMATIN DOMAINS

Knowledge of the distribution of acetylated histones in chromatin is central to the formulation of hypotheses on their role in chromatin function. It has recently become possible to examine this distribution directly by immunofluorescence microscopy using antisera specific for acetylated H4. This approach has been used by Allis and co-workers to monitor developmental changes in the level and distribution of acetylated H4 in the macronuclei and micronuclei of Tetrahymena and by Pfeffer and colleagues to show a striking increase in H4 acetylation in erythrocyte nuclei activated by fusion with transcriptionally active cultured cells. These experiments have confirmed the correlation between increased levels of histone acetylation and transcription. More recently, the immunomicroscopical approach has been used to address the role of site-specific H4 acetylation; that is, to ask whether H4 isoforms acetylated at different lysine residues have particular functional roles.

The yeast work has demonstrated functional differences between the different histones and has shown that substitution of particular acetylatable lysines with neutral amino acids can have specific phenotypic effects. However, direct evidence for the importance of site-specific acetylation has come from the use of antibodies that can distinguish H4 isoforms acetylated at lysines 5, 8, 12, or 16. If these isoforms do indeed have different functions, then one might expect them to have different locations along the genome and to be associated with functionally different chromatin domains. It is not practical to try and test this by microscopical analysis of interphase, diploid cells in which the chromosomes are tightly packaged in a small nucleus. However, it can be done in the giant, polytene chromosomes found in certain insect larvae. These chromosomes have been most widely studied in the larvae of the geneticists favorite organism, the fruit fly *Drosophila*. The cells containing these chromosomes have gone through several rounds of DNA replication without intervening cell divisions, resulting in the formation of polytene chromosomes in which 1,000 or more strands of DNA are aligned in exact register. Differential packaging of DNA along these chromosomes results in a characteristic and reproducible banding pattern that allows one to navigate along the chromosomes and locate specific genes and genetically defined regions. Immunolabelling of such polytene chromosomes with site-specific antibodies to acetylated H4 has shown that isoforms acetylated at particular residues have their own characteristic patterns of distribution through the genome. Of particular significance is the finding that, in two instances, the distribution of acetylated H4 isoforms has been correlated with particular aspects of genomic function. Firstly, H4 in condensed, genetically inactive (heterochromatic) regions is consistently underacetylated. Secondly, H4 acetylated at lysine 16 is found predominantly (though not exclusively) on the X chromosome in male cells. The significance of this latter finding lies in the fact that in *Drosophila*, dosage compensation (i.e., the equalization of X chromosome gene products between XY males and XX females) is achieved by doubling the transcriptional activity of X chromosome genes in male cells. Thus, a single chromosome (the male X) that shows a particular functional characteristic (i.e., doubling the rate of transcription of its genes) is marked by a specific acetylated isoform of H4 (H4Ac16).

The existence of such an X chromosome marker is a potentially important element in the mechanism of dosage compensation in *Drosophila*. Four genes that are essential for dosage compensation in male *Drosophila* have been identified through their mutants. Immunostaining of polytene chromosomes with antisera to the protein products of two of these genes, MLE and MSL-1, by Kuroda and colleagues at Baylor College of Medicine, shows that both are localized almost exclusively on the X chromosome in male cells and co-localize along the chromosome in many discrete bands. Double immunolabelling experiments with rabbit antiserum to H4Ac16 and goat antiserum to MSL-1 have shown that the two proteins are, with only rare exceptions, located in exactly the same chromatin regions. This finding supports the proposition that H4Ac16, MLE, and MSL-1 are all components of a multi-subunit complex required to increase the transcriptional activity of genes on the male X chromosome. The functional relationships between the components of this hypothetical complex remain to be worked out, as does the mechanism by which H4 lysine 16 is acetylated predominantly on the male X chromosome.

Unfortunately mammals do not have polytene chromosomes, so analyzing the distribution of acetylated histones through the interphase genome requires more complex experimental approaches. However, the chromosomes of mammalian cells are accessible microscopically in metaphase cells and, while not offering the resolution achievable with *Drosophila* polytene chromosomes, can be used to define, in broad terms, patterns of distribution. Immunolabelling of metaphase chromosomes from human and mouse cells with antisera to acetylated H4 initially revealed two properties that are consistent with the idea that in mammals, as in other eukaryotes, acetylated H4 serves as a genomic marker. First, H4 in constitutive, centric heterochromatin is underacetylated at all four lysines. The fact that H4 in *Drosophila* heterochromatin is also underacetylated suggests that this is a conserved property of heterochromatin. Perhaps, in higher eukaryotes, as in yeast, H4 acetylation can prevent formation of the multi-subunit protein complexes required for chromatin condensation. Second, those regions of the chromosome arms that label most strongly with antisera to acetylated H4 correspond, in general terms, to regions known to be enriched in coding DNA, and identified by conventional chromosome staining procedures as R-bands.

But what is the significance of this for gene regulation? Heterochromatin contains few if any genes, so its underacetylation may be more concerned with packaging in the nucleus than transcriptional control. That this is not the case has been shown by the finding that H4 is underacetylated not only in the genetically inert, constitutive heterochromatin found at and around the centromeres of mammalian chromosomes, but also in facultative heterochromatin, i.e., chromatin that, in some cells, is condensed in order to switch off the genes it contains.

The most dramatic example of facultative heterochromatin in mammals is the inactive X chromosome in female cells. In mammals, unlike fruit flies, dosage compensation is achieved by the genetic silencing of one of the two X chromosomes in female cells. This occurs early in development when the X to be inactivated is selected at random and then packaged into heterochromatin. The inactive X remains condensed through interphase, often taking up a position adjacent to the nuclear envelope, and replicates late in S-phase. The decision of which X to inactivate is, with only rare exceptions, irreversible, with the same X remaining inactive in all the descendants of any given cell. Immunofluorescent staining of metaphase chromosomes from various mammalian cell types with site-specific antibodies to acetylated H4 has shown that the inactive X (Xi) in mammalian females is marked by very low acetylation of H4 at all four lysines.

SUMMARY

It has been established over a number of years that histone acetylation is correlated with the transcriptional state of chromatin. Recent experiments have begun to analyze this correlation in more detail and have established some important facts.

1) Histone acetylation is not dependent on, or a result of, the transcription process. In chicken erythrocytes, histones on the β-globin gene were found to be acetylated in both early and late embryos, even though the gene is transcribed only in the latter. This result suggests that histone acetylation may define transcribable genes rather than those that are actually being transcribed.

2) The experiments with yeast mutants have shown that different histones have different roles *in vivo* and their acetylation may have different functional effects. These experiments also raise the possibility that the individual lysine residues acetylated are significant and that H4 lysine 16 may be particularly important in determining whether some genes are stably switched on or off.

3) The use of antibodies that can distinguish H4 molecules acetylated at different lysine residues has established the functional significance of acetylation of different lysines in the fruit fly *Drosophila*. H4 acetylated at

lysine 16 marks the transcriptionally hyperactive X chromosome in male cells.

4) Underacetylation of H4 at all sites (with the exception of lysine 12 in *Drosophila*) is characteristic of genetically inactive heterochromatin in most, though not all species and cell types so far studied. Perhaps deacetylation of H4 facilitates the assembly of the protein complex that establishes a compact and genetically silent chromatin state.

Despite recent advances, many unanswered questions remain. It would be particularly useful to know when in the cell cycle, and how, the pattern of histone acetylation characteristic of a particular cell is set and how this pattern is faithfully transmitted from one cell generation to the next. This will require a much more detailed understanding of the enzymology of the histone acetylation/deacetylation cycle. It will also be essential to establish by what mechanism(s) histone acetylation can influence chromatin function. Is it by regulating the assembly of multi-protein complexes that have either positive or negative effects on gene expression? If so, what are the proteins that bind to the histone amino terminal domains and how is this binding regulated by acetylation? Finally, although most attention to date has been focused on histones H3 and H4, H2A and H2B are also frequently acetylated. What is the functional significance of acetylation of these histones? The list of unanswered questions could go on and will inevitably lengthen as new results throw up more mysteries and unanticipated puzzles.

RECOMMENDED READINGS

General

Alberts, B., Bray, D., Lewis, J., Raff, M., Roberts, K., & Watson, J.D. (1989). In: Molecular Biology of the Cell, 2nd. ed., Garland, New York.

Paranjape, S.M., Kamakaka, R.T., & Kadonaga, J.T. (1994). Role of chromatin structure in the regulation of transcription by RNA polymerase II. Ann. Rev. Biochem. 63, 265–297.

Turner, B.M., & O'Neill, L.P. (1995). Histone acetylation in chromatin and chromosomes. Sem. Cell Biol. 6, 229–236.

Turner, B.M. (1991). Histone acetylation and control of gene expression. J. Cell Sci. 99, 13–20.

van Holde, K.E. (1988). In: Chromatin. Springer-Verlag, New York.

Tail Demains and Chromatin Structure

Garcia-Ramirez, M., Dong, F., & Ausio, J. (1992). Role of the histone "tails" in the folding of oligonucleosomes depleted of histone H1. J. Biol. Chem. 267, 19587–19595.

Norton, V.G., Imai, B.S., Yau, P., & Bradbury, E.M. (1989). Histone acetylation reduces nucleosome core particle linking number change. Cell 57, 449–457.

Whitlock, J.P., & Simpson, R.T. (1977). Localisation of the sites along nucleosomal DNA which interact with NH_2-terminal histone regions. J. Biol. Chem. 252, 6516–6520.

The Dynamics of Histone Acetylation

Covault, J., & Chalkley, R. (1980). Identification of distinct populations of acetylated histone. J. Biol. Chem. 255, 9110–9116.

Perry, M., & Chalkley, R. (1982). Histone acetylation increases the solubility of chromatin and occurs sequentially over most of the chromatin. J. Biol. Chem. 255, 9110–9116.

Transcription and Transcription Factor Binding

Braunstein, M., Rose, A.B., Holmes, S.G., Allis, C.D., & Roach, J.R. (1993). Transcriptional silencing in yeast is associated with reduced histone acetylation. Genes. Dev. 7, 592–604.

Hayes, J.P., & Wolffe, A.P. (1992). The interaction of transcription factors with nucelosomal DNA. BioEssays 14, 597–603.

Hebbes, T.R., Thorne, A.W., Clayton, A.L., & Crane-Robinson, C. (1992). Histone acetylation and globin gene switching. Nuc. Acids Res. 20, 1017–1022.

Lee, D.Y., Hayes, J.L., Pruss, D., & Wolffe, A.P. (1993). A positive role for histone acetylation in transcription factor access to nucleosomal DNA. Cell 72, 1–20.

O'Neill, L.P., & Turner, B.M. (1995). Histone H4 acetylation distinguishes coding regions of the human genome from heterochromatin in a differentiation-dependent but transcription-indenpendent manner. EMBO J. 14, 3946–3957.

Workman, J.L., & Buchman, A.R. (1993). Multiple functions of nucleosomes and regulatory factors in transcription. Trends Biological Sci. 18, 90–95.

Tail Domains and Transcription

Hecht, A., Laroche, T., Strahl-Bosinger, S., Gasser, S.M., & Grunstein, M. (1995). Histone H3 and H4 N-termini interact with the Silent Information Regulators Sir3 and Sir4: A molecular model for the formation of heterochromatin in yeast. Cell 80, 583–592.

Johnson, L.M., Kayne, P.S., Kahn, E.S., & Grunstein, M. (1990). Genetic evidence for an interaction between SIR3 and histone H4 in the repression of the silent mating loci in Saccharomyces cerevisiae. Proc. Natl. Acad. Sci. USA 87, 6286–6290.

Mann, R.K., & Grunstain, M. (1992). Histone H3 N-terminal mutations allow hyperactivation of the yeast *GAL1* gene *in vivo*. EMBO J. 11, 3297–3306.

Intrachromosomal Distribution of Acetylated H4 in *Drosophila* and Mammals

Bone, J.R., Lavender, J., Richman, R., Palmer, M.J., Turner, B.M., & Kuroda, M.I. (1994). Acetylated histone H4 on the male X chromosome is associated with dosage compensation in *Drosophila* Genes Devel. 8, 96–104.

Jeppesen, P., & Turner, B.M. (1993). The inactive X chromosome in female mammals is distinguished by a lack of histone H4 acetylation, a cytogenetic marker for gene expression. Cell 74, 281–289.

Turner, B.M., Birley, A.J., & Lavender, J. (1992). Histone H4 isoforms acetylated at specific lysine residues define individual chromosomes and chromatin domains in *Drosophila* polytene nuclei. Cell 69, 375–384.

Chapter 5

Protein-Nucleic Acid Recognition and Interactions

ALICE VRIELINK and PAUL S. FREEMONT

INTRODUCTION

The interactions between protein molecules and DNA are fundamental to the viability of all cells. These interactions mediate important biological processes such as transcriptional control, cellular differentiation, development, and replication. An example of these interactions is in the control of transcription, a process which

Principles of Medical Biology, Volume 5
Molecular and Cellular Genetics, pages 85–115.
Copyright © 1996 by JAI Press Inc.
All rights of reproduction in any form reserved.
ISBN: 1-55938-809-9

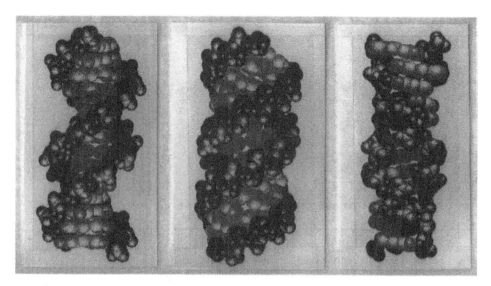

Figure 1. Space filling representations showing the structures of (**a**) B-DNA, (**b**) A-DNA, and (**c**) Z-DNA. Reproduced with permission from Saenger (1984).

makes RNA from genes and is fundamental to all living cells. Essential to this process is the recognition of a particular sequence of nucleotides, by a specific protein molecule, an event which is mediated through the three-dimensional structures of both the DNA and the protein. Such recognition events are commonly seen in the binding of a protein or enzyme to its specific ligand or substrate. In DNA-protein interactions many of the same principles apply; however, a major contributing feature is hydrogen bonding interactions between specific protein amino acid side chains and nucleotide bases which make up the recognition sequence of the DNA. In the case where proteins do not recognize a specific nucleotide sequence, but bind nonselectively to DNA, recognition is often mediated through interactions between the protein and the sugar phosphate DNA backbone. In both cases, however, local conformations of the DNA play a key role in protein-DNA recognition and binding.

A large number of structures of DNA binding proteins have been determined both crystallographically and by nuclear magnetic resonance (NMR). These structures reveal a number of specific protein motifs which are important in DNA interaction. These motifs which are seen in many eukaryotic and mammalian regulatory proteins include the helix-turn-helix, the zinc finger, the ribbon-helix-helix, the basic region leucine zipper motif, and the basic region helix-loop-helix motif. In this chapter we will present examples of each of these classes of DNA binding proteins and show the features of the motif which are essential for DNA interaction.

DNA STRUCTURE

DNA double helices have been observed in a number of structural forms based on their environmental conditions and oligonucleotide sequence. These include the right-handed A and B types of DNA and, the left-handed form, Z-DNA (for review see Saenger, 1984). The most commonly occurring form of DNA, which has been observed both in aqueous solution and in numerous crystal structures, is B-DNA (Figure 1a). B-DNA is characterized by a helical turn consisting of 10 nucleotides with an axial rise of 3.3 Å per nucleotide, giving rise to a distance between phosphate groups of approximately 7 Å. B-DNA also has good helical stacking and, unlike A-DNA, the major and minor grooves of B-DNA are approximately equal in size (Figure 1a). Numerous crystal structures of B-DNA of various oligomeric lengths have shown that the DNA helix can bend, although these distortions could be caused by crystal packing forces (e.g., Wing et al., 1980). However, in general small modulations in structure are observed both in the crystal and in solution which are sequence-dependent and could provide local structural variations to facilitate specific protein-nucleic acid recognition.

In contrast, the structure of A-DNA which is similar to that of A-RNA (Fuller et al., 1965) consists of 11 nucleotides per helical turn, with an axial rise of 2.6 Å per nucleotide, resulting in a phosphate-phosphate distance of approximately 6 Å (Figure 1b). In A-DNA, the base pairs are tilted relative to the helix axis producing an open central channel if viewed down the helical axis. A-DNA is also more compact than B-DNA and has a deep major groove but a shallow minor groove (Figure 1b). However, it is important to note that A-DNA has never been observed in aqueous solution and the structure of A-DNA results from dehydration of B-DNA. All the A-DNA structures to date have been obtained from crystals grown in the presence of alcohol which acts as a dehydrating agent inducing a B-to-A transition or in dehydrating conditions in solution (Zimmerman and Pheiffer, 1979). In addition, oligomers containing a GG sequence are thought to be more stable in the A form (Wang et al., 1982). An interesting feature of both A-DNA and B-DNA is their ability to form DNA-RNA hybrids. DNA-RNA hybrids are observed biologically when DNA bases are transcribed into complementary messenger RNA by RNA polymerase (Wang et al., 1977). Another example of DNA-RNA hybrid formation is during the production of DNA from phage or viral RNA by the enzyme reverse transcriptase. However, there is no evidence to date which supports the existence of A-DNA *in vivo*.

An unusual form of DNA which is also only produced in dehydrating conditions is Z-DNA. Z-DNA consists of a left-handed double helical structure made up of alternating and repeating sequences of dG-dC (Figure 1c) (Wang et al., 1979; Drew et al., 1980). Studies have shown that B-DNA is able to transform to Z-DNA without complete strand separation and that this transition moves along the helix via a co-operative mechanism (Wang et al., 1979). The Z-DNA structure shows that twelve Watson-Crick type G-C base pairs are needed to complete a turn of helix. The phosphate groups in left-handed DNA are located at two different alternating radii and result in neighboring sugar units which point in opposite directions. This gives a zigzag line for the connecting phosphate groups, hence the name Z-DNA (Figure 1c). Another unique feature of Z-DNA is the existence of only a deep and narrow minor groove with the major groove filled by C_5 of cytosine and N_7 and C_8 of guanine (Figure 1c). This is a result of the fact that the G-C base pairs are not symmetrically located near the helical axis but are shifted radially outwards causing a convex, rather than a concave, major groove. Z-DNA structures are also stabilized by methylation of cytosine residues at C_5. Interestingly, sequences containing methylated cytosines are frequently observed in eukaryotic DNA as a result of DNA methylation (Razin and Riggs, 1980), suggesting that Z-DNA may exist under physiological conditions. In addition, Z-DNA has been suggested to exist in *Drosophila* chromosomes (Santella et al., 1981) and may be involved in the regulation of supercoiling (Nickol et al., 1982).

From these numerous crystallographic and NMR studies on a variety of different DNA oligomers, it is apparent that DNA can adopt a number of different conformations according to the experimental conditions. However, in aqueous solution, and presumably *in vivo*, oligonucleotides generally show the presence of B-like conformations. Furthermore, different DNA sequences can show differing local geometries. This leads to the idea that such local variations might be important in the recognition process or in the binding specificity, although this requires that the conformations of target sites differ significantly from other pieces of DNA. However, at present the importance of such small local variations of DNA conformation on protein specificity and recognition is unclear. It is now well established that DNA can be highly flexible over long ranges as illustrated by the ability of chromatin DNA to supercoil and to wrap around the histone octamer to form the core particle. This bendability and flexibility is sequence dependent and has been attributed to strong base-base interactions and the formation of non-Watson-Crick hydrogen bonds along the helix axis (for review see Travers, 1993). This leads to the proposal that the intrinsic flexibility of different DNA sequences may be important for biological specificity and function. Given that globular proteins tend to be stiffer than DNA, it is likely that the DNA will deform more than the protein on forming specific protein-DNA complexes, as observed in a number of protein-DNA complex structures. A striking example of this is the catabolite gene activator protein (CAP) repressor-DNA complex which shows B-like DNA bent by almost 90° around the protein dimer (Schultz et al., 1991).

HELIX-TURN-HELIX MOTIF

From the early crystal structures of the dimeric λ phage *Cro* repressor (Anderson et al., 1981) and *Escherichia coli* CAP (McKay and Steitz, 1981), it was noted that a structural motif consisting of two α-helices from each monomer was separated by 34 Å equivalent to one turn of duplex B-DNA and was related by the two-fold symmetry of the dimeric protein. It was proposed that one α-helix from each pair would lie in successive major grooves, with the two-fold symmetry of the protein dimer coincident with the two-fold symmetry of the DNA operator sequence (for review see Brennan, 1992). This structure of two α-helices (Figure 2), now termed the helix-turn-helix (HTH) DNA binding motif, contains a tight β-turn resulting in an interhelical angle of nearly 90° and an amino acid sequence preference which has been extended to other DNA binding proteins. One of the α-helices (Figure 3) lies in the major groove providing sequence-specific DNA interactions and is termed the recognition helix. The N-termini of both helices point towards the phosphate backbone using the positive helix dipoles for the correct positioning of the recognition helix. The amino acid side chains jutting out of the recognition helix

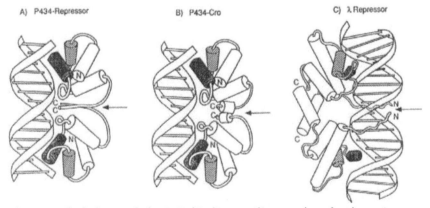

Figure 2. The helix-turn-helix DNA-binding motif in a number of prokaryotic repressors. Schematic representation of P434, P434-*cro* and λCl repressor-operator complexes as derived from their crystal structures, with α-helices as cylinders. The helix-turn-helix motif is shaded with the recognition helix interacting with base pairs in the major groove. The molecular two-fold of each complex is indicated by an arrow and lies coincident with the diad axis of the operator sequence. The DNA structures are represented as ribbons. (**A**) P434 repressor-operator complex (adapted from Aggarwal et al., 1988). It should be noted a short C-terminal helix similar to that in the P434-*cro* repressor was observed on refinement (Aggarwal and Harrison, 1988). (**B**) P434 *cro* repressor-operator complex (adapted from Wolberger et al., 1991). (**C**) λCl repressor-operator complex (adapted from Jordan and Pabo, 1988). From Freemont et al. (1991) with permission.

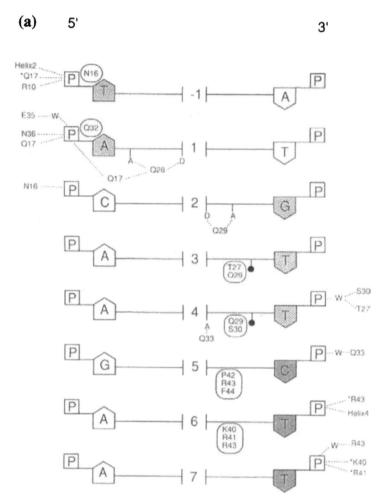

Figure 3. Schematic drawing of the specific amino-acid-base contacts responsible for repressor-operator recognition. The side chains protrude from the recognition helix of the helix-turn-helix motif which lies in the major groove. (a) P434 repressor-operator complex (Harrison and Aggarwal, 1990). (b) P434 *cro* repressor-operator complex (Wolberger et al., 1991) (c) λCl repressor-operator complex (Jordan and Pabo, 1988). Amino acids are indicated by the single letter code. Amino acid interactions are shown as dotted lines. The phosphate (☐P☐), sugar (⌂) and base (—|) make up one nucleotide. Hydrogen bond donors (D) and acceptors (A) are indicated on the bases involved in specific amino acid interactions. Non-polar interactions (☐) are also shown. Amino acids which form hydrogen bonds through main chain atoms are flagged with an asterisk (*). Interactions which occur in the major and minor groove are indicated by hatched and stippled shading of the specific bases, respectively. The base numbering between the nucleotides is taken from the original references.

(b) 5'   3'

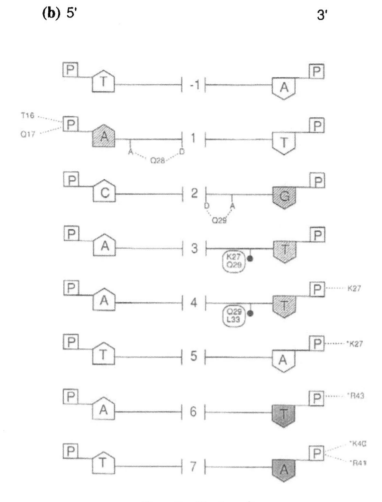

Figure 3. (Continued)

are able to make sequence-specific interactions with exposed hydrogen bonding groups in the major groove of the DNA. Although the HTH motif forms a compact stable structural unit and is highly conserved, the orientation of the motif relative to the DNA binding site is highly variable. The HTH structural motif has been found in many prokaryotic repressors. The structures of several of these repressors bound to their DNA target sequences have now been elucidated and include the *E. coli trp* (Otwinowski et al., 1988) and CAP repressor-operator complexes (Schultz et al., 1991), bacteriophage λ CI (Jordan and Pabo, 1988) and *cro* repressor-operator complexes (Brennan et al., 1990), bacteriophage P434 (Aggarwal et al., 1988) and

(c)   5'   3'

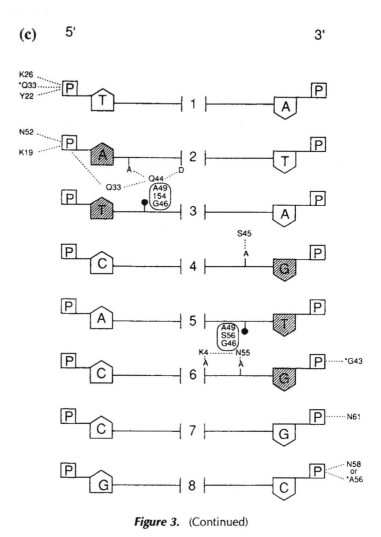

Figure 3. (Continued)

P434 *cro* (Mondragon and Harrison, 1991) repressor-operator complexes. All of these structures show that amino acids from the recognition helix make specific contacts with base pairs in the major groove through either direct hydrogen bonding or even via ordered water molecules (Figure 3). It is clear that specific operator recognition relies not only on specific amino acid-base contacts, but also on a network of nonspecific phosphate backbone contacts which surround the recognition helix orienting the helix to maximize favorable contacts. Another feature which emerged from these related prokaryotic repressor-operator complexes was the possible dependence of recognition specificity on local sequence-dependent variations in DNA geometry. It is now generally accepted that the ability of certain DNA sequences to adopt more flexible conformations contributes to specific recognition. Furthermore, specific recognition is a concerted property of the entire protein/DNA interface, as illustrated in the structures of 434 *cro* repressor bound to three related operator sites (Rodgers and Harrison, 1993).

A number of variations of the HTH motif have now been characterized in a number of eukaryotic transcription factor proteins. These include the homeodomain proteins from *Drosophila* which are involved in embryonic development and contain a 60 amino acid DNA-binding motif. The motif comprises three α-helices and an amino terminal arm and is structurally related to the HTH motif, although the equivalent recognition helix is significantly longer in the homeodomain (Figure 4). Structures of the *engrailed* and *antennapedia* homeodomains complexed with their target sequences have been solved (Kissinger et al., 1990; Otting et al., 1990) as has the yeast MATα2 homeodomain-DNA complex (Wolberger et al., 1991). All three complexes show that the third longer recognition helix is inserted into the major groove making base-specific contacts analogous to the prokaryotic repressors (Figure 5). Further HTH variants have recently been discovered where the HTH motif definition has been loosened to permit loops instead of turns between the

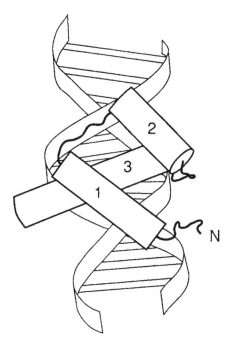

Figure 4. The helix-turn-helix variant motif as found in homeodomains. Cartoon of the *engrailed* homeodomain protein-DNA complex showing a longer recognition helix (3) positioned in the major groove. From Freemont et al. (1991) with permission.

5'   3'

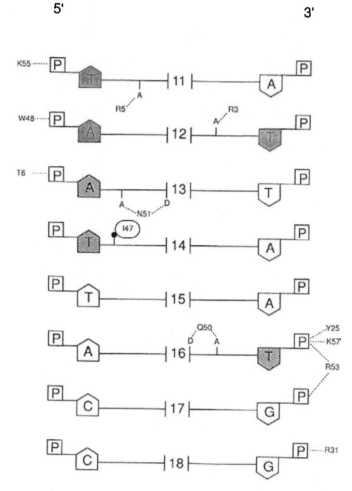

Figure 5. Specific amino acid-base contacts observed in the *engrailed* homeodomain protein DNA complex (Kissinger et al., 1990). See Figure 3 legend for details of symbols. From Freemont et al. (1991) with permission.

α-helices. These variants include a number of eukaryotic transcription factors, namely the third repeat of c-Myb (Ogata et al., 1992), the POU-specific (POUs) domain of Oct-1 (Assa-Munt et al., 1993; Dekker et al., 1993), an atypical homeodomain from hepatocyte nuclear factor I (LFB1/HNF1; Letting et al., 1993; Ceska et al., 1993), and the fork head domain of hepatocyte nuclear factor (HNF)-3γ-DNA complex (Clark et al., 1993). The variations of the motif include a one amino acid longer turn in c-Myb, a two-amino acid longer turn and extension at the C-terminus of the first helix in POUs, and most extreme, an insertion of 21 residues into the HTH motif for HNF1. The fork head HNF-3γ-DNA complex shows that the recognition helix lies in the major groove making base-specific contacts in a manner similar to that of the monomeric eukaryotic homeodomain proteins, and analogous to bacterial repressors. Interestingly, the structure of the HNF-3γ fork head domain is very similar to that of histone GH5, which also contains a HTH motif (Ramakrishnan et al., 1993), showing that the HTH motif can be found in very diverse proteins with little sequence similarity (Figure 6). This probably reflects the functional stability and potential variability that the motif can provide with DNA recognition achieved by specific amino acid-base interactions from a single α-helix placed within the major groove.

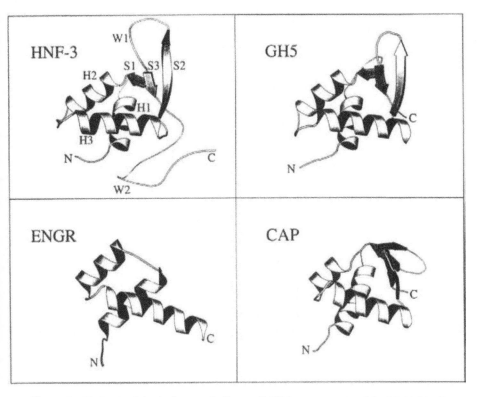

Figure 6. Variants of the helix-turn-helix motif. Ribbon cartoons of the DNA-binding domains of HNF-3λ (Clark et al., 1993), histone H5 (Ramakrishnan et al., 1993), *engrailed* homeodomain (ENGR; Kissinger et al., 1990), and catabolite activator protein (CAP; Schultz et al., 1991). All of the proteins are viewed in the same orientation with the helix-turn-helix motif indicated in the HNF-3 structure as H2 and H3. From Clark et al. (1993) with permission.

ZINC FINGER MOTIFS

Zinc fingers are autonomously folded protein domains which require zinc for stabilization through tetrahedral coordination via cysteine or histidine amino acid

(a)

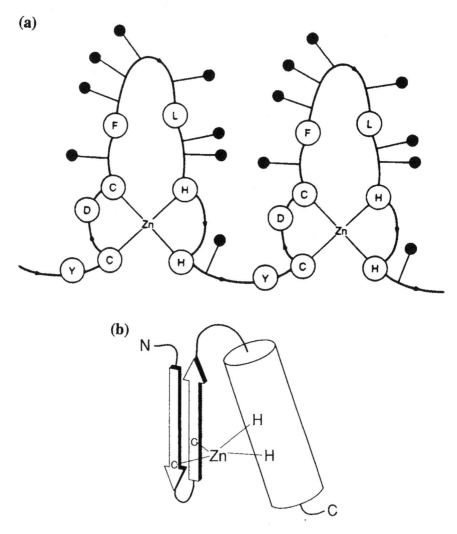

(b)

Figure 7. Schematic drawing of the Cys_2-His_2 zinc finger motif as discovered in the sequence of the *Xenopus* transcription factor TFIIIA. (**a**) The domain is centered on a tetrahedral arrangement of zinc ligands with the conserved hydrophobic residues responsible for stabilizing the secondary structure indicated. (**b**) Schematic model of a single zinc finger showing two anti-parallel β-strands and an α-helix tethered together by a zinc atom.

side chains. The archetypal zinc finger comprises 30 amino acids and was originally identified in the frog transcription factor TFIIIA as a tandemly repeated sequence motif (Miller et al., 1985). The spacing between the conserved cysteines and histidines allows a single zinc atom to coordinate four residues to form a fingerlike structure if drawn out on paper (Figure 7). It is now accepted that the motif generally constitutes a specific nucleic acid binding motif which binds zinc and DNA with

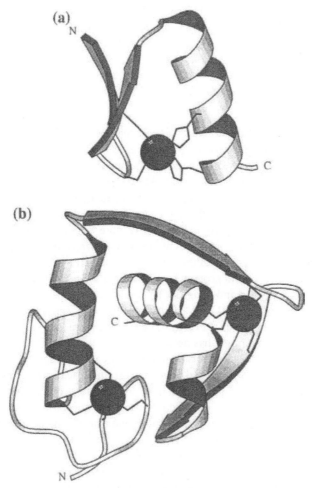

(continued)

Figure 8. Representative structures of zinc and DNA binding domains from the various zinc finger families. Zinc is shown as a black sphere and the zinc ligands are drawn in black lines. (a) The Cys$_2$-His$_2$ TFIIIA type zinc finger (Pavletich and Pabo, 1991). (b) The glucocorticoid hormone receptor DNA binding domain (Luisi et al., 1991). (c) The GAL4 two zinc cluster (Marmorstein et al., 1992). Reproduced from Current Opinion in Structural Biology with permission.

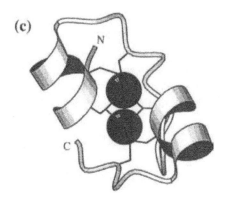

Figure 8. (Continued)

high affinity. Several NMR structures have been determined of either single or double zinc finger domains, and crystal structures of three zinc fingers and five zinc fingers bound specifically to DNA have also been described (Paveltich and Pabo, 1991, 1993; reviewed by Kaptein, 1992). All of the structures show that a single zinc finger domain comprises an antiparallel two-stranded β-sheet and an α-helix held together by a zinc atom which is bound tetrahedrally by the conserved cysteines and histidines, as predicted earlier (Figure 8a). There are some small structural variations between the different zinc fingers which have been studied to date, including differing lengths of the β-strand and α-helix, and an extra β-strand as observed in the *Drosophila* protein Tramtrack, but the overall structural frame-work is the same (Schwabe and Klug, 1994). The sequence-specific DNA interactions occur at the N-terminal end of the α-helix with each finger generally recognizing three base pairs as previously proposed (Figure 9a).

A second class of zinc finger has been described in the steroid/nuclear receptor family of proteins. This motif differs from the classical zinc finger in that it binds two zinc atoms to form a single folded domain of 60 amino acids, with four cysteine ligands for each zinc atom (Figure 8b). The crystal structure of the glucocorticoid receptor zinc finger bound to its specific DNA target sequence (Luisi et al., 1991) is very similar to that of the estrogen receptor zinc finger domain, as determined both by X-ray crystallography and 2D-NMR (Schwabe et al., 1993a,b), and shows that the receptor binds to its DNA target sequence as a dimer (Figure 9b), unlike the TFIIIA-like zinc fingers which bind as separate monomeric units. Sequence-specific DNA interactions occur through an α-helix, which lies in the major groove of the DNA (Figure 9b), similar to the Zif268 zinc finger structure and previously observed in the HTH -DNA complexes.

Another cysteine-rich motif which differs significantly from that of the TFIIIA and steroid/nuclear zinc fingers has been described in the yeast transcription factor GAL4. The motif comprises six cysteines and forms the entire DNA-binding

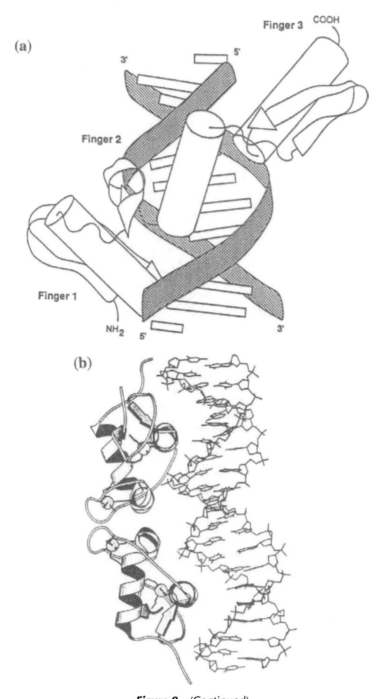

Figure 9. (Continued)

(c)

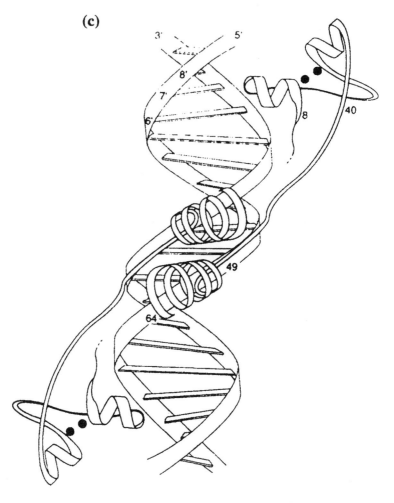

Figure 9. (Continued) Sequence-specific DNA binding by zinc finger domains. (**a**)
Schematic representation of the Zif269 three zinc finger-DNA complex with α-helices
as cylinders and β-strands as ribbons. The three fingers wrap around the major groove
with base-specific contacts formed between residues near the N-termini of each helix
and a three base pair subsite (adapted from Pavletich and Pabo, 1991). (**b**) Structure
of the dimeric glucocorticoid receptor DNA-binding domain bound to DNA (Luisi et
al., 1991). α-Helices from each monomer lie in successive major grooves as observed
in bacterial repressor-operator complexes. (**c**) Structure of the dimeric GAL4 DNA
binding domain bound to DNA (Marmorstein et al., 1992). Zinc atoms are shown as
small black spheres. The dimerization interface (49-64) comprises a coiled-coil. Parts
b and **c** are reproduced from Current Opinion in Structural Biology with permission.

domain of GAL4 (Figure 8c). Two zinc atoms are bound by the motif which, together with a coiled-coil dimerization domain, allows specific binding to a palindromic DNA target sequence. The domain is twofold symmetric and comprises two α-helices held together by two zinc atoms with one α-helix lying in the major groove forming sequence-specific DNA interactions (Figure 9c; Marmorstein et al., 1992).

Interestingly, the only common feature between the above described zinc finger structural motifs is that all three use either cysteine and/or histidine as zinc ligands. All three structures display different tertiary frameworks which are stabilized by bound zinc atoms that surprisingly do not always require four unique ligands, as shown in GAL4 where two zinc atoms are shared by two cysteine ligands. Due to the redox potential of the cell, zinc atoms bound to cysteine residues can substitute for disulfide bonds which are only found extracellularly. Therefore, by using cysteines and histidines as specific zinc ligands, a potentially large number of different tertiary templates can be formed, each stabilized by bound zinc (Figure 8). On this theme, a number of recent zinc fingerlike DNA/RNA-binding structures have been elucidated, most notably the eukaryotic transcription factors GATA-1 and TFIIS (for review see Schwabe and Klug, 1994).

β-SHEET MOTIF (RIBBON-HELIX-HELIX)

Although most DNA binding structures use α-helical motifs for DNA recognition, a number of recently determined structures are found to use antiparallel β-sheets for interaction with the DNA bases to mediate specific sequence recognition. These include the prokaryotic regulatory proteins, MetJ and Arc repressors; both structures have been determined bound to an operator half site (Figure 10a,b; Somers and Phillips, 1992; Raumann et al., 1994a) and the Mnt repressor, the primary sequence of which is approximately 40% homologous to that of Arc (Sauer et al., 1983). The *E. coli* MetJ repressor is involved in the control of methionine biosynthesis by binding cooperatively, as dimers, to tandem repeats of six specific operator sites (Phillips et al., 1989). Arc and Mnt repressors are encoded in the *immunity* I region of the *Salmonella* bacteriophage P22 and both are autoregulatory (Knight et al., 1990). The binding motif of these proteins consists of a $(\beta\alpha\alpha)_2$ structure with the β sheets of each monomer arranged antiparallel in the major groove of the DNA half site (Figure 10c). They bind to operator sites of 18–21 base pairs as tetramers with dimers binding to each half of the operator making cooperative interactions with the neighboring dimer (reviewed by Raumann et al., 1994b). The $(\beta\alpha\alpha)_2$ motif contacts the DNA through extensive phosphate backbone contacts which also serve to orient the β-sheet in the major groove. Many of the phosphate contacts involve hydrogen bonds donated by NH groups mainly from helix B of each protein monomer and from a turn region preceding the β-strand (Figure 10c). Differences between the MetJ-DNA and Arc-DNA complexes include the spacing between the half sites in the operators and the cooperative binding interactions between the two

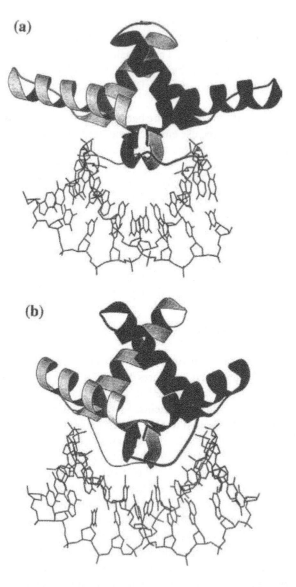

Figure 10. Major groove DNA recognition by β-sheets. (a) Ribbon representation of the MetJ dimer bound to an operator half site (Somers and Phillips, 1992). The two antiparallel β-strands lie in the major groove making DNA sequence-specific contacts. (b) Arc dimer bound to an operator half site viewed in the same orientation as the MetJ complex (Raumann et al., 1994b). (c) Schematic cartoon showing the position of the two antiparallel β-strands in the major groove. The conserved phosphate contacts observed in the Arc and MetJ complexes by helix B and a tandem turn are indicated by arrows. Reproduced from Current Opinion in Structural Biology with permission.

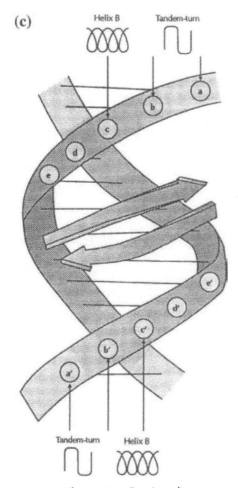

Figure 10. (Continued)

neighboring dimers. In both complexes the DNA bends by approximately 50° from its linear conformation in order to allow the two dimers to interact cooperatively. Other differences in the DNA structure for each of the complexes, for example, variations in the helical repeat and the groove width, are most likely specific to each of the repressors and cannot be considered general observations for the class of $(\beta\alpha\alpha)_2$ proteins.

Recently, sequence patterns have been found in the eukaryotic auxin induced plant proteins which suggest the presence of a $(\beta\alpha\alpha)_2$ motif (Abel et al., 1994). These sequence patterns appear to be significantly similar to the $(\beta\alpha\alpha)_2$ DNA binding domains of Arc and MetJ. The sequence similarity is further supported by a comparison of the pattern of hydrophobicity between the prokaryotic $(\beta\alpha\alpha)_2$ domains of Arc and MetJ and that of the plant protein and the amphipathic nature of their secondary structure elements. These sequence and proposed structural similarities suggest the presence of the $(\beta\alpha\alpha)_2$ domain in eukaryotic proteins as a means for conveying specificity for DNA binding and heterodimerization.

Yet another example of a protein which interacts with DNA through β-sheet structure is the TATA-box-binding protein TBP-2. TBP-2 plays a central role in

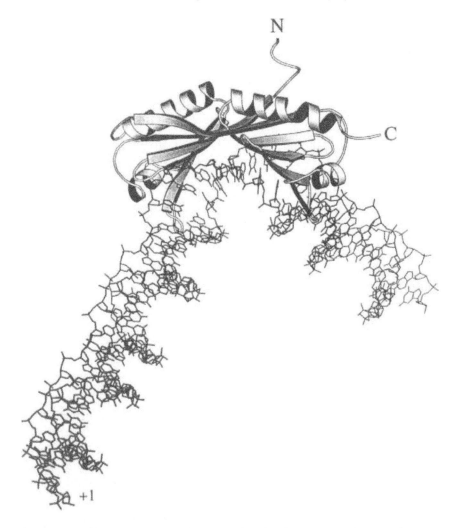

Figure 11. Ribbon diagram of the TATA box-binding protein from *Arabidopsis* complexed with a TATA element. The protein forms a "saddle" which sits on the DNA. Reproduced from Current Opinion in Structural Biology with permission.

transcription where, upon binding to the TATA box, it causes the formation of a transcriptional pre-initiation complex through establishing protein-protein contacts with other transcriptional elements. The structure forms a dimer shaped like a saddle (see Figure 11) (Nikolov et al., 1992). The concave surface of the saddle comprises an eight stranded β-sheet with two further β-strands on each edge of the sheet forming the spurs. The convex surface of the protein is covered by four α-helices and exposes predominantly hydrophobic residues which make direct contact to the DNA. Two crystal structures of complexes between TBP-2 and two different DNA sequences have been determined characterizing the interactions between the protein and DNA (Kim et al., 1993a,b). Both structures show major distortions in the DNA structure with the TBP binding to the minor groove of the DNA (see Figure 11).

BASIC REGION LEUCINE ZIPPER COILED-COIL MOTIF

DNA binding proteins which contain the basic region leucine zipper coiled-coil motif (bZIP) can form homo- or heterodimers, as well as recognize specific DNA target sequences (reviewed by Kerppola and Curan, 1991). The bZIP motif is found in a variety of eukaryotic transcription factors and can be divided into two subdomains: C-terminal leucine zipper region which functions in dimerization and an N-terminal basic region which contacts the DNA target site. The dimerization region consists of a two-stranded parallel coiled-coil arrangement of α-helices. The primary sequence of leucine zippers consists of a heptad repeat $(abcdefg)_n$ of 30–40 residues with leucines situated at position a and hydrophobic residues such as Val or Ile found at position d of the heptad (Figure 12). This positioning of these residues results in a hydrophobic region along one surface of each helix. This surface induces the helices to form dimers which are stabilized by the packing of knobs, formed by

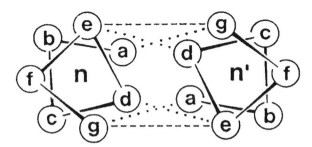

Figure 12. Schematic diagram of the α-helical interactions in a hypothetical helical coil-coil motif. The two α-helices are shown in a parallel alignment, although in reality they wrap around each other to form a left-handed superhelix. Residues at position a and d are generally hydrophobic (leucines in leucine zippers), whereas position e and g are generally charged amino acids.

(a)

(b)

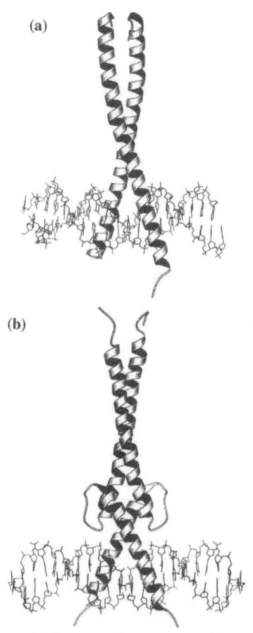

Figure 13. Structures of the basic region leucine zipper and basic region helix-loop-helix DNA binding domains. (a) GCN4 bZIP dimer-DNA complex showing the basic helices from each monomer positioned in adjacent major grooves (Ellenberger et al., 1992). (b) The Max bHLH homodimer-DNA complex (Ferre-D'Amare et al., 1993). Reproduced from Current Opinion in Structural Biology with permission.

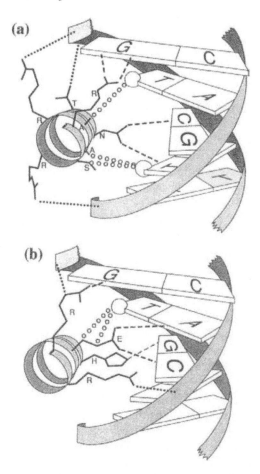

Figure 14. Schematic diagram of the specific amino acid-base contacts in the basic regions of GCN4 and Max. (**a**) GCN4/AP-1 complex with specific hydrogen bonds shown as dashed lines (Ellenberger et al., 1992). (**b**) Max DNA complex (Ferre-D'Amare et al., 1993). The sequence specificity for the base at the center of the hexamer DNA target site is governed by an arginine residue shown at the top. Reproduced from Current Opinion in Structural Biology with permission.

hydrophobic residues of one helix, into holes, formed by the spaces between the hydrophobic side chains of the neighboring helix (Crick, 1953). The N-terminal basic region contains approximately 30 residues and is rich in conserved arginines and lysines, as well as other conserved residues. The basic region also adopts a helical conformation which lies in the major groove of the DNA target sequence. Many of the conserved residues are involved in contacts with both the phosphodiester backbone and the nucleotide bases (for review see Ellenberger, 1994).

An example of a protein containing basic region leucine zipper coiled-coil motif is the yeast transcriptional activator GCN4. Two crystal structures have been determined of the GCN4 homodimer basic region in complex with different DNA sites, one with adjacent half sites (Konig and Richmond, 1993) and one with overlapping half sites (Ellenberger et al., 1992). Both structures reveal a pair of continuous α-helices that form a parallel coiled-coil over their carboxy-terminal dimerization region and diverge toward their basic amino terminal DNA binding region (Figure 13a). The divergence from the coiled-coil facilitates the amino termini to pass through the major groove of the DNA binding site, making extensive contacts with the DNA. The main contacts to the DNA made by the protein are with the edges of base pairs and the phosphate backbone of one half site (Figure 14a).

BASIC REGION HELIX-LOOP-HELIX MOTIF

Proteins which comprise a basic region helix-loop-helix motif (bHLH) are similar to those which form leucine zipper motifs, in that they can form homo- or heterodimers. However, rather than having two distinct α-helices as in the bZIP family, this class of DNA binding proteins has four distinct α-helices (Figure 13b). The monomeric unit of these proteins comprises two α-helices, with the longer C-terminal helix forming a leucine zipper as in the bZIP family (see above). The N-terminal helix forms the basic region of each monomer and contains residues along one surface which contact base pairs in the major groove of the DNA binding site, as well as residues adjacent to this helical surface which contact the phosphodiester backbone (for review see Ellenberger, 1994).

An example of a protein containing the basic region helix-loop-helix motif is the Max homodimer. Max is an important member of the bHLH family as it heteroligomerizes with the Myc oncoproteins mediating specific DNA interaction *in vivo*. The oncogenic activity of c-Myc requires dimerization with Max and it appears that Max can function as both a suppressor and activator of Myc *in vivo*. The crystal structure of the Max homodimer in complex with a DNA target sequence shows that the helix-loop-helix adopts a left-handed parallel four helix bundle (Ferre-D'Amare et al., 1993; Figure 13b). The two basic α-helices are in contact with the major groove of the DNA binding site. The loop region adopts an extended strand conformation connecting the two helices of each monomer and is stabilized by contacts with both the DNA backbone and with the DNA binding helix from the second monomer (Figure 12b). Most bHLH proteins bind hexameric sequences of -CANNTG- in which the central bases (NN) are -CG- or -GC- (Blackwell et al., 1993). The Max homodimer makes numerous contacts with the conserved base pairs including important hydrogen bonds between the cytosine and adenine bases and a glutamic acid residue which is conserved within the family (Figure 13b). In addition a highly conserved histidine residue, located one helical turn away from the conserved glutamic acid, also hydrogen bonds to the conserved guanine residue of the DNA target sequence (Figure 14b).

DNA MODIFICATION AND CLEAVAGE ENZYMES

A number of structures of DNA modification and cleavage enzymes either bound or unbound to specific DNA sequences have now been determined. Among these are bacterial restriction endonucleases which are examples of dimeric proteins that recognize small (4–8 base pairs) specific DNA target sequences and cleave phosphodiester bonds between particular bases with the requirement of Mg^{2+} as co-factor (for review see Anderson, 1993). These enzymes have revolutionized modern molecular biology and, in particular, DNA cloning technology as they provide exquisite sequence specificity and cleavage. There are three structures to date of restriction endonucleases, namely BamHI (Newman et al., 1994), EcoR1 (Kim et al., 1990), and EcoRV (Winkler et al., 1993). Interestingly, BamH1 and EcoR1 are structurally related despite little sequence homology between them, and all three enzymes share conserved active site residues and geometry (Newman et al., 1994). EcoR1 binds to the major groove side of its DNA target sequence using a bundle of four α-helices and a short region of extended chain to make base-specific contacts (Rosenberg, 1991). EcoRV, however, approaches its recognition sequence from the minor groove side and contacts both the major and minor grooves using four short loops (Winkler et al., 1993). Although both enzymes bind to their respective target sequences differently, they share a common structural motif for phosphodiester cleavage, namely two acidic residues and a basic residue.

The most commonly occurring group of DNA modification enzymes are DNA methylases. These enzymes act as monomers and transfer a methyl group from the co-factor S-adenosyl-methionine to one of the bases within the specific recognition sequences (Wilson and Murray, 1991). In bacteria, there are three main classes of DNA methylases classified according to which atom the methyl group is transferred to (e.g., m5C-methylases transfer the methyl group to the C5 atom of cytosine bases). Eukaryotic methylases, on the other hand, methylate only the C5 atom of cytosines in CG sequences (Bird, 1992). Among the different classes of methylases, there exist significant sequence similarities including motifs specific to all methylases (Anderson, 1993). Recently, the structure of the bacterial cytosine C5 methylase HhaI has been determined bound to its DNA target sequence (Klimasauskas et al., 1994). The structure is remarkable in that it shows the cytosine base which is methylated, swung out of the DNA double helix, and positioned next to the co-factor for methyl transfer. The enzyme comprises two domains, a larger catalytic domain and a smaller domain, between which there is a cleft where the DNA binds. Two glycine rich loops contributed by the small domain contact the DNA from the major groove and are responsible for all the sequence specific contacts. What is extraordinary about this complex is that it highlights the enormous potential conformational flexibility of DNA, in that a nucleotide base can be swung out of the double helix where necessary, without drastic effects on the overall conformation of the DNA.

DNA REPLICATION AND RECOMBINATION PROTEINS

The recombination and template-directed replication of DNA are among the most complex and difficult biological processes known. To design suitable protein structural frameworks to perform such complex tasks would indeed be a daunting prospect. However, the structures of an increasing number of bacterial proteins which perform some of these functions are now available, allowing detailed insights into the mechanisms of such processes. The most notable of these is the Klenow fragment of DNA polymerase I, the structure of which has increased our under- standing of DNA replication and fidelity (for review see Joyce, 1991). DNA polymerase I has two major roles in *E. coli*, namely the repair of damaged duplex DNA and the processing of Okazaki fragments, and is comprised of three domains. The Klenow fragment contains two domains, the DNA polymerase domain and the 3'-5' exonuclease domain, and catalyzes the template directed DNA replication, making sure that the correct nucleotides are incorporated and editing out mis- matched bases if they occur. The structure of the Klenow fragment clearly shows the two domains forming a large U-shaped cleft of appropriate dimensions to bind duplex DNA (Ollis et al., 1985). Numerous biochemical and crystallographic studies have led to a detailed understanding of the polymerase and editing activities and a structural mechanism has been proposed for DNA polymerization (for review see Steitz, 1993; Beese et al., 1993). Recently, several structures of polymerase domains have been determined including HIV-1 reverse transcriptase (Kohlstaedt et al., 1992), T7 RNA polymerase (Sousa et al., 1993), and rat DNA polymerase β (Davies et al., 1994; Pelletier et al., 1994). Interestingly, there are distinct topologi- cal differences between the structures of the different polymerase domains suggest- ing that although they catalyze the same nucleotide transfer reaction, they have evolved separately. However, a number of carboxylate residues in the active sites are conserved as is the binding of divalent metal ions.

The inversion and deletion of DNA is of primary importance in the propagation and rearrangement of genetic material in both prokaryotes and eukaryotes. There are a number of prokaryotic enzymes which catalyze site-specific recombination reactions that have been the focus of several structural analyses. Most notable is γδ-resolvase from *E. coli* which catalyzes a site-specific strand exchange reaction resolving the cointegrate molecule formed as an intermediate in the transposition of the transposon (for review see Grindley, 1994). Dimers of γδ-resolvase bind to three DNA target sequences within a 115 base pair site called *res*, forming half of the synaptic complex which comprises 12 resolvase monomers and two *res* sites. γδ-resolvase consists of two domains, a catalytic domain and a DNA binding domain which contains a helix-turn-helix motif, as seen previously in bacterial repressors. The structure of the dimeric catalytic domain reveals a flexible mole- cule, possibly reflecting the requirement of the 12 molecules within the synaptic complex to perform differing tasks (Rice and Steitz, 1994). Another prokaryotic recombinase which has received considerable attention is Hin invertase from

Salmonella. Hin has significant sequence homology with γδ-resolvase and also comprises a catalytic domain and a DNA binding domain which contains an HTH motif. Recently the structure of the DNA binding domain of Hin bound to a DNA target sequence has been determined and shows that DNA contacts are made primarily in the major groove by a recognition helix (Feng et al., 1994a,b). The overall fold of the Hin domain is similar to that of the eukaryotic homeodomains, although the contacts with DNA between both domains are quite different. A model based on the DNA binding domain of Hin and the dimeric catalytic domain of γδ-resolvase has been proposed, suggesting that the DNA binding domains grip the DNA on the opposite side to the catalytic domains (Feng et al., 1994a,b).

SUMMARY

The application of protein crystallography and two-dimensional NMR to study proteins which interact with DNA has provided detailed information about protein-DNA recognition and specificity. Numerous structural motifs have evolved to allow a diverse variety of interactions between proteins and DNA. These include motifs such as the helix-turn-helix, zinc finger, basic region leucine zipper, and ribbon-helix-helix, some of which are found in both prokaryotes and eukaryotes. DNA recognition is usually achieved through specific protein amino acid-DNA base pair interactions which can be mediated via water molecules, in either the major or minor grooves of the DNA. However, interactions between protein side chains and the phosphodiester backbone is a common feature of non-sequence-specific protein-DNA recognition. The flexibility and existence of DNA in a number of different conformations contributes significantly to the overall process of protein-DNA recognition. The enormous variety of protein structures and motifs which can interact with DNA probably reflects the diversity of biological function which requires the formation of protein-DNA complexes.

ACKNOWLEDGMENTS

We are very grateful to Dr. David Weedon of Current Science Ltd. (London) for providing previously published figures and to all the authors who gave permission for their figures to be reproduced. We also thank Dr. Kathy Borden and Professor John Fothergill for helpful comments on the manuscript.

REFERENCES

Abel, S., Oeller, P.W., & Theologis, A. (1994). Early auxin-induced genes encode short-lived nuclear proteins. Proc. Natl. Acad. Sci. USA 91, 326–330.

Aggarwal, A.K., Rodgers, D.W., Drottar, M., Ptashne, M., & Harrison, S.C. (1988). Recognition of a DNA operator by the repressor of phage 434: A view at high resolution. Science 242, 899–907.

Anderson, J.E. (1993). Restriction endonucleases and modification methylases. Curr. Opin. Struct. Biol. 3, 24–30.

Anderson, W.F., Ohlendorf, D.H., Takeda, Y., & Matthews, B.W. (1981). Structure of the *cro* repressor from bacteriophage λ and its interaction with DNA. Nature 290, 754–758.

Assa-Munt, N., Mortishire-Smith, R.J., Aurora, R., Herr, W., & Wright, P.E. (1993). The solution structure of the Oct-1 POU-specific domain reveals a striking similarity to the bacteriophage λ repressor DNA-binding domain. Cell 73, 193–205.

Beese, L.S., Derbyshire, V., & Steitz, T.A. (1993). Structure of DNA polymerase I Klenow fragment bound to duplex DNA. Science 260, 352–355.

Bird, A. (1992). The essentials of DNA methylation. Cell 70, 5–8.

Blackwell, T.K., Huang, J., Ma, A., Kretzner, L., Alt, F.W., Eisenman, R.N., & Weintraub, H. (1993). Binding of myc proteins to canonical and noncanonical DNA sequences. Mol. Cell. Biol. 13, 5216–5224.

Brennan, R.G. (1992). DNA recognition by the helix-turn-helix motif. Curr. Opin. Struct. Biol. 2, 100–108.

Brennan, R.G., Roderick, S.L., Takeda, Y., & Matthews, B.W. (1990). Protein-DNA conformational changes in crystal structure of a λ *cro*-operator complex. Proc. Natl. Acad. Sci. USA 87, 8165–8169.

Ceska, T.A., Lamers, M., Monaci, P., Nicosia, A., Cortese, R., & Suck, D. (1993). The X-ray structure of a atypical homeodomain present in the rat liver transcription factor LFB1/HNF1 and implication for DNA binding. EMBO J. 12, 1805–1810.

Clark, K.L., Halay, E.D., Lai, E., & Burley, S.K. (1993). Co-crystal structure of the HNF-3/*fork head* DNA-recognition motif resembles histone H5. Nature 364, 412–420.

Crick, F.H.C. (1953). The packing of α-helices: simple coiled-coil. Acta Cryst. 6, 689–697.

Davies, J.F., Almassy, R.J., Hostomska, Z., Ferre, A.R., & Hostomsky, Z. (1994). 2.3 Å crystal structure of the catalytic domain of DNA polymerase β. Cell 76, 1123–1133.

Dekker, N., Cox, M. Boelens, R., Verrijzer, C.P., van der Vliet, P.C., & Kaptein, R. (1993). Solution structure of the POU-specific DNA-binding domain of Oct-1. Nature 362, 852–855.

Drew, H., Takano, T., Tanaka, S., Itakura, K., & Dickerson, R.E. (1980). High-salt d(CpGpCpG): A left-handed Z-DNA double helix. Nature 286, 567–573.

Ellenberger, T. (1994). Structures of the bZIP and bHLH DNA-binding domains. Curr. Opin. Struct. Biol. 4,12–21.

Ellenberger, T.E., Brandl, C.J., Struhl, K., & Harrison, S.C. (1992). The GCN4 basic region leucine zipper binds DN as a dimer of uninterrupted α-helices: Crystal structure of the protein-DNA complex. Cell 71, 1223–1237.

Feng, J.A., Johnson, R.C., & Dickerson, R.E. (1994a). Hin recombinase bound to DNA: The origin of specificity in major and minor groove interactions. Science 263, 348–355.

Feng, J.A., Dickerson, R.E., & Johnson, R.C. (1994b). Proteins that promote DNA inversion and deletion. Curr. Opin. Struct. Biol. 4, 60–66.

Ferre-D'Amare, A.R., Prendergast, G.C., Ziff, E.B., & Burley, S.K. (1993). Recognition by max of its cognate DNA through a dimeric b/HLH/Z domain. Nature 363, 38–45.

Freemont. P.S., Lane, A.N., & Sanderson, M.R. (1991). Structural aspects of protein-DNA recognition. Biochem. J. 278, 1–23.

Fuller, W., Wilkins, M.H.F., Hamilton, H.R., & Arnott, S. (1965). The molecular configuration of deoxyribonucleic acid. IV. X-ray diffraction study of the A-form. J. Mol. Biol. 12, 60–80.

Grindley, N.D.F. (1994). Resolvase mediated site specific recombination. In: Nucleic Acids and Molecular Biology, Vol. 8 (Eckstein, F. & Lilley, D.M.J., eds.). Springer-Verlag, Berlin and New York.

Joyce, C.M. (1991). Can DNA polymerase I (Klenow fragment) serve as a model for other polymerases? Curr. Opin. Struct. Biol. 1, 123–129.

Jordan, S.R., & Pabo, C.O. (1988). Structure of the lambda complex at 2.5 Å resolution: Details of the repressor-operator interactions. Science 242, 893–899.

Kaptein, R. (1992). Zinc finger structures. Curr. Opin. Struct. Biol. 2, 109–115.

Kerpolla, T.K., & Curran, T. (1991). Transcription factor interactions: Basics on zippers. Curr. Opin. Struct. Biol. 1, 71–79.

Kim, Y. Garble, J.C., Love, R., Greene, P.J., & Rosenberg, J.M. (1990). Refinement of EcoR1 endonuclease crystal structure: A revised chain tracing. Science 249, 1307–1309.

Kim, Y., Geiger, J.H., Hahn, S., & Sigler, P.B. (1993a). Crystal structure of a yeast TBP/TATA box complex. Nature 365, 512–520.

Kim, J.L., Nikolov, D.B., & Burley, S.K. (1993b). Co-crystal structure of TBP recognizing the minor groove of a TATA element. Nature 365, 520–527.

Kissinger, C.R., Liu, B., Martin-Blanco, E., Korberg, T.B., & Pabo, C.O. (1990). Crystal structure of an engrailed homeodomain/DNA complex at 2.8 Å resolution: A framework for understanding homeodomain/DNA interactions. Cell 63, 579–590.

Klimasauskas, S., Kumar, S., Roberts, R.J., & Cheng, X. (1994). HhaI methyltransferase flips its target base out of the DNA helix. Cell 76, 357–369.

Knight, K.L., Bowie, J.U., Vershon, A.K., Kelley, R.D., & Sauer, R.T. (1990). The arc and mnt repressors: A new class of sequence-specific DNA binding proteins. J. Biol. Chem. 264, 3639–3642.

Konig, P., & Richmond, T.J. (1993). The X-ray structure of GCN4 bZIP bound to ATF/CREB site DNA shows the complex depends on DNA flexibility. J. Mol. Biol. 233, 139–154.

Kohlstaedt, L.A., Wang, J., Friedman, J.M., Rice, P.A., & Steitz, T.A. (1992). Crystal structure at 3.5 Å resolution of HIV-1 reverse transcriptase complexed with inhibitor. Science 256, 1783–1790.

Letting, B., DeFrancesco, R., Tomei, L., Cortese, R., Otting, G., & Wüthrich, K. (1993). Three-dimensional NMR-solution structure of the polypeptide fragment 195-286 of the LFB1/HNF1 transcription factor from rat liver comprises a non-classical homeodomain. EMBO J. 12, 1797–1803.

Luisi, B.F., Xu, W.X., Otwinowski, Z., Freedman, L.P., Yamamoto, K.R., & Sigler, P. B. (1991). Crystallographic analysis of the interaction of the glucocorticoid receptor with DNA. Nature 352, 497–505.

Marmorstein, R., Carey, M., Ptashne, M., & Harrison, S.C. (1992). DNA recognition by GAL4: Structure of a protein-DNA complex. Nature 356, 408–414.

McKay, D.B., & Steitz, T.A. (1981). The structure of catabolite gene activator protein at 2.9 Å resolution suggests binding to left handed B-DNA. Nature 290, 744–749.

Miller, J., McLachlan, A.D., & Klug, A. (1985). A repetitive zinc-binding domain in the protein transcription factor IIIA from *Xenopus* oocytes. EMBO J. 4, 1609–1614.

Mondragon, A., & Harrison, S.C. (1991). The phage 434 *cro*/Or1 complex at 2.5 Å resolution. J. Mol. Biol. 219, 321–334.

Newman, M., Strzelecka, T., Dorner, L.F., Schildkraut, I., & Aggarwal, A.K. (1994). Structure of restriction endonuclease *Bam* HI and its relationship to *Eco*RI. Nature 368, 660–664.

Nickol, J., Behem M., & Felsenfeld, G. (1982). Effect of the B-Z transition in poly(dG-m^{5}dC) poly(dG-m^{5}dC) on nucleosome formation. Proc. Natl. Acad. Sci. USA 79, 1771–1775.

Nikolov, D.B., Hu, S.-H., Lin, J., Gasch, A., Hoffmann, A., Horikoshi, M., Chua, N.-H., Roeder, R.G., & Burley, S.K. (1992). Crystal structure of TFIID TATA-box binding protein: A central transcription factor. Nature 360, 40–46.

Ogata, K., Hojo, H., Aimoto, S, Nakai, T., Nakamura, H., Sarai, A., Ishii, S., & Nishimura, Y. (1992). Solution structure of a DNA-binding unit of myb: A helix-turn-helix motif with conserved tryptophans forming a hydrophobic core. Proc. Natl. Acad. Sci. USA 89, 6428–6432.

Ollis, D.L., Brick, P., Hamlin, R. Xuong, N.G., & Steitz, T.A. (1985). Structure of the large fragment of *Escherichia coli* DNA polymerase I complexed with dTMP. Nature 313, 762–766.

Otting, G., Qian, Y.-Q., Billeter, M., Müller, M., Affolter, M., Gehring, W.J., & Wüthrich, K. (1990). Protein-DNA contacts in the structure of a homeodomain-DNA complex determined by NMR spectroscopy in solution. EMBO J. 9, 3085–3092.

Otwinowski, Z., Schevitz, R.W., Zhang, R.G., Lawson, C.L., Joachimiak, A., Marmorstein, R.Q., Luisi, B.F., & Sigler, P.B. (1988). Crystal structure of *trp* repressor/operator complex at atomic resolution. Nature 355, 321–329.

Pavletich, N.P., & Pabo, C.O. (1991). Zinc finger-DNA recognition: Crystal structure of a Zif268-DNA complex at 2.1 Å. Science 252, 809–817.

Pavletich, N.P., & Pabo, C.O. (1993). Crystal structure of a five-finger GLI-DNA complex: New perspectives on zinc fingers. Science 261, 1701–1707.

Pelletier, H., Sawaya, M.R., Kumar, A., Wilson, S.H., & Kraut, J. (1994). Structures of ternary complexes of rat DNA polymerase β, a DNA template-primer and ddCTP. Science 264, 1891–1903.

Phillips, S.E., Manfield, I. Parsons, I., Davidson, B.E., Rafferty, J.B., & Somers, W.S. (1989). Cooperative tandem binding of the *E. coli* methionine repressor. Nature 341, 711–715.

Ramakrishnan, V., Finch, J.T., Graziano, V., Lee, P.L., & Sweet, R.M. (1993). Crystal structure of globular domain of histone H5 and its implications for nucleosome binding. Nature 362, 219–223.

Raumann, B.E., Rould, M.A., Pabo, C.O., & Sauer, R.T. (1994a). Arc repressor-operator complex at 2.6 Å resolution: New perspectives on β-sheet-DNA interactions. Nature 367, 754–757.

Raumann, B.E., Brown, B.M., & Sauer, R.T. (1994b). Major groove DNA recognition by β-sheets. Curr. Opin. Struct. Biol. 4, 36–43.

Razin, A., & Riggs, A.D. (1980). DNA methylation and gene function. Science 210, 604–610.

Rice, P.A., & Steitz, T.A. (1994). Refinement of γδ resolvase reveals a strikingly flexible molecule. Structure 2, 371–384.

Rodgers, D.W., & Harrison, S.C. (1993). The complex between phage 434 repressor DNA binding domain and operator site Or3; structural differences between consensus and non-consensus half-sites. Structure 1, 227–240.

Rosenberg, J.M. (1991). Structure and function of restriction endonucleases. Curr. Opin. Struct. Biol. 1, 104–113.

Saenger, W. (1984). In: Principles of Nucleic Acid Structure. Springer Verlag, Berlin.

Sauer, R.T., Krovatin, W., De Anda, J., Yourerian, P., & Susskind, M.M. (1983). Primary structure of the *Imm*I immunity region of bacteriophage P22. J. Mol. Biol. 168, 699–713.

Santella, R.M., Grunberger, D., Weinstein, I.B., & Rich, A. (1981). Induction of the Z conformation in poly(dG-dC)·poly(dG-dC) by binding of N-2-acetylaminofluorene to guanine residues. Proc. Natl. Acad. Sci. USA 78, 1451–1455.

Schultz, S.C., Shields, G.C., & Steitz, T.A. (1991). Crystal structure of a CAP-DNA complex: The DNA is bent by 90°. Science 253, 1001–1007.

Schwabe, J.W.R., & Klug, A. (1994). Zinc mining for protein domains. Nature Struct. Biol. 1, 345–349.

Schwabe, J.W.R., Chapman, L., Finch, J.T., & Rhodes, D. (1993a). The crystal structure of the estrogen receptor DNA-binding domain bound to DNA: How receptors discriminate between their response elements. Cell 75, 567–578.

Schwabe, J.W.R., Chapman, L., Finch, J.T., Rhodes, D., & Neuhaus, D. (1993b). DNA recognition by the estrogen receptor: From solution to crystal. Structure 1, 187–204.

Somers, W.S., & Phillips, S.E. (1992). Crystal structure of the Met repressor-operator complex at 2.8 Å resolution reveals DNA recognition by β-strands. Nature 359, 387–393.

Sousa, R., Chung, Y.J., Rose, J.P., & Wang, B.C. (1993). Crystal structure of bacteriophage T7 RNA polymerase at 3.3 Å resolution. Nature 364, 593–599.

Steitz, T.A. (1993). DNA- and RNA-dependent DNA polymerases. Curr. Opin. Struct. Biol. 3, 31–38.

Travers, A. (1993). DNA structure. In: DNA-Protein Interactions, pp. 1–26, Chapman and Hall, London.

Wang, J.C., Jacobsen, J.H., & Saucier, J.M. (1977). Physiochemical studies on interactions between DNA and RNA polymerase. Unwinding of the DNA helix by *Escherichia coli* RNA polymerase. Nucl. Acids. Res. 4, 1225–1241.

Wang, A.H.J., Quigley, G.J., Kolpak, F.J., Crawford, J.L., van Boom, J.H., van der Marel, G., & Rich, A. (1979). Molecular structure of a left-handed double helical DNA fragment at atomic resolution. Nature 282, 680–686.

Wang, A.H.J., Fujii, S., van Boom, J.H., & Rich, A. (1982). Molecular structure of the octamer of d(GGCCGGCC): Modified A-DNA. Proc. Natl. Acad. Sci. USA 79, 3968–3972.

Wing, R., Drew, H., Takano, R., Broka, C., Tanaka, S., Itakura, K., & Dickerson, R.E. (1980). Crystal structure analysis of a complete turn of DNA. Nature 287, 755–758.

Wilson, G.G., & Murray, N.E. (1991). Restriction and modification systems. Ann. Rev. Genet. 25, 585–627.

Winkler, F.K., Banner, D.W., Oefner, C., Tsernoglou, D., Brown, R.S., Heathman, S.P., Bryan, R.K., Martin, P.D., Petratos, K., & Wilson, K.S. (1993). The crystal structure of EcoRV endonuclease and of its complexes with cognate and non-cognate DNA fragments. EMBO J. 12, 1781–1795.

Wolberger, C., Vershon, A.K., Liu, B., Johnson, A.D., & Pabo, C.O. (1991). Crystal structure of a Mat α2 homeodomain-operator complex suggests a genetic model for homeodomain-DNA interactions. Cell 67, 517–528.

Zimmermann, S.B., & Pheiffer, B.H. (1979). A direct demonstration that the ethanol-induced transition of DNA is between the A and B forms: An X-ray diffraction study. J. Mol. Biol. 135, 1023–1027.

Chapter 6

Transcription Factors

DAVID S. LATCHMAN

Principles of Medical Biology, Volume 5
Molecular and Cellular Genetics, pages 117–133.
Copyright © 1996 by JAI Press Inc.
All rights of reproduction in any form reserved.
ISBN: 1-55938-809-9

INTRODUCTION

Human gene regulation is a highly regulated process as should be obvious to anyone who has ever dissected a human body. The various different tissues and organs differ dramatically from one another and they all synthesize different proteins: immunoglobulins in B lymphocytes, myosin in muscle, insulin in the pancreas, and so on. Moreover, with few exceptions, all these different cells have the same sequence of DNA encoding these proteins as was present in the initial single-celled zygote from which they arose during embryonic development. Clearly some process of gene regulation operates to decide which genes within the DNA will be active in producing proteins in each cell type.

In most cases, this process operates at the level of transcription by selecting which genes will be copied into the primary RNA transcript (for review see Latchman, 1995). Once this has been achieved, it will be followed by all the other stages which process this RNA into the mRNA that is translated into protein by the ribosomes. Thus the regulation of transcription has a critical role in the process of gene regulation. In turn transcription is controlled by the presence of short DNA sequences within gene regulatory regions which act as the binding sites for specific proteins known as transcription factors.

TRANSCRIPTION FACTORS AND THEIR REGULATION

Function of Transcription Factors

The primary function of transcription factors is to bind to specific sequences within the gene regulatory region and then to interact with one another and with the RNA polymerase enzyme itself in order to regulate the rate of transcription (for detailed review see Latchman, 1995). The nature of the transcription factors which bind to a specific gene plays a critical role in determining its pattern of transcription. Thus, if the gene binds predominantly transcription factors which are active in all cell types it will be expressed in all cells, whereas if it binds factors which are synthesized or are active in only one cell type, the gene will be expressed in a cell type-specific manner. Hence, in many cases, the tissue-specific activity of a particular gene will be dependent upon its binding of factors which are capable of activating its expression only in that tissue. It is therefore necessary to consider how the activity of transcription factors is regulated so that they are functional only in a specific tissue or cell type. Two basic methods are used to achieve this. These are regulation of synthesis and regulation of activity (Figure 1).

Regulation of Transcription Factor Synthesis

A classical case of a gene which is expressed only in one specific cell type is provided by the immunoglobulin genes which encode the proteins of the antibody

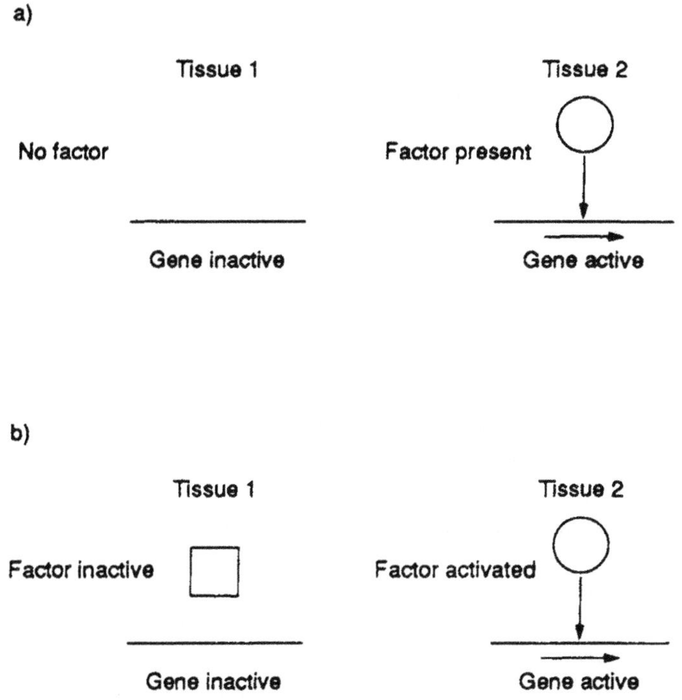

a)

Tissue 1   Tissue 2

No factor   Factor present

Gene inactive   Gene active

b)

Tissue 1   Tissue 2

Factor inactive   Factor activated

Gene inactive   Gene active

Figure 1. Transcription factors can activate transcription in a particular tissue if (a) they are synthesized only in that tissue or (b) are present in an active form only in that tissue.

molecule and which are transcribed only in B lymphocytes. These genes contain a binding site for the octamer binding transcription factor Oct-2 in their regulatory region or promoter upstream of the start site for transcription. The Oct-2 factor is synthesized only in B lymphocytes and binds only to the immunoglobulin gene promoter in B cells resulting in the transcription of the immunoglobulin genes only in antibody producing B cells. Similarly, genes expressed only in muscle cells, such as the creatine kinase gene, contain binding sites for the MyoD transcription factor which is synthesized only in skeletal muscle cells. This case is even more dramatic, however, because the artificial expression of MyoD in nonmuscle cells such as fibroblasts is sufficient to convert them into muscle cells, indicating that MyoD activates transcription of all the genes whose protein products are necessary to produce a differentiated muscle (Olson, 1990).

In many cases, such regulated synthesis of a transcription factor is achieved by regulating the transcription of the gene encoding the factor. Thus, like the Oct-2 protein, the Oct-2 mRNA is absent from non-B cells suggesting the existence of transcriptional control mechanisms acting on the gene encoding Oct-2, while the gene encoding the liver-specific factor C/EBP has been directly shown to be transcribed only in the liver by using a nuclear run-on assay.

It is clear that such regulation of transcription only sets the problem of gene regulation one stage further back, requiring mechanisms to activate the transcription of the gene encoding the transcription factor itself. It is not surprising, therefore, that the synthesis of specific transcription factors is often modulated by posttranscriptional control mechanisms. Such posttranscriptional regulation is observed, for example, in the case of the yeast factor GCN4 which activates the genes encoding the enzymes of amino acid biosynthesis in response to amino acid starvation. In this case the increased synthesis of GCN4 following amino acid starvation is mediated by increased translation of its specific mRNA (Fink, 1986).

Regulation of Transcription Factor Activity

Given the obvious advantages of regulating transcription factor synthesis at a posttranscriptional level, it is not surprising that in other transcription factors this process has been taken further with such factors being present in all tissues and becoming activated posttranslationally in response to a particular signal or in a specific tissue. A number of different mechanisms by which this can be achieved have been observed (Figure 2).

A simple example of such modulation occurs in the yeast transcription factor ACE1 which activates transcription of the metallothionein gene in response to copper. In this case the protein undergoes a conformational change in the presence of copper which allows it to bind to regulatory sites in the metallothionein gene and activate transcription (Figure 2a).

A different example in which activation of the transcription factor depends on the dissociation of an inhibitory factor (Figure 2b) is observed in B lymphocytes. Thus, unlike the case of MyoD (see above), the expression of Oct-2 alone is not sufficient to produce differentiated B cells. This is because other transcription factors that are specifically active in B cells are also involved in producing the expression of genes specific to B cells such as those encoding the immunoglobulins. One such factor is NF kappa B, which binds to a DNA sequence in the regulatory region of the immunoglobulin kappa light chain gene. Interestingly, unlike Oct-2, the NF kappa B protein is present in all cell types. In most cells, however, it is present in an inactive form in which it is complexed with an inhibitory protein, resulting in it being restricted to the cell cytoplasm. In mature B cells, however, NF kappa B is released from the inhibitory protein and moves to the nucleus, where it can bind to its DNA target sequence and activate the transcription of the immunoglobulin kappa light-chain gene. Interestingly, this activation of NF kappa B also occurs when resting T lymphocytes are activated by antigenic stimulation and is the main reason for the improved growth of the human immunodeficiency virus in activated (compared with resting) T cells since NF kappa B can bind to two sites

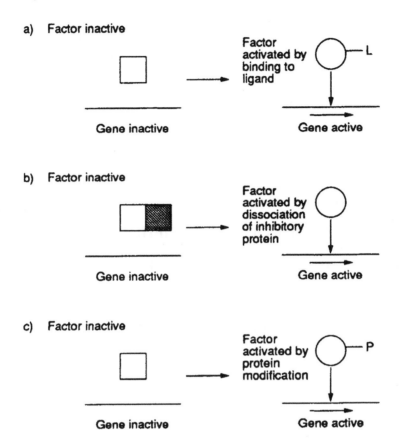

Figure 2. Activation of transcription factors by (**a**) ligand binding (**b**) dissociation of an inhibitory protein, and (**c**) protein modification.

within the HIV promoter and activate viral transcription (Lenardo and Baltimore, 1989).

In addition to protein-protein interaction, activation of transcription factors can also be achieved by protein modification, providing a direct means of activating a particular factor in response to a specific signal (Figure 2c). One example of this is provided by the CREB transcription factor which mediates the activation of several cellular genes following cyclic AMP treatment. Thus, cyclic AMP is known to stimulate the protein kinase A enzyme; in turn, this enzyme phosphorylates CREB stimulating the ability of the factor to activate transcription following DNA binding. Hence, stimulation of gene expression by cyclic AMP is directly mediated via its stimulation of protein kinase A and the consequent phosphorylation of CREB. Interestingly, the activation of NF kappa B activity by antigenic stimulation in lymphocytes (see above) also involves phosphorylation. In this case, however, it is the inhibitory protein associated with NK kappa B (see above) which is phosphorylated, causing it to dissociate from the NF kappa B protein, allowing NF kappa B to bind to DNA and activate transcription. Although phosphorylation is obviously an important means of achieving activation by protein modification, transcription factors are also modified in other ways, for example, by glycosylation, and activation could therefore take place by alteration of the levels of these modifications.

A variety of mechanisms involving both increased synthesis and protein activation by ligand binding, protein modification, or disruption of protein-protein interaction therefore exist to allow specific factors to become active in response to a particular signal or in a particular cell type.

DNA BINDING BY TRANSCRIPTION FACTORS

Modular Structure of Transcription Factors

Having considered the manner in which transcription factors become activated, it is necessary to consider the features of these factors which allow them to modulate the rate of transcription. Detailed studies of these factors have revealed that they

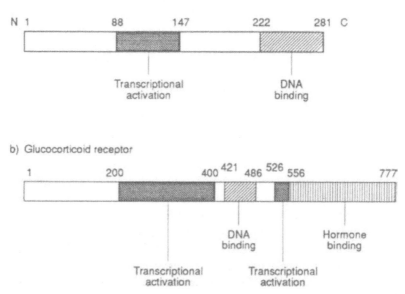

Figure 3. Domain structure of the yeast GCN4 transcription factor (**a**) and the mammalian glucocorticoid receptor (**b**). Note the distinct domains which are active in DNA binding or transcriptional activation.

Table 1. Transcription Factor Domains

Domain	Role	Factors Containing Domain	Comments
Homeobox	DNA binding	Numerous *Drosophila* homeotic genes, related genes in other organisms	DNA binding mediated via helix-turn-helix motif
Cysteine-histidine zinc finger	DNA binding	TFIIIA, Kruppel, Sp1, etc.	Multiple copies of finger motif
Cysteine-cysteine zinc finger	DNA binding	Steroid-thyroid hormone receptor family	Single pairs of fingers, related motifs in adenovirus E1A and yeast GAL4, etc.
Basic element	DNA binding	C/EBP, c-*fos*, c-*jun*, GCN4	Often found in association with leucine zipper
Leucine zipper	Protein dimerization	C/EBP, c-*fos*, c-*jun*, GCN4, c-*myc*	Mediates dimerization which is essential for DNA binding by adjacent domain
Helix-loop-helix	Protein dimerization	c-*myc*, *Drosophila* daughterless MyoD, E12, E47	Mediates dimerization which is essential for DNA binding by adjacent domain
Amphipathic acidic α-helix	Gene activation	Yeast GCN4, GAL4, steroid-thyroid receptors, etc.	Probably interacts directly with TFIID
Glutamine-rich region	Gene activation	SP1	Related regions in Oct-1, Oct-2, AP2, etc.
Proline-rich region	Gene activation	CTF/NF1	Related regions in AP2, c-*jun*, Oct-2

have a modular structure in which distinct regions of the protein mediate each of its different functions (Figure 3; see Johnson and McKnight, 1989). In the case of the glucocorticoid receptor, a transcription factor which mediates the induction of specific genes in response to glucocorticoid, different regions of the protein mediate its ability to bind to DNA, its ability to activate gene transcription following such binding, and its ability to bind glucocorticoid hormones.

Hence, specific regions of the transcription factor protein mediate its ability to bind to DNA and its ability to activate transcription by interacting with other factors following such binding. A number of different motifs which can mediate DNA binding have been identified in different transcription factors and have been used to classify these factors into families with a common related DNA binding domain (see Table 1 and Figure 4). Each of the major classes of DNA binding domain will be discussed in turn (Struhl, 1989).

The Zinc Finger

One of the first transcription factors to be purified and cloned was TFIIIA which plays a critical role in the transcription of the 5S ribosomal RNA genes by RNA

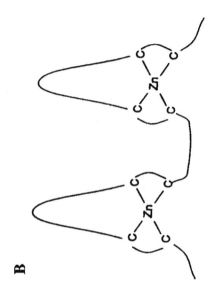

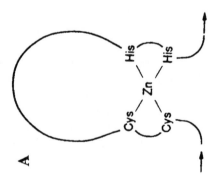

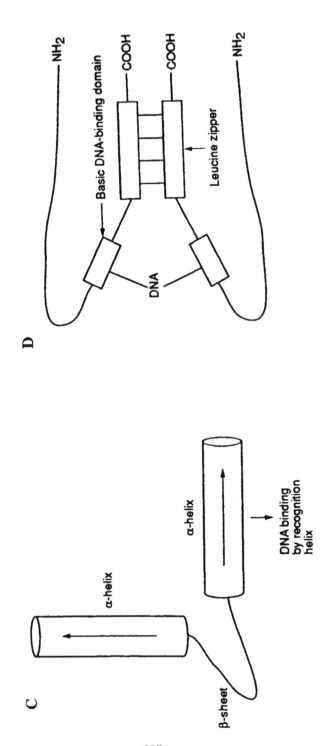

Figure 4. DNA binding domains. (**A**) The cysteine-histidine zinc finger. (**B**) The multi-cysteine zinc finger. (**C**) The helix-turn-helix motif. (**D**) The leucine zipper and adjacent basic DNA binding domain following dimerization.

polymerase III. The DNA binding region of this factor contains nine repeats of a 30 amino acid sequence of the form Tyr/Phe-X-Cys-X-Cys-$X_{2,4}$-Cys-X_3-Phe-X_5-Leu-X_{-2}-His-$X_{-3,4}$-His-X_5 where X is a variable amino acid. Each of these repeats therefore contains two invariant pairs of cysteine and histidine residues which coordinate a single atom of zinc. This results in a fingerlike structure (Figure 4A) in which the conserved phenylalanine and leucine residues and several basic residues in the finger project from the surface of the protein (Evans and Hollenberg, 1988).

Multiple examples of this zinc finger motif have subsequently been identified in a number of transcription factors for genes transcribed by RNA polymerase II including Spl, the *Drosophila* kruppel protein, the yeast ADRI protein and many others. Interestingly, a single mutation in one of the zinc finger motifs of the kruppel protein which replaces a cysteine by a serine that cannot bind zinc, results in a mutant fly whose appearance is exactly identical to that produced by complete deletion of the gene. Hence the ability to bind zinc is essential for DNA binding activity and, therefore, for the functioning of the protein as a transcription factor.

A similar zinc binding motif is also found in the DNA binding regions of members of the steroid/thyroid hormone receptor family which bind specific steroid hormones and either activate or repress gene expression by binding to specific DNA sequences in target genes. In this case, however, the binding region consists of two fingers each of which contains four cysteine residues coordinating the zinc atom rather than two each of cysteine and histidine, and it also lacks the conserved phenylalanine and leucine residues found in the other type of finger (Figure 4B). In addition the two finger element is present only once in each receptor, as opposed to the multiple fingers ranging from two to 37 found in genes having cysteine-histidine fingers. The two types of fingers may not therefore be evolutionarily related.

The Helix-Turn-Helix Motif

A number of homeotic genes which dramatically affect the development of the fruit fly *Drosophila melanogaster* have been described. These genes contain a common highly conserved 60 amino acid region which is known as the homeobox or homeodomain and which mediates the DNA binding ability of these proteins (Affolter et al., 1990). Structure predictions of this region indicated that it could form a helix-turn-helix motif in which an α-helical region is followed by a β-turn and then another α-helical region (Figure 4C).

Following its original identification in the *Drosophila* homeotic genes, similar homeoboxes containing the helix-turn-helix motif have also been identified in the yeast mating type transcriptional regulatory proteins, in a variety of amphibian and mammalian transcription factors, and in plants such as Antirrhinum. More recently, another class of regulatory proteins has been identified in which the homeobox forms one part of a larger conserved domain known as the POU domain that also includes another POU-specific region. These proteins, which include the octamer binding proteins Oct-1 and Oct-2, the pituitary specific protein Pit-1, and the nematode gene unc-86, all use the POU (Pit-Oct-Unc) domain to bind to DNA. The relative contribution of the homeodomain and the POU-specific domain differs between the different proteins however, with the homeodomain being sufficient for sequence specific binding of Pit-1 while both the homeodomain and the POU-specific domain are necessary in the case of Oct-1 (for review see Herr et al., 1988).

The Leucine Zipper and the Basic DNA Binding Domain

Another element found in several transcription factors such as the liver specific transcription factor C/EBP, the yeast factor GCN4, and the proto-oncogene proteins Fos and Jun is the leucine zipper (Abel and Maniatis, 1989). In this structure leucine residues occur every seven amino acids in an α-helical structure such that the leucines occur every two turns on the same side of the helix.

Rather than acting directly as a DNA binding motif, however, the zipper facilitates the dimerization of the protein by interdigitation of two leucine-containing helices on different molecules. In turn such dimerization results in the correct protein structure for DNA binding by the adjacent region which in C/EBP, Fos, and Jun is a highly basic region, distinct from those discussed so far, that can interact directly with the acidic DNA.

Both the Fos and Jun proteins bind to sequences in DNA known as AP-1 sites which mediate gene induction following growth factor or phorbol ester treatment. Interestingly, however, whereas the Jun protein can bind specifically to this sequence as a homeodimer, the Fos protein can only do so after formation of a heterodimer with Jun. This difference is directly due to a difference in the leucine zipper motif of the two proteins which prevents Fos homodimer formation. Thus substitution of the Fos leucine zipper region with that of Jun allows the chimeric protein to bind to DNA through the basic region of Fos. The requirement for dimer formation of these proteins prior to DNA binding thus introduces another potential regulatory point in the control of gene expression.

Although originally identified in leucine zipper-containing proteins, the basic DNA binding domain has also been identified by homology comparisons in a number of other transcriptional regulatory proteins including the muscle regulatory protein MyoD. In this case, however, the basic domain is associated with an adjacent region that can form a helix-loop-helix structure in which two amphipathic helices (containing all the charged amino acids on one side of the helix) are separated by an intervening non-helical loop. Although originally thought to be the DNA binding domain of MyoD, this helix-loop-helix motif is now believed to play a similar role to the leucine zipper in mediating protein dimerization and facilitating DNA binding by the adjacent basic DNA binding motif. Hence, both the leucine zipper and the helix-loop-helix motif play a critical role in allowing dimerization of the factors which contain them. This dimerization is critical for subsequent DNA binding mediated by the basic DNA binding domain (Figure 4D).

Other DNA Binding Motifs

Although the majority of DNA-binding domains analyzed so far fall into the three classes we have discussed, not all do so. Thus the DNA binding domains of the transcription factors AP2, CTF/NF1, and SRF are distinct from the known motifs and from each other. As more factors are studied, DNA-binding motifs similar to those of these proteins are likely to be identified and they will become founding members of new families of DNA-binding motifs. Indeed this process is already under way; the similarity between the mammalian factor HNF-3 and the *Drosophila* fork head factor, for example, having led to the identification of a new family of proteins containing the so-called fork head DNA binding motif.

It is clear, therefore, that a number of different structures exist which can mediate sequence-specific DNA binding. Several of these are common to a number of different transcription factors, with differences in the precise amino acid sequence of the motif in each factor controlling the precise DNA sequence which it binds and hence the target genes for the factor.

MODULATION OF TRANSCRIPTION

Although binding to DNA is obviously a necessary prerequisite for a factor to affect transcription, it is not in itself sufficient. Thus following binding the factor must interact with other factors or the RNA polymerase itself in order to modulate transcription. Although such an interaction very often results in the activation of transcription, a number of cases have now been described in which factor binding can result in transcriptional repression. Activation and repression of transcription will be discussed in turn.

Activation

In order to activate transcription following DNA binding the transcription factor must interact with other factors or the RNA polymerase itself to stimulate transcription. Such interaction is dependent on specific activation domains in the molecule which are normally distinct from the region which mediates DNA binding (Ptashne, 1988).

When the activating regions of a number of transcription factors were compared, it was found that, although they did not share amino acid homology, they possessed a high proportion of acidic amino acids. These acidic amino acids were arranged in such a way that they formed an amphipathic α-helix in which all the negative changes were displayed along one surface of the helix. In agreement with the critical role of this structure in transcriptional activation, a peptide which can form an acidic amphipathic helix can activate transcription when linked to the DNA binding domain of the yeast transcription factor GAL4, whereas the same amino acids placed in a random order could not do so.

Although this acidic activation domain is common to a number of transcription factors from yeast to man, other nonacidic activation domains have also been described (see Table 1). Thus the activation domain of the transcription factor Spl contains a glutamine rich region, while that of CTF/NFl is very proline rich. The finding of similar glutamine or proline rich regions in other transcription factors indicates that activation domains of these types are not confined to a single protein.

It is likely that the different activation domains act by interacting with other protein factors in order to facilitate transcription. Although this may occur by direct interaction with the RNA polymerase itself, at least in the case of the acidic activation domain, it seems more likely that its effect is mediated either via the TATA box binding factor TFIID or via another transcription factor TFIIB (Sharp, 1991). Both these factors are components of the basal transcriptional complex which forms on the gene promoter prior to transcription and which interacts with RNA polymerase II itself. By stimulating the assembly or stability of this complex, activation domains act to enhance the rate of gene transcription. Thus specific regions of transcription factors can activate transcription following DNA binding, by interacting with another bound factor and facilitating the assembly of a stable transcriptional complex.

Repression

Although the majority of transcription factors act in a positive manner, a number of cases have now been described in which a transcription factor exerts an inhibiting effect on transcription, and several possible mechanisms by which this can be achieved have been described (Levine and Manley, 1989; Goodbourn, 1990; see Figure 5).

The simplest means of achieving repression is seen in the β-interferon promoter where the binding of two positively acting factors is necessary for gene activation. Another factor acts negatively by binding to this region of DNA and simply preventing the positively acting factors from binding (Figure 5a). In response to viral infection, the negative factor is inactivated, allowing the positively acting factors to bind and transcription occurs. In a related phenomenon (Figure 5b) repression is achieved by formation of a complex between the activator and the repressor in solution, preventing the activator binding to the DNA. This is seen in the case of the inhibitory factor Id which dimerizes with the muscle determining factor MyoD via its helix-loop-helix motif and prevents it binding to DNA and activating transcription.

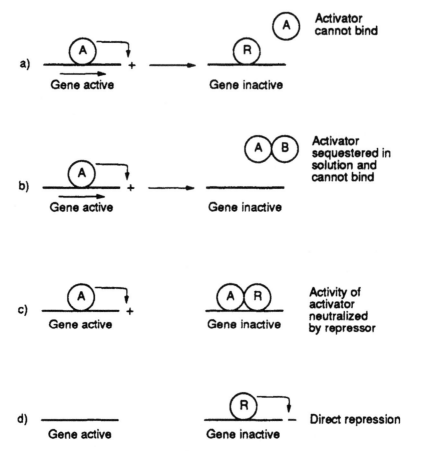

Figure 5. Repression of transcription by (a) competition for binding, (b) inhibition of binding, (c) quenching of activity, or (d) direct repression.

In addition to inhibiting DNA binding, a negative factor can also act by interfering with the activation of transcription mediated by a bound factor in a phenomenon known as quenching (Figure 5c). Thus the negatively acting yeast protein GAL80 prevents gene activation by the GAL4 protein by binding to it and masking the activation domain of GAL4. In response to the presence of galactose, however, GAL80 dissociates allowing GAL4 to activate the genes required for the metabolism of galactose.

In these cases the negative factor exerts its inhibiting effect by neutralizing the action of a positively acting factor by preventing either its DNA binding or its activation of transcription. It is likely, however, that some factors directly inhibit transcription by means of inhibitory domains, analogous to activation domains but with the opposite effect on the formation or stability of the basal transcriptional complex (Figure 5d).

MALREGULATION OF TRANSCRIPTION FACTORS IN DISEASE

Failure of Transcription Factor Function

In view of the critical importance of transcription factors, it is not surprising that a number of human diseases are due to alterations in transcription factor function. One type of congenital severe combined immunodeficiency is caused by a failure of HLA class II gene transcription, resulting in the absence of these proteins. In turn this failure of transcription is dependent on the lack of a specific transcription factor necessary for the transcription of these genes (Reith et al., 1988). Similarly cases of human dwarfism involving pituitary deficiency have been shown to involve mutations in the gene encoding the POU family transcription factor Pit-1 which normally activates gene expression in the pituitary gland (Radovick et al., 1992).

Proto-Oncogenes

The absence or inactivity of a particular transcription factor can result in a failure to express a specific gene or genes leading to disease. In addition, however, transcription factors can also cause disease if they are expressed or become active in the wrong place or at the wrong time, resulting in the inappropriate expression of the genes dependent upon their activity.

This form of malregulated gene expression is central to the development of certain cancers. Thus it is now clear that most, if not all, human cancers are caused by the mutation or overexpression of certain specific cellular genes known as proto-oncogenes, which results in their conversion into cellular oncogenes capable of causing cancer. While many of these proto-oncogenes encode genes whose protein products act as growth factors or their receptors, several, such as *erbA*, *fos*, *jun*, *myb*, and *myc*, encode cellular transcription factors that are involved in regulating the expression of specific genes. After the conversion of these proto-oncogenes into oncogenes by mutation or overexpression, corresponding altera-tions occur in the expression of the genes which they regulate, resulting in cancer.

One example of this is provided by the related Fos and Jun proteins both of which encode cellular transcription factors (for review see Curran and Franza, 1988). As described above both Fos and Jun are synthesized following treatment with growth factors and then activate the transcription of specific genes whose protein products are necessary for cellular growth. Normally, however, Fos and Jun are synthesized only transiently in response to exposure to the growth factor, resulting in only a transient activation of gene expression and thereby producing the controlled growth factor regulated proliferation characteristic of normal cells. If for any reason, however, Fos and Jun are continually synthesized, either owing to a mutation, resulting in their continual overexpression, to infection with expression, or to infection with a virus expressing one or other of them. The cell is stimulated to grow continually even in the absence of growth factors. Such continuous uncontrolled growth is characteristic of the cancer cell.

Hence the *fos* and *jun* genes are proto-oncogenes whose products have a critical role in the growth of normal cells but which can be converted into oncogenes capable of transforming cells. Moreover, in contrast to the other diseases we have discussed, in this case malregulation of gene expression and disease is caused not by failure of transcription factor function but rather by failure to correctly regulate the activity of transcription factors. Hence as with other cellular processes, the synthesis and activity of transcription factors are subject to complex regulatory mechanisms, the failure of which can be as disastrous as the failure of the basic process itself.

SUMMARY

This chapter describes the properties of transcription factors which allow them to bind to DNA and influence the rate of transcription either positively or negatively. When combined with the synthesis or activation of many of these factors only in specific cell types, these properties allow transcription factors to play a critical role both in the basic process of transcription itself and in its regulation in specific cell types.

REFERENCES

Abel, T., & Maniatis, T. (1989). Action of leucine zippers. Nature 341, 24–25.

Affolter, M., Schier, A., & Gehring, W.J. (1990). Homeodomain proteins and the regulation of gene expression. Cur. Op. Cell Biol. 2, 485–495.

Curran, T., & Franza, B.R. (1988). Fos and Jun: The AP-1 connection. Cell 55, 395–397.

Evans, R.M., & Hollenberg, S.M. (1988). Zinc fingers: Guilt by association. Cell 52, 1–3.

Fink, G.R. (1986). Translational control of transcription in eukaryotes. Cell 45, 155–156.

Goodbourn, S. (1990). Negative regulation of transcriptional initiation in eukaryotes. Biochim. Biophys. Acta 1032, 53–77.

Herr, W., Sturm, R.A., Clerc, R.G., Corcoran, L.M., Baltimore, D., Sharp, P.A., Ingraham, H.A., Rosenfeld, M.G., Finney, M., Ruvkun, G., & Horvitz, H.R. (1988). The POU domain: A large conserved region in the mammalian pit-1, Oct-1 Oct-2 and *Caenorhabditis elegans* unc-86 gene products. Genes Devel. 2, 1513–1516.

Johnson, P.F., & McKnight, S.L. (1989). Eukaryotic transcriptional regulatory proteins. Ann. Rev. Biochem. 58, 799–839.

Latchman, D.S. (1995). In: Gene Regulation: A Eukaryotic Perspective, 2nd edn. Chapman and Hall, London.

Latchman, D.S. (1995). In: Eukaryotic Transcription Factors, 2nd edn. Academic Press, San Diego.

Lenardo, M.J., & Baltimore, D. (1989). NF-kappa B: A pleiotropic mediator of inducible and tissue-specific gene control. Cell 58, 227–229.

Levine, M., & Manley, J.L. (1989). Transcriptional repression of eukaryotic promoters. Cell 59, 405–408.

Olson, E.N. (1990). MyoD family: A paradigm for development? Genes Devel. 4, 1454–1461.

Ptashne, M. (1988). How eukaryotic transcriptional activator work. Nature 335, 683–689.

Radovick, S., Nations, M., Du, Y., Berg, L.A., Weintraub, B.D., & Wondisford, F.E. (1992). A mutation in the POU-homeodomain of Pit-1 responsible for combined pituitary hormone deficiency. Science 257, 1115–1118.

Reith, W., Satola, S., Sanchey, C.H., Amaldi, I., Lisowska-Grospiere, B., Griscelli, C., Hadam, M.R., & Mach, B. (1988). Congenital immunodeficiency with a regulatory defect in MHC Class II gene expression lacks a specific HLA-DR promoter binding protein RF-X. Cell 53, 897–906.

Sharp, P.A. (1991). TFIIB or not TFIIB? Nature 351, 16–18.

Struhl, K. (1989). Helix-turn-helix, zinc finger, and leucine zipper motifs for eukaryotic transcriptional regulatory proteins. Trends Biochem. Sci. 14, 137–140.

RECOMMENDED READINGS

Latchman, D.S. (1995). In: Gene Regulation, 2nd edn. Chapman, and Hall, London.

Latchman, D.S. (1995). In: Eukaryotic Transcription Factors, 2nd edn. Academic Press, London.

Chapter 7

Alternative RNA Splicing

DAVID S. LATCHMAN

INTRODUCTION

Perhaps the major fundamental discovery to come from the current era of gene cloning was the finding that for the vast majority of eukaryotic genes, the protein coding regions are not present as one uninterrupted block within genomic DNA. Rather they exist as short segments, known as exons, which are separated by intervening sequences, known as introns (for review see Breathnach and Chambon,

Principles of Medical Biology, Volume 5
Molecular and Cellular Genetics, pages 135–147.
Copyright © 1996 by JAI Press Inc.
All rights of reproduction in any form reserved.
ISBN: 1-55938-809-9

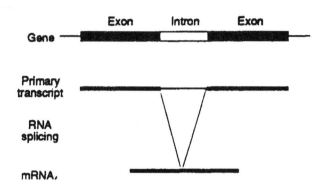

Figure 1. Removal of an intervening sequence from the primary RNA transcript by RNA splicing.

1981). Following the transcription of both the exons and intervening introns into a large primary RNA transcript the intervening sequences are removed by the process of RNA splicing in which the exons are joined to produce a mature mRNA molecule that can be translated into protein (Figure 1; for reviews see Sharp, 1987; Lamond, 1991).

Following the discovery of RNA splicing there was much speculation that this process might represent a major point of gene regulation. Thus, in theory, an RNA species transcribed in several tissues might be correctly spliced to yield a functional RNA in one tissue and remain unspliced in another tissue. Such an unspliced RNA would either be rapidly degraded within the nucleus or, if transported to the cytoplasm, would be unable to produce a functional protein due to the interruption of the protein coding regions.

However, while a number of such processing versus discard decisions have been described in the fruit fly *Drosophila* (for review see Bingham et al., 1988), they

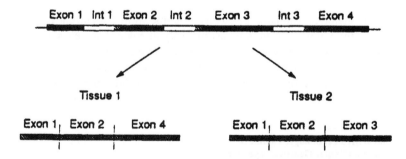

Figure 2. Alternative splicing of the same primary transcript to yield two different mRNAs.

have not been observed in mammals. Another form of regulation at the level of RNA splicing is widely used in both mammals and other organisms and forms the subject of this chapter. In this process known as alternative splicing (for reviews see Leff et al., 1986; Latchman, 1990) the primary intron-containing RNA transcript of a gene is spliced in two or more different ways in different tissues resulting in different combinations of exons within the resulting messenger RNA (Figure 2).

FUNCTION OF ALTERNATIVE SPLICING

In many cases the two or more different mRNAs produced by alternative splicing will produce distinct but related proteins with different functions. Thus, alternative splicing in the immunoglobulin heavy-chain gene produces two mRNAs which encode two distinct forms of the heavy-chain protein either with or without a transmembrane region which allows the protein containing it to insert in the cell membrane rather than being secreted. Both of these protein forms are otherwise identical and following association with an appropriate immunoglobulin light-chain can bind antigen with identical specificity. Hence, in this case, alternative splicing is used to control the progression of the immune response to a particular antigen, with the first antibody produced being capable of interacting with antigen only when bound to the surface of the antibody-producing B cell. Such an interaction results in the proliferation of the antibody-bearing B cell which then produces a secreted form of the antibody capable of neutralizing the antigen in solution.

A similar role for alternative splicing in producing distinct but related proteins is seen in the case of the major sarcomeric muscle proteins with different forms of myosin, troponin, and tropomyosin all being produced by different types of muscle cells. In an extreme example of this phenomenon, alternative splicing in the gene encoding troponin is used to generate up to 64 distinct forms of the protein with different properties (Breitbart et al., 1987).

Interestingly, in some cases of alternative splicing the distinct but related proteins produced by alternative splicing can actually be processed further by proteolytic cleavage to produce two entirely distinct peptides with different properties. This is well illustrated in the case of the gene encoding the calcium regulatory hormone calcitonin. Thus, when this gene was first cloned it was found that the calcitonin mRNA was encoded by an mRNA produced by the joining of four exons (1 to 4 in Figure 3). This RNA was then translated into a protein precursor which was cleaved to yield mature calcitonin. Interestingly, however, this gene also produces an alternatively spliced mRNA in which the fourth calcitonin exon is omitted and exons 1 to 3 are joined to two additional downstream exons (5 and 6 in Figure 3). This results in the production of a distinct mRNA which is translated into a protein precursor that is related to but distinct from that producing calcitonin. Moreover, this precursor is cleaved to produce a peptide completely distinct from calcitonin, known as calcitonin gene related peptide (CGRP) (Leff et al., 1987). This peptide has an entirely different protein sequence to that of calcitonin since

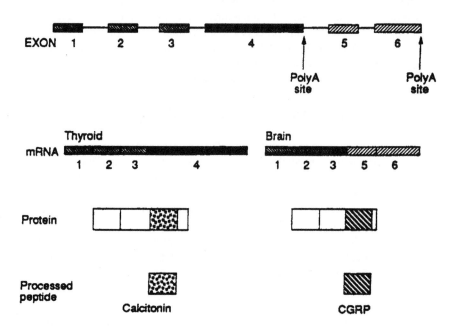

Figure 3. Alternative splicing of the calcitonin/CGRP transcript followed by proteolytic cleavage results in the production of calcitonin in the thyroid gland and CGRP in the brain.

exons 1 to 3—which are common to both mRNAs—encode a common region of the two protein precursors but do not contribute to the mature peptides. Hence, alternative splicing produces two peptides with completely distinct properties from a single gene. Both of these peptides are important for normal body function, calcitonin having a critical role in calcium homeostasis while CGRP (which was only discovered through the study of the calcitonin gene) is the most potent known vasodilator.

Hence, alternative splicing represents a highly flexible means of generating protein diversity from a single gene producing distinct but related proteins with different properties or which can be processed by proteolytic cleavage to yield completely unrelated peptides. In view of the utility of this process, it is not surprising that it is widely used, being involved in many different processes ranging from the control of sex determination in *Drosophila* (for review see Baker, 1989) to the production of neuropeptides in the mammalian brain (Kitamura et al., 1983). A representative selection of cases of alternative splicing chosen to illustrate the many different systems in which it is used is given in Table 1.

As indicated in Table 1, alternative splicing is a highly regulated process with the outcome of a specific splicing decision being different in different cell types.

Thus the calcitonin/CGRP gene is transcribed in both thyroid C cells and in neuronal cells. However, alternative splicing produces only the calcitonin mRNA in thyroid cells and only the CGRP mRNA in the brain (Leff et al., 1987). Similarly, as discussed above, production of the membrane bound form of the immunoglobulin heavy chain protein occurs early in the immune response with subsequent production of the secreted form later in the response.

Table 1. Cases of Alternative Splicing Which Are Regulated Developmentally or Tissue Specifically

Protein	Species	Nature of Transcripts Which Undergo Alternative Splicing	Cell Types Carrying Out Alternative Splicing
Immune system			
Immunoglobulin heavy chain IgD, IgE, IgG, IgM	Mouse	3' end differs	B cells
Lyt-2	Mouse	Same transcript	T cells
Enzymes			
Alcohol dehydrogenase	*Drosophila*	5' end differs	Larva and adult
Aldolase A	Rat	5' end differs	Muscle and liver
α-Amylase	Mouse	5' end differs	Liver and salivary gland
(2'5') oligo A synthetase	Human	3' end differs	B cells and monocytes
Muscle			
Myosin light chain	Rat/mouse/human/ chicken	5' end differs	Cardiac and smooth muscle
Myosin heavy chain	*Drosophila*	3' end differs	Larva and adult muscle
Tropomyosin	Mouse/rat/human/ *Drosophila*	Same transcript	Different muscle cell types
Troponin T	Rat/quail/chicken	Same transcript	Different muscle cell types
Nerve cells			
Calcitonin/CGRP	Rat/human	3' end differs	Thyroid C cells or neural tissue
Myelin basic protein	Mouse	Same transcript	Different glial cells
Neural cell adhesion molecule	Chicken	Same transcript	Neural development
Preprotachykinin	Bovine	Same transcript	Different neurons
Others			
Fibronectin	Rat/human	Same transcript	Fibroblasts and hepatocytes
Early retinoic acid induced gene 1	Mouse	Same transcript	Stages of embryonic cell differentiation
Thyroid hormone receptor	Rat	Same transcript	Different tissues

In this manner, distinct but related proteins can be produced in different cell types, in response to specific stimuli or at different stages of development, from only a single gene. Such a process is clearly more economical than having a distinct gene encoding each of these forms. Moreover, as shown in Table 1, alternative splicing is especially prevalent in terminally differentiated non-dividing cells such as nerve and muscle. In these non-dividing cells, it may be particularly difficult to alter gene expression at the transcriptional level. Thus such a transcriptional response may require the opening up of chromatin regions containing the genes which are being activated and will therefore require a reprogramming of chromatin structure which is not possible without a cell division (for discussion see Latchman, 1995). Hence, alternative splicing offers a particularly valuable means of regulating gene expression and responding to specific stimuli.

REGULATION OF ALTERNATIVE SPLICING

Cases of alternative splicing can be divided into three classes depending on the nature of the transcripts which are differentially processed in different tissues (see Table 1, and Leff et al., 1986). These three types will be discussed in turn since different mechanisms are used for their regulation.

Situations Where the 5′ End of the Differentially Processed Transcripts Are Different

In these cases, as in the case of the mouse α-amylase gene (Young et al., 1981), two distinct primary transcripts which differ at their 5′ ends are produced in different tissues due to the use of two distinct promoters for the initiation of transcription. These two transcripts are then spliced differently to produce two different mRNAs. Such cases can be explained simply on the basis that differences in the sequences and structure of the two primary transcripts and the sequences they contain control the pattern of splicing. Hence these cases really represent cases of transcriptional control in which the alternative splicing decision is dependent on which promoter is selected in each tissue. They are thus dependent on the same processes of transcriptional regulation that control which genes are transcribed in each tissue (for discussion of transcriptional regulation see Latchman, 1995).

Situations Where the 3′ End of the Differentially Processed Transcripts Are Different

In the case of the membrane bound and secreted forms of the immunoglobulin heavy-chain discussed above, two distinct transcripts are spliced to produce the mRNAs encoding the two forms. Unlike the case of α-amylase, however, these two transcripts differ not at the 5′ end but rather at the 3′ end where they are polyadenylated at two different points. Interestingly, if the polyadenylation site used in the production of the shorter secreted immunoglobulin RNA is deleted, the expected decrease in the production of secreted immunoglobulin is paralleled by a corresponding increase in the use of the downstream polyadenylation site and consequent synthesis of the membrane bound form. Hence the choice of polyadenylation site determines which pattern of splicing is followed. As with the cases of alternative promoter usage, therefore, the regulated pattern of splicing which is observed is dependent upon some other process, in this case the regulated choice of polyadenylation site.

Situations Where Both the 5′ and 3′ Ends of the Differentially Processed Transcripts Are Identical

Primary regulation at the stage of transcription or polyadenylation followed by consequent differences in the pattern of splicing cannot explain the third form of alternative splicing. Numerous cases have been described in which the same promoters and polyadenylation sites are used in different tissues resulting in the production of two identical RNA transcripts which are then spliced in different ways. Examples of this type include the human fibronectin gene, the *ras* oncogene and, most dramatically, the rat skeletal muscle troponin T gene discussed above where the same RNA can be spliced in up to 64 different ways in different muscle cell types. In these cases the process of splicing itself must be highly regulated resulting in two different splicing patterns in different cell types.

Moreover, these differences in splicing can be reproduced when mini-gene constructs containing only the alternatively spliced exons and introns are introduced into two different cell types which normally splice the gene transcript in two different ways. Hence these effects cannot be dependent upon subtle (and undetected) differences in the nature of the transcripts produced in different tissues which produce the observed differences in the splicing pattern. Rather they must be dependent upon the existence of specific *trans*-acting splicing factors which are differentially expressed in different tissues and which interact with the RNA to control the pattern of alternative splicing.

Such factors may also play a critical role in the regulation of alternative splicing even in some cases where different polyadenylation sites are used. In the case of the calcitonin/CGRP gene, alternative splicing events produce the mRNAs encoding the calcium modulatory hormone calcitonin and the neuropeptide CGRP. This alternative splicing takes place on RNAs which have different 3′ ends due to the use of different polyadenylation sites (Figure 3). In this case, however, unlike the case of immunoglobulin discussed above, deletion of the upstream (calcitonin) polyadenylation site does not result in the production of CGRP in a calcitonin producing tissue (Figure 3) but only in the accumulation of unspliced precursor (Leff et al., 1987).

Hence, tissue specific splicing factors present in the tissues producing CGRP must be necessary for CGRP-specific splicing to occur. Indeed even in the immunoglobulin case discussed above where splicing factors specific to B cells producing membrane bound immunoglobulin are clearly not absolutely required for membrane-specific splicing, injection of nuclear extracts from such cells can enhance the splicing of membrane bound immunoglobulin in *Xenopus* oocytes, indicating that factors specific to such cells may increase the efficiency of the splicing events required to produce membrane bound immunoglobulin (Peterson and Perry, 1989). These findings indicate that specific alternative splicing factors play a critical role in regulating a number of alternative splicing decisions. In the next section we will consider the nature and mode of action of these factors.

ALTERNATIVE SPLICING FACTORS

Mode of Action of Alternative Splicing Factors

In an attempt to investigate the manner in which an alternative splicing factor might act, Rosenfeld and colleagues (Crenshaw et al., 1987) expressed a calcitonin/CGRP mini-gene in all the cells of a transgenic mouse and then monitored the pattern of splicing. In these experiments, calcitonin-specific splicing occurred in a very wide variety of tissue types. In contrast, CGRP-specific splicing occurred only in brain and heart of all mouse tissues. These findings strongly suggest that the constitutively expressed RNA splicing machinery, which is present in all tissues and cell types, can carry out calcitonin-specific splicing. However, a tissue-specific splicing factor, expressed in only a few cell types, is required in addition for CGRP-specific splicing.

The existence of factors necessary for one particular pattern of splicing clearly begs the question of how such factors act. In many cases, even when multiple alternative splice sites exist in a single RNA, the regulatory factor may control only one of the alternative splicing decisions. Once this decision has been made, the remaining splicing decisions will follow naturally due to the structure of the RNA intermediate produced. In the case of the rat β-tropomyosin gene exons 5, 6, and 8 are joined in fibroblasts whereas exons 5, 7, and 8 are joined in skeletal muscle. As expected, HeLa cells are capable only of carrying out the fibroblast-specific splicing pattern on transfection. If exon 7 is joined to exon 8 prior to transfection, however, these cells can produce the muscle-specific splice of exons 5 and 7, indicating that the joining of exon 8 to either exon 6 or 7 is the critical regulated event in this system (Helfman et al., 1988).

Alternative splicing factors are likely to act by affecting the outcome of one critical splicing decision with other secondary splicing events following as a consequence of this initial decision. Although the tissue specific expression of such *trans*-acting factors is responsible for determining the nature of this decision, they

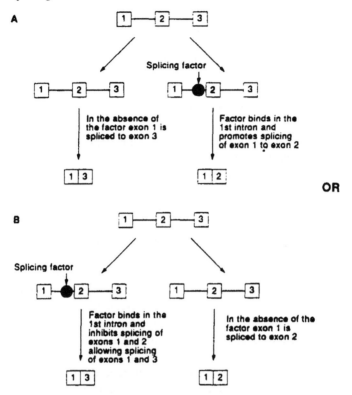

Figure 4. Possible models by which an alternative splicing factor can affect splicing by binding to a *cis*-acting sequence. In (**A**), the factor acts by binding to the weaker of the two potential splicing sites promoting its use, while in (**B**) it acts by binding to the stronger of the two sites and inhibiting its use so that the other, weaker, site is used.

must act by recognizing *cis*-acting sequences within the RNA transcript itself. Clearly the interaction of these factors with such sequences could produce alternative splicing either by promoting splicing at the site of the *cis*-acting sequence at the expense of the alternative splice site (Figure 4A) or by inhibiting splicing at the site of binding and thereby promoting the use of the alternative splice site (Figure 4B).

Both of these mechanisms appear to operate in different cases. Thus in the case of the calcitonin/CGRP gene, the removal of sequences between exon 3 and the calcitonin-specific exon 4 (see Figure 3) results in calcitonin-specific splicing in cells normally producing CGRP (Emeson et al., 1989). Hence the tissue-specific splicing factor which promotes CGRP-specific splicing is likely to do so by binding to these sequences and inhibiting the calcitonin-specific splicing of exons 3 and 4, thereby promoting the CGRP-specific splicing of exons 3 and 5. In contrast, in the *Drosophila* dsx transcript which is differentially spliced in males and females, mutations which affect the binding of an alternative splicing factor active only in females map in the intron adjacent to the female specific exon, indicating that the alternative splicing factor binds here and promotes the female-specific splice (reviewed by Baker, 1989).

Nature of Alternative Splicing Factors

Although studies on the effects of specific mutations have enhanced our understanding of the manner in which alternative splicing is achieved, further advances in this area are likely to be dependent on the isolation of the alternative splicing factors themselves.

In the fruit fly *Drosophila*, which is genetically very well characterized, the approach to this has been a genetic one and a number of genes whose disruption by mutation affects alternative splicing have been characterized. Some of these, such as Sxl and tra-2, have been studied at the DNA level and their corresponding proteins shown to contain one or more copies of a ribonucleoprotein (RNP) consensus sequence. This sequence is also found in a wide variety of RNA binding proteins including the U1 70kd RNA binding protein which plays a key role in the basic splicing process itself and is likely to constitute an RNA binding domain (review by Bandziulis et al., 1989). Interestingly, a similar RNA binding domain is also found in the constitutively expressed mammalian splicing factor ASF-1/SF2 whose concentration has been shown to influence the outcome of alternative splicing of the Simian virus 40 RNA in an *in vitro* extract (Kramer et al., 1991).

On the basis of those currently characterized, alternative splicing factors appear to have features in common with the factors which control the basic process of splicing itself, allowing them to act by binding to the RNA and influencing its splicing. This similarity between alternative splicing factors and those mediating constitutive splicing has prompted a search for tissue-specific forms of the small nuclear U series RNAs and proteins of the spliceosome which mediate the basic process of splicing itself (for review see Maniatis and Reed, 1987). Thus, while the majority of these RNAs and proteins are present in all tissues paralleling the requirement for the basic process of splicing itself in all cell types, tissue specific variants of them would represent excellent candidates for a key role in alternative splicing.

Such investigations have revealed the existence of tissue-specific forms of the U1 and U4 RNAs in *Xenopus* (Lund and Dahlberg, 1987) and of the U1 RNA in the mouse (Lund et al., 1985). The relationship of these variants to the constitutively expressed snRNAs suggests that they could associate with the spliceosome and regulate alternative splicing. Interestingly, the U1-specific A protein has been shown to bind more strongly to a U1a-containing ribonucleoprotein particle than to a U1b-containing RNP, indicating one means by which snRNA variants may influence splicing. In no case, however, has the expression of a specific snRNA variant in particular tissues been correlated with a particular pattern of RNA splicing in these tissues.

However, such a correlation has been reported in the case of a tissue-specific variant of a protein component of the spliceosome. Both our laboratory and others' have reported the existence of a mammalian protein, SmN which is closely related to the constitutively expressed SmB and SmB' proteins of the spliceosome. Unlike SmB and B', however, this protein is expressed in only a limited range of tissues and cell types such as brain, EC cells, and the neuronal cell line PC12 (McAllister et al., 1988; Sharpe et al., 1990). Interestingly the cell types which express SmN are precisely those which are capable of correctly splicing the calcitonin/CGRP transcript to produce CGRP mRNA (Crenshaw et al., 1987; Leff et al., 1987), whereas all the cell types in these experiments that could produce only the calcitonin mRNA also lack SmN.

It is possible therefore that SmN may be the tissue-specific splicing factor required for CGRP-specific mRNA splicing, as well as for other specific splicing events occurring in brain and heart tissue. However, direct evidence that this is the case is lacking. Indeed when SmN was artificially expressed in a fibroblast cell type which normally lacks it, it did not influence the splicing of the calcitonin/CGRP gene (Delsert and Rosenfeld, 1992). Hence, if SmN is involved in regulating CGRP-specific splicing it must require another factor expressed only in neuronal cells in order to function.

Whatever its involvement in alternative splicing of the calcitonin/CGRP gene, SmN does represent the first example of a tissue-specific factor related to the constitutive splicing proteins. It is therefore likely to play an important role in regulating splicing decisions in neuronal cells. Such a vital role for SmN in neuronal cells is supported by the finding that the expression of SmN appears to be inactivated in human patients with Prader-Willi syndrome, suggesting that a failure of SmN function may be the cause of the neurological deficiencies observed in these patients (Ozcelik et al., 1992).

SUMMARY

It is clear that alternative splicing represents a very widely used mechanism of gene regulation allowing distinct but related proteins to be produced in different cell types. Initial progress in characterizing many such cases of alternative splicing has now been followed up by more detailed characterization of the effects of specific mutations on this process and of the alternative splicing factors themselves. More work is required to elucidate exactly how alternative splicing factors act to modulate the outcome of the splicing process.

REFERENCES

Baker, B.S. (1989). Sex in flies: The splice of life. Nature 340, 521–524.

Bandziulis, R.J., Swanson, M.S., & Dreyfuss, G. (1989). RNA-binding proteins as developmental regulators. Genes Devel. 3, 431–437.

Bingham, P.M., Chou, T.-B., Mims, I., & Zachar, Z. (1988). On/off regulation of gene expression at the level of gene splicing. Trends Gen. 4, 134–138.

Breathnach, R., & Chambon, P. (1981). Organization and expression of eukaryotic split genes coding for proteins. Ann. Rev. Biochem. 50, 349–383.

Breitbart, R.E., Andreadis, A., & Nadal-Ginard, B. (1987). Alternative splicing: A ubiquitous mechanism for the generation of multiple protein isoforms from different genes. Ann. Rev. Biochem. 56, 467–495.

Crenshaw, E.B., Russo, A.F., Swanson, L.W., & Rosenfeld, M.G. (1987). Neuron-specific alternative RNA processing in tranagenic mice expressing a metallothionein-calcitonin fusion gene. Cell 49, 389–398.

Delsert, C.D., & Rosenfeld, M.G. (1992). A tissue specific small nuclear ribonucleoprotein and the regulated splicing of the calcitonin/calcitonin gene-related protein transcript. J. Biol. Chem. 267, 14573–14579.

Emeson, R.B., Hedjran, F., Yeakley, J.M., Guise, J.W., & Rosenfeld, M.G. (1989). Alternative production of calcitonin and CGRP mRNA is regulated at the calcitonin-specific splice acceptor. Nature 341, 76–80.

Helfman, D.M., Ricci, W.M., & Finn, L.A. (1988). Alternative splicing of tropomyosin pre-mRNAs *in vitro* and *in vivo*. Genes Devel. 2, 1627–1638.

Kitamura, N., Tagagaki, Y., Furuto, S., Tanaka, T., Nawa, H., & Nakanishi, S. (1983). A single gene for bovine high molecular weight kininogens. Nature 305, 545–549.

Kramer, A.R., Mayeda, A., Kozak, D., & Binns, G. (1991). Functional expression of cloned human splicing factor SF2: Homology to RNA-binding proteins, U1 70k and Drosophila splicing regulators. Cell 66, 383–394.

Lamond, A.I. (1991). Nuclear RNA processing. Curr. Opin. Cell Biol. 3, 493–500.

Latchman, D.S. (1990). Cell-type specific splicing factors and the regulation of alternative RNA splicing. New Biol. 2, 297–303.

Latchman, D.S. (1995). Gene Regulation: A Eukaryotic Perspective. 2nd edn., Chapman and Hall, London.

Leff, S.E., Evans, R.M., & Rosenfeld, M.G. (1987). Splice commitment dictates neuron-specific alternative RNA processing in calcitonin/CGRP gene expression. Cell 48, 517–524.

Leff, S.E.. Rosenfeld, M.G., & Evans, R.M. (1986). Complex transcriptional units: Diversity in gene expression by alternative RNA processing. Ann. Rev. Biochem. 55, 1091–1117.

McAllister, G., Amara, S.G., & Lerner, M.R. (1988). Tissue-specific expression and cDNA cloning of small nuclear ribonucleoprotein associated polypeptide. N. Proc. Natl. Acad. Sci. USA 85, 5296–5300.

Maniatis, T., & Reed, R. (1987). The role of small nuclear ribonucleoprotein particles in pre-mRNA splicing. Nature 325, 673–678.

Ozcelik, T., Leff, S., Robinson, W., Donlon, T., Lalande, M., Sanjines, E., Schinzel, A., & Francke, U. (1992). Small nuclear ribonucleoprotein polypeptide N (SNRPN), an expressed gene in the Prader-Willi syndrome critical region. Nature Genet. 2, 265–269.

Peterson, M.L., & Perry, R.P. (1989). The regulated production of Um and Us mRNA is dependent on the relative efficiencies of Us polyA site usage and trhe CU4 to M1 splice. Mol. Cell. Biol. 9, 726–738.

Sharp, P.A. (1987). Splicing of messenger RNA precursors. Science 235, 766–771.

Sharpe, N.G., Williams, D.G., & Latchman, D.S. (1990). Regulated expression of small nuclear ribonucleoprotein particle protein SmN in embryonic stem cell differentiation. Mol. Cell. Biol. 10, 6817–6820.

RECOMMENDED READING

Latchman, D.S. (1995). Gene Regulation. 2nd edn. Chapman and Hall, London.

Latchman, D.S. (1990). Cell type specific splicing factors and the regulation of alternative RNA splicing. New Biol. 2, 297–303.

Leff, S.E., Rosenfeld, M.G., & Evans, R.M. (1986). Complex transcriptional units: Diversity in gene expression by alternative RNA processing. Ann. Rev. Biochem. 55, 1091–1117.

Chapter 8

Antisense RNA and DNA as Potential Therapeutic Agents

LINDA J. VAN ELDIK

INTRODUCTION

The rapid explosion of knowledge resulting from recombinant DNA and gene cloning technology has led to exciting new developments in the treatment of

Principles of Medical Biology, Volume 5
Molecular and Cellular Genetics, pages 149–162.
Copyright © 1996 by JAI Press Inc.
All rights of reproduction in any form reserved.
ISBN: 1-55938-809-9

disease. One such development is the use of antisense approaches for selectively deactivating individual genes. The antisense methodology is based on the known specificity of base pairing between complementary nucleic acids that occurs during the process of gene expression. In principle, this allows the possibility of selectively interfering with the expression of any target gene by using a complementary, or "antisense," sequence that will hybridize to the target sequence and block expression of that gene. It should be possible, in theory, to use antisense strategies on any gene for which nucleic acid sequence information is available, and on any cell or organism into which the antisense agent can be introduced. In practice, however, there are a number of problems and challenges that need to be addressed in attempts to transfer current research findings to the clinic, including questions of specificity, potency, delivery, and toxicity. Antisense inhibition of gene expression is a rapidly expanding field of interest that has gone from abstract idea to real possibility in a very short time, and millions of dollars are being spent to bring this technology to the clinical arena. It is likely that antisense therapeutics will be available for the physician of the 21st century. Therefore, it is imperative to understand both the fundamental assumptions of antisense approaches, as well as the important unanswered questions that are currently being addressed.

It is beyond the scope of this review to cover in depth all the possible antisense methodologies and agents that have been employed, nor could one chapter do justice to this rapidly expanding research area. The reader is referred to a number of excellent articles on various aspects of antisense technology, listed at the end of this chapter. What will be emphasized in this chapter are: (a) a brief summary of the various classes of antisense reagents and the general principles of their action; (b) some practical considerations, concerns, and challenges that must be addressed in current and future attempts to develop antisense therapeutics; and (c) a set of case-study examples that demonstrate the potential of utilizing antisense strategies in human medicine, particularly in the areas of viral and cancer therapy.

GENERAL CLASSES, PRINCIPLES, AND MECHANISMS OF ANTISENSE REAGENTS

In order to understand how antisense technology works, it is useful to review briefly some of the fundamentals of gene expression. Figure 1 shows a schematic diagram of the major steps in the pathway from gene to protein product. During gene expression, the DNA transcriptional unit is transcribed into a primary RNA transcript, which is then processed within the nucleus into the mature messenger RNA (mRNA) form(s). The mRNA is then transported to the cytoplasm, where it is translated into the protein product(s). Even from this simplified overview of gene expression, it is obvious that there are a number of potential points in the pathway where antisense nucleic acids could exert their inhibitory functions; e.g., at the level of transcription, splicing, mRNA transport from nucleus to cytoplasm, or transla-

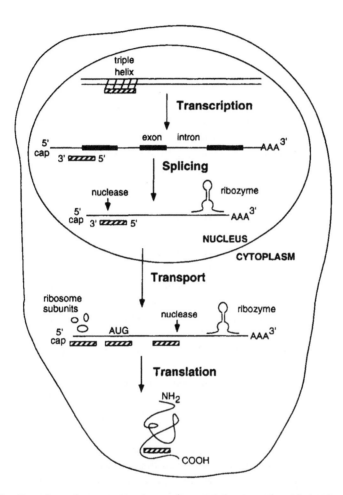

Figure 1. Overview of gene expression and potential points of regulation by antisense reagents. The mechanism of inhibition by antisense agents can be by physically blocking important interactions, stimulating degradation by nucleases, or by direct cleavage by ribozymes. Antisense agents can interfere with gene expression at various points along the information pathway from DNA transcriptional unit to protein: (a) with transcription by triple helix formation, (b) with splicing by binding at intron-exon junctions or interfering with capping or polyadenylation, (c) with transport of mRNA from nucleus to cytoplasm, (d) with translation by inhibiting binding of initiation factors or ribosomal subunits or by direct cleavage of the mRNA, or (e) with protein function by direct binding of oligonucleotide to protein. Antisense agents are depicted by slashed lines.

tion. There are examples in the literature of antisense inhibition at each of these points in the information pathway from DNA to RNA to protein.

One of the present limitations of antisense technology is that the precise molecular mechanisms by which an antisense agent inhibits gene expression are not completely defined. It has been found that gene expression may be blocked by one of several possible mechanisms, including hindering other molecules from interacting with the target sequence and thereby interrupting the expression pathway, stimulating nucleases specific for the double-stranded hybrid, or directly inactivating the target by intercalation or cleavage.

Antisense agents can be grouped broadly into antisense RNA (including catalytic RNAs or ribozymes) and antisense DNA. A brief outline of the development and mechanisms of action of these agents is given below.

Antisense RNA

The principle of gene inhibition by standard antisense RNA methods is simple: RNA that is complementary (or antisense) to a target mRNA will form specific RNA:RNA hybrids by base pairing, leading to inhibition of translation from the target mRNA. It has been known for some time that certain viruses and bacteria regulate the expression of some genes by making just such antisense RNAs. The theory that this natural antisense mechanism in bacteria and viruses could be applied to eukaryotic gene expression was first demonstrated by Izant and Weintraub (1984). They designed an expression vector that would transcribe the "wrong" strand of the thymidine kinase (TK) gene, thus producing antisense RNA. The antisense RNA was postulated to be able to bind to the mRNA, and prevent protein translation by one of several means, such as interference with ribosome interactions, degradation by nucleases, interfering with transport from nucleus to cytoplasm, or base modifications. This is what happened: synthesis of the TK enzyme from a sense expression vector was blocked when the antisense expression vector was introduced into mouse cells. These results showed that antisense RNA could inhibit the function of cloned genes. Later studies by a number of investigators confirmed that antisense RNA could also inhibit endogenous cellular genes. In addition, an antisense expression vector was not required for the success of this technology. For example, antisense RNA injected directly into cells can inhibit translation (see Melton, 1985; Green et al., 1986; Weintraub, 1990).

An extension of these classical antisense RNA strategies makes use of the discovery by Cech and Altman that certain RNAs can act as enzymes to catalyze self-cleavage or cleavage of other RNA molecules with which they hybridize (for reviews, see Cech and Bass, 1986; Haseloff and Gerlach, 1988; Symons, 1992; Castanotto et al., 1994). These RNA enzymes, or ribozymes, occur naturally in a variety of biological systems and show specificity in the RNA sequences that they recognize and cleave. The advantage of using ribozymes as antisense agents is that they act catalytically; i.e., they dissociate after cleavage of the target RNA sequence and are free to bind to and cleave additional RNA molecules. Because of this catalytic activity, the ribozyme concentrations necessary for gene inhibition should theoretically be lower than with conventional antisense RNAs, thus making them an attractive possibility for engineering into antisense RNA reagents. There are a variety of types of ribozymes with differing specificities and functional requirements. The main classes can be divided into: (a) the Tetrahymena (group 1 intron) ribozymes, (b) RNase P, (c) "hammerhead" ribozymes, and (d) "hairpin" ribozymes. The hammerhead and hairpin ribozymes have been studied the most extensively for development of antisense reagents that will recognize and cleave specific RNA sequences. Therefore, only these two ribozyme types with their generalized structure and cleavage sites are illustrated schematically in Figure 2.

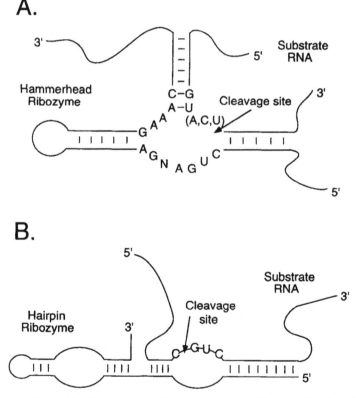

Figure 2. Generalized structure and cleavage sites of hammerhead and hairpin ribozymes. (**A**) Schematic diagram of a hammerhead ribozyme with the cleavage site of the substrate RNA marked with an arrow (see Haseloff and Gerlach, 1988 for more details). (**B**) Schematic diagram of a hairpin ribozyme with its consensus cleavage site marked with an arrow (see Hampel et al., 1990 for more details).

References at the end of this chapter should be consulted for more details on the mechanism of action of ribozymes. A specific example of the use of a ribozyme for inhibition of HIV is given in the section on case-study examples of antisense inhibition.

Antisense DNA

The second major type of antisense reagents are based on using antisense oligodeoxynucleotides, i.e., antisense DNA. The basic strategy is that a short sequence of single-stranded DNA complementary to the target sequence is synthesized *in vitro* and delivered to the cell, where it binds to and inhibits the target sequence. One of the first examples of using antisense DNA to inhibit gene expression was the work of Zamecnik and Stephenson in 1978, who synthesized an oligodeoxynucleotide complementary to a region of the sequence of Rous sarcoma virus, and showed that this antisense DNA could inhibit the translation (Stephenson and Zamecnik, 1978) and replication (Zamecnik and Stephenson, 1978) of the virus in cultured chick embryo fibroblasts. Since those pioneering studies, there have been numerous examples of the use of antisense DNA to inhibit specific genes. There is also an increasing literature on new and improved strategies for oligonucleotide design that employ DNA modified at particular linkages or terminal groups to improve binding affinity, cellular uptake, targeting, or resistance to degradation by nucleases. Studies concerning the mechanisms by which antisense DNA can shut down gene expression have shown that a number of different mechanisms can come into play. Most of the early studies of antisense DNA have targeted specific sequences in the mRNA. The resulting oligonucleotide-mRNA complex can prevent translation of the mRNA by either a steric hindrance mechanism to prevent ribosome binding or other important interactions required for translation, or by stimulation of mRNA degradation by RNase H which cleaves the RNA component of RNA:DNA hybrids. Antisense DNA has also been targeted to splice sites on pre-mRNA to prevent maturation of the RNA. For excellent general reviews on mechanisms of antisense DNA inhibition, see Helene and Toulme, 1990; Uhlmann and Peyman, 1990; Ghosh and Cohen, 1992; Murray, 1992.

Although the standard target for antisense DNA is a complementary sequence in the mRNA of interest, more recent antisense strategies have targeted protein or DNA sequences. Direct targeting of a protein with a specific antisense oligonucleotide is a strategy under active investigation for DNA binding proteins such as transcription factors, polymerases, etc. Antisense oligonucleotides can also be designed to interfere directly with gene transcription by binding to double-stranded DNA to form a triple helix structure through Hoogsteen base pairing. Hoogsteen base pairing is a pattern of hydrogen bonding in DNA where a thymine (T) can bind to an adenine-thymine (A-T) pair, or where a protonated cytosine (C^+) can bind to a guanine-cytosine (G-C) pair to form T.A-T and C^+.G-C triplets. A

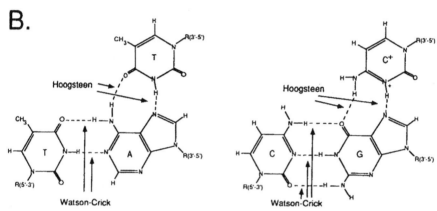

Figure 3. Generalized structure of triple helix DNA. (**A**) Schematic diagram of the binding of a homopyrimidine oligonucleotide (black) to double-stranded DNA helix. (**B**) Triple helix formation occurs through Hoogsteen hydrogen bonding of thymine (T) and protonated cytosine (C^+) to Watson-Crick A.T and G.C base pairs, respectively (see Helene and Toulme, 1990 for more details).

schematic diagram of the structure of triple helix DNA is shown in Figure 3. The formation of triple helix DNA can lead to inhibition of gene expression by interfering with binding of transcription factors, methylases, or other proteins important for transcription. These antisense oligonucleotides that target duplex DNA have been called anti-gene oligonucleotides (see Helene and Tulme, 1990).

The ability to utilize anti-gene oligonucleotides to form triple helices for regulating gene expression has potential advantages over the standard antisense DNA approach of inhibiting at the mRNA level. Because multiple copies of mRNA are transcribed from each copy of DNA, a much smaller number of molecules need to be inhibited when DNA rather than RNA is the antisense target, and much lower doses of antisense DNA should be needed. However, studies with anti-gene oligonucleotides are in their infancy, and the potential efficiency of anti-gene oligonucleotides compared to classical antisense oligonucleotides remains to be determined. Recent studies are also evaluating the prospect of designing triple-helix forming oligonucleotides that are modified by addition of cross-linking, alkylating, or cleavage reagents in attempts to generate irreversible inhibition of genes.

CONSIDERATIONS, CONCERNS, AND CHALLENGES IN THE USE OF ANTISENSE AGENTS

It is evident from this brief survey of antisense methodologies that it is possible to inhibit gene expression by using a variety of antisense agents that target different parts of the information pathway from DNA to RNA to protein in eukaryotic cells. There is now a sizeable literature employing antisense methodologies, with more than 2,000 publications appearing since 1992. However, in very few cases have the precise mechanisms of the antisense inhibition been elucidated and results with antisense agents are not always consistent or predictable. Thus, rational development of effective antisense agents is challenging at the present time. Nevertheless, there are some general considerations to keep in mind when designing an antisense reagent for inhibition of a specific gene of interest. These are discussed briefly below.

Specificity

In developing a new antisense reagent for deactivating a specific gene, it is difficult to predict with certainty what sites on the target gene or what part of the gene expression pathway will be most effectively regulated by the antisense compound. Ideally, the target sequence needs to be unique to the desired gene and not occur in other genes, and the antisense reagent should not have nonspecific biological effects. There is no generally applicable optimal length for antisense agents, and there is a trade-off between designing an antisense compound that is of the length and sequence to bind specifically to the target, but not so long that nonspecific binding or secondary structure problems occur. There is also no way at the present time to predict *a priori* what sequence on the gene is best to target for antisense inhibition. Many investigators have had success with antisense agents targeted to the 5' region of genes (e.g., the cap site, ribosome binding site, initiation codon, splice sites), but there are currently no definitive guidelines for the "best" design of an antisense construct and much of the development of an antisense reagent is still a trial and error proposition.

Potency and Efficacy

It has been found for both antisense RNA and DNA that it is not easy to turn off gene expression completely; rather, the extent of inhibition generally is proportional to the dose of antisense reagent used. In addition, the maximal level of inhibition achieved can vary with the particular gene targeted or antisense compound used. Antisense strategies are most useful for applications where a complete shutoff of the gene is not necessary for a therapeutic benefit. Because of their exquisite selectivity and discriminating capability, antisense reagents are also attractive when a disease is characterized by a unique nucleic acid sequence or difference in genetic information from the nondiseased state (e.g., virus or parasitic infection, mutated or translocated gene).

Delivery

Another important challenge for antisense strategies is designing methods for more efficient delivery of the antisense agent to its specific target gene in the cell. There are several strategies being explored in this regard. For example, encapsulating antisense RNA into liposomes that carry antibodies directed against particular cell surface antigens have the potential to allow targeting and interaction with a specific cell of interest. Other modifications for more efficient delivery are being developed, such as conjugation of antisense oligonucleotides to lipophilic groups (cholesterol) or to polycations. In addition, recent studies with *in vivo* administration of antisense oligonucleotides into rodent brain have demonstrated that antisense agents can be effective in the whole animal. The challenge, of course, is to deliver the agent specifically to the target gene with minimal nonspecific interactions.

Stability and Toxicity

Another challenge in designing an antisense compound is to be able to develop an agent with a good therapeutic concentration range, and minimal side effects. An extensive amount of research is being done on antisense reagents that are modified in structure to increase their stability and effectiveness. For example, a number of oligonucleotide analogues with backbone modifications (methylphosphonates, phosphorothioate, α-oligomers) or added terminal groups (cholesterol, polylysine, acridine rings, alkylating reagents) have been developed in order to improve activity, cell uptake, or resistance to nucleases. An interesting class of compounds that is being investigated for its potential as antisense reagents is a peptide nucleic acid (PNA) analogue of DNA, where the deoxyribose phosphate backbone of DNA is replaced by a peptidelike polymer of 2-aminoethylglycine units to which bases (thymine, cytosine, adenine, or guanine) are attached (Nielsen et al., 1993). PNAs appear to have a high affinity for RNA and DNA, are quite resistant to nucleases and proteases, and are relatively easy to synthesize and modify chemically in attempts to develop improved PNA derivatives.

These studies on development of modified antisense reagents are also being complemented by continued efforts to reduce potential nonspecific toxicity, immunogenicity, or mutagenic potential of the modified derivatives. Investigation of the pharmacokinetics of oligopharmaceuticals is also an active area of research (see Vlassov and Yakubov, 1991). Finally, successful first-generation antisense therapies will require major advances in issues dealing with the synthesis, characterization, and quality control for large-scale production of antisense agents in order to lower the cost of the reagents to realistic levels for therapeutic applications.

CASE-STUDY EXAMPLES OF ANTISENSE INHIBITION

Antisense Inhibition of Oncogenes:
Antisense Oligonucleotides Against Ha-*ras*

Mammalian p21 *ras* proteins are GTP/GDP binding proteins with intrinsic GTPase activity. Although the precise mechanism of action of p21 *ras* proteins are not well-defined, they are thought to play an important role in regulation of signal transduction pathways critical for control of cell proliferation (for review, see Barbacid, 1987). It has been found that many human tumors are characterized by point mutations in the coding region of the *ras* genes, which lead to the production of mutated p21 *ras* proteins. For example, point mutations in the 12th or 61st codon of the Ha-*ras* gene result in single amino acid changes in the Ha-*ras* protein product. This single amino acid difference converts the normal, regulated *ras* proto-oncogene into the mutated Ha-*ras* oncogenic protein that is constitutively active and has transforming activity. In addition to these naturally occurring mutations in *ras* oncogenes associated with human neoplasms, mutations in *ras* genes have also been found in a number of chemical- and radiation-induced tumors.

It is obvious that methods to selectively inhibit the expression of the mutated *ras* oncogene without affecting the normal *ras* gene (which is necessary for normal cell proliferative functions) would be of great value. In this regard, an antisense strategy is appealing because selective targeting of the mutated mRNA could theoretically be achieved through the use of properly designed antisense reagents.

In fact, this theoretical possibility has now been demonstrated experimentally for mutations at either codon 12 or codon 61 of Ha-*ras*. Saison-Behmoaras et al. (1991) designed 9-residue antisense oligodeoxynucleotides modified with an ac-ridine intercalating agent and/or a hydrophobic substituent to enhance their activity and their uptake by cells. The antisense oligonucleotides were designed to inhibit selectively the Ha-*ras* oncogene mutated at codon 12 (G to T mutation). The data demonstrated selective inhibition of the mutant *ras* mRNA and inhibition of the growth of bladder carcinoma cells (which contain the activated Ha-*ras*), with no effect on the expression of normal *ras* mRNA and no inhibition of normal mammary cell growth.

Monia et al. (1992) tested a series of phosphorothioate-modified antisense oligonucleotides ranging in length from 5 to 25 bases, each targeted to the Ha-*ras* codon 12 G to T point mutation. They showed selective targeting of the point mutation by the antisense oligonucleotides, with selectivity being critically depend-ent on oligonucleotide length and concentration.

Chang et al. (1991) targeted the Ha-*ras* codon 61 A to T point mutation with 11-residue antisense oligonucleoside methylphosphonates and their psoralen derivatives. In mixed cell cultures expressing both the mutated and normal *ras* p21 proteins, the antisense oligonucleotides were shown to selectively inhibit the mutated p21 *ras* while leaving the normal p21 virtually unaffected.

These data demonstrate the feasibility of using antisense oligonucleotides to selectively target an oncogene that contains a single-base point mutation without affecting the normal gene, and suggest that continued effort should be expended on developing antisense treatment strategies for those oncogene-mediated neoplasms that result from gene mutations. Antisense strategies are also being explored for modulation of expression of other oncogenes (for reviews, see Neckers et al., 1992; Neckers and Whitesell, 1993; Millian et al., 1994).

Antisense Inhibition of Viruses: Ribozymes Against HIV

Antisense reagents are currently being tested in human clinical trials for the treatment of acquired immune deficiency syndrome (AIDS). Since the identification of human immunodeficiency virus (HIV) as a causative agent for AIDS, there has been a concerted effort to develop antisense strategies that inhibit HIV replication. A large number of antisense reagents (both antisense RNA and DNA) have been shown to successfully impair HIV activity. One antisense approach is the use of ribozymes targeted against specific regions of HIV (for reviews, see Rossi and Sarver, 1992; Yu et al., 1994; Smythe and Symonds, 1995).

One of the first reports of ribozymes as potential anti-HIV therapeutic agents was by Sarver et al. (1990), who designed a hammerhead ribozyme targeted to the HIV-1 gag structural gene. They showed that cells expressing the ribozyme were significantly protected from HIV infection. Other areas of the HIV genome that have since been shown to be able to be inhibited by ribozymes include the env structural gene, the tat regulatory gene, the integrase gene, and a region of the 5' leader sequence.

The ribozyme targeted to the 5' leader sequence is particularly interesting. Wong-Staal and colleagues designed a hairpin ribozyme to recognize and cleave an RNA sequence near the beginning of the HIV mRNA that is conserved in diverse strains of HIV (Ojwang et al., 1992; Yu et al., 1993). The ribozyme was found to be highly effective to reduce the amount of virus produced by HIV-infected cells. Based on these results, clinical trials utilizing an HIV-1 leader sequence ribozyme were approved and initiated. The experimental paradigm is to isolate the patient's T lymphocytes, insert the ribozyme-producing retroviral vector into the cells, and then return the cells to the patient in hopes that the ribozyme will inactivate the HIV mRNA (see Barinaga, 1993; Yu et al., 1994). As of early 1996, this study is in phase I clinical trial. This study is just one example of numerous ongoing clinical trials utilizing HIV antisense therapies, illustrating the rapid progress being made in this field.

Additional areas of intense current investigation include designing trafficking signals into the ribozyme in order to direct the molecule to specific routes in the cell where HIV transcripts are localized, or linking together several ribozymes targeted to different parts of the HIV genome. This latter approach of using multi-site ribozymes is being done in hopes of minimizing the effect of HIV's ability to mutate and potentially evade a ribozyme directed against only a single site. Although a number of challenges remain to be overcome concerning ribozyme efficiency, specificity, safety, and delivery, the expectation is high that the antisense approach to HIV will lead to an effective antiviral therapy.

SUMMARY

It is evident that antisense technology has practical potential in terms of biotechnological and pharmacological applications. Several biotechnology companies have antisense strategies as a primary focus, and several established pharmaceutical companies have initiated antisense research and development programs. Antisense technology is already proving extremely useful to the biotechnology industry interested in commercial plant development; e.g., tomatoes that ripen more slowly and tobacco plants that are more resistant to virus infections. However, it is becoming increasingly clear that antisense strategies are not as straightforward or as easily applicable to the whole animal or human as had been suggested in early experiments with cultured cells. Part of the problem is that the exact mechanisms of gene inhibition are not known in molecular terms for each class of antisense agents. Therefore, the specificity and extent of gene inhibition are difficult to predict. A number of other problems related to specificity, drug delivery, and pharmacokinetics will also need to be overcome to allow the rational design of antisense therapeutic strategies. Nevertheless, the rapidly expanding database of knowledge concerning the molecular mechanism of action of antisense inhibition of gene expression, and the continual development of new classes of selectively modified antisense compounds hold great promise that antisense therapeutics for human disease will be an important component of 21st century medicine.

ACKNOWLEDGMENTS

These studies were supported in part by NIH grant AG11138. I am grateful to Drs. D. Martin Watterson and John Franks for encouragement and critical reading of the manuscript.

REFERENCES

Barbacid, M. (1987). ras genes. Ann. Rev. Biochem. 56, 779–827.
Barinaga, M. (1993). Ribozymes: Killing the messenger. Science 262, 1512–1514.

Castanotto, D., Rossi, J.J., & Sarver, N. (1994). Antisense catalytic RNAs as therapeutic agents. Adv. Pharm. 25, 289–317.

Cech, T.R., & Bass, B.L. (1986). Biological catalysis by RNA. Ann. Rev. Biochem. 55, 599–629.

Chang, E.H., Miller, P.S., Cushman, C., Devadas, K., Pirollo, K.F., Ts'o, P.O.P., & Yu, Z.P. (1991). Antisense inhibition of ras p21 expression that is sensitive to a point mutation. Biochemistry 30, 8283–8286.

Ghosh, M.K., & Cohen, J.S. (1992). Oligodeoxynucleotides as antisense inhibitors of gene expression. Prog. Nucl. Acid Res. and Mol. Biol. 42, 79–126.

Green, P.J., Pines, O., & Inouye, M. (1986). The role of antisense RNA in gene regulation. Ann. Rev. Biochem. 55, 569–597.

Hampel, A., Tritz, R., Hicks, M., & Cruz, P. (1990). "Hairpin" catalytic RNA model: Evidence for helices and sequence requirement for substrate RNA. Nucleic Acids Res. 18, 299–304.

Haseloff, J., & Gerlach, W.L. (1988). Simple RNA enzymes with new and highly specific endonuclease activities. Nature 334, 585–591.

Helene, C., & Toulme, J.-J. (1990). Specific regulation of gene expression by antisense, sense and antigene nucleic acids. Biochim. Biophys. Acta 1049, 99–125.

Izant, J.G., & Weintraub, H. (1984). Inhibition of thymidine kinase gene expression by antisense RNA: A molecular approach to genetic analysis. Cell 36, 1007–1115.

Melton, D.A. (1985). Injected anti-sense RNAs specifically block messenger RNA translation in vivo. Proc. Natl. Acad. Sci. USA 82, 144–148.

Milligan, J.F., Jones, R.J., Froehler, B.C., & Matteucci, M.D. (1994). Development of antisense therapeutics. Implications for cancer gene therapy. Ann. N.Y. Acad. Sci. 716, 228–241.

Monia, B.P., Johnston, J.F., Ecker, D.J., Zounes, M.A., Lima, W.F., & Freier, S.M. (1992). Selective inhibition of mutant Ha-ras mRNA expression by antisense oligonucleotides. J. Biol. Chem. 267, 19954–19962.

Murray, J.A.H. (1992). Antisense RNA and DNA. In: Modern Cell Biology (Murray, J.A.H., ed.), Vol. 11. Wiley-Liss, New York.

Neckers, L., & Whitesell, L. (1993). Antisense technology: biological utility and practical considerations. Amer. J. Physiol. 9, L1–L12.

Neckers, L., Whitesell, L., Rosolen, A., & Geselowitz, D.A. (1992). Antisense inhibition of oncogene expression. Crit. Rev. Oncogenesis 3, 175–231.

Nielsen, P.E., Egholm, M., Berg, R.H., & Buchardt, O. (1993). Peptide nucleic acids (PNAs): Potential antisense and anti-gene agents. Anticancer Drug Des. 8, 53–63.

Ojwang, J.O., Hampel, A., Looney, D.J., Wong-Staal, F., & Rappaport, J. (1992). Inhibition of human immunodeficiency virus type 1 expression by a hairpin ribozyme. Proc. Natl. Acad. Sci. USA 89, 10802–10806.

Rossi, J.J., & Sarver, N. (1992). Catalytic antisense RNA (ribozymes): Their potential and use as anti-HIV-1 therapeutic agents. In: Innovations in Antiviral Development and the Detection of Virus Infection (Block, T., Jungkind, D., Crowell, R.L., Denison, M., & Walsh, L.R., eds.), pp. 95–109, Plenum Press, New York.

Saison-Behmoaras, T., Tocque, B., Rey, I., Chassignol, M., Thuong, N.T., & Helene, C. (1991). Short modified antisense oligonucleotides directed against Ha-*ras* point mutation induce selective cleavage of the mRNA and inhibit T24 cells proliferation. EMBO J. 10, 1111–1118.

Sarver, N., Cantin, E.M., Chang, P.S., Zaia, J.A., Ladne, P.A., Stephens, D.A., & Rossi, J.J. (1990). Ribozymes as potential anti-HIV-1 therapeutic agents. Science 247, 1222–1225.

Smyth, J.A., & Symonds, G. (1995). Gene therapeutic agents: the use of riboszymes, antisense, and RNA decoys for HIV-1 infection. Inflamm. Res. 44, 11–15.

Stephenson, M.I., & Zamecnik, P.C. (1978). Inhibition of Rous sarcoma viral RNA translation by a specific oligodeoxyribonucleotide. Proc. Natl. Acad. Sci. USA 75, 285–288.

Symons, R.H. (1992). Small catalytic RNAs. Ann. Rev. Biochem. 61, 641–671.

Ulhmann, E., & Peyman, A. (1990). Antisense oligonucleotides: A new therapeutic principle. Chem. Rev. 90, 543–584.

Vlassov, V.V., & Yakubov, L.A. (1991). Oligonucleotides in cells and in organisms: Pharmacological considerations. In: Prospects for Antisense Nucleic Acid Therapy of Cancer and AIDS (Wickstrom, E., ed.), pp. 243–266, Wiley-Liss, Inc., New York.

Weintraub, H.M. (1990). Antisense RNA and DNA. Scientific American 262, 40–46.

Yu, M., Ojwang, J., Yamada, O., Hampel, A., Rapapport, J., Looney, D., & Wong-Staal, F. (1993). A hairpin ribozyme inhibits expression of diverse strains of human immunodeficiency virus type 1. Proc. Natl. Acad. Sci. USA 90, 6340–6344.

Yu, M., Poeschla, E., & Wong-Staal, F. (1994). Progress towards gene therapy for HIV infection. Gene Therapy 1, 13–26.

Zamecnik, P.C., & Stephenson, M.I. (1978). Inhibition of Rous sarcoma virus replication and cell transformation by a specific oligodeoxyribonucleotide. Proc. Natl. Acad. Sci. USA 75, 280–284.

RECOMMENDED READINGS

Castanotto, D., Rossi, J.J., & Sarver, N. (1994). Antisense catalytic RNAs as therapeutic agents. Adv. Pharm. 25, 289–317.

Ghosh, M.K., & Cohen, J.S. (1992). Oligodeoxynucleotides as antisense inhibitors of gene expression. Prog. Nucl. Acid Res. and Mol. Biol. 42, 79–126.

Helene, C., & Toulme, J.-J. (1990). Specific regulation of gene expression by antisense, sense and antigene nucleic acids. Biochim. Biophys. Acta 1049, 99–125.

Murray, J.A.H., & Crockett, N. (1992). Antisense techniques: An overview. In: Antisense RNA and DNA. Modern Cell Biology, Vol. 11, pp. 1–49. Wiley-Liss, New York.

Weintraub, H.M. (1990). Antisense RNA and DNA. Scientific American 262, 40–46.

Chapter 9

Regulation of Gene Expression and Gene Amplification

JOHN D. HAWKINS and IAN R. PHILLIPS

Principles of Medical Biology, Volume 5
Molecular and Cellular Genetics, pages 163–199.
Copyright © 1996 by JAI Press Inc.
All rights of reproduction in any form reserved.
ISBN: 1-55938-809-9

INTRODUCTION

All of an animal's somatic cells contain the whole of its genome. However, the various cell types are specialized so that each cell produces only a limited number of proteins, using just part of the total genome. In addition, some proteins are produced only at defined stages of development or at certain times during the cell cycle. A major current interest in molecular biology and genetics is the search for the mechanisms by which only certain genes are selected for transcription in the different tissues of the body or at different times. Most of the information that has been acquired so far is concerned with positive control, which involves the binding of protein(s) to the DNA in the vicinity of a gene, thereby stimulating its transcription. However, examples are increasingly coming to light of negative control in which proteins, binding to DNA, inhibit the transcription of nearby genes.

Not all proteins are produced in equal amounts, and there are several stages in their formation where control over their final amount can be exercised. These include the rate of transcription of their genes, the processing of the pre-mRNAs so formed and the rate of transport of the mature mRNAs out of the nucleus. In the cytoplasm the binding of mRNAs to the ribosomes and their subsequent rate of degradation are also points at which control can be exerted.

A number of mRNAs and proteins concerned with the general functioning of all cells are found in most tissues. These molecules are encoded by genes known as housekeeping genes, which exhibit features that differ from those of other genes.

Prokaryotes have special mechanisms for controlling the transcription of their genes, particularly in response to changes in their environment. The presence of a particular substrate may lead to the activation of a whole set of adjacent genes directing the production of the enzymes necessary for the metabolism of that substance. Such a set of genes, under the control of a single promoter, is known as an operon. Conversely, when an essential metabolite that the organism normally synthesizes is present, transcription of the genes in the operon encoding the enzymes involved in this synthetic process will be inhibited. This mechanism does not seem to operate in eukaryotes, though there are situations in which their cells can respond to certain metabolites by synthesizing some or all of the enzymes required to metabolize them. For example, the genes encoding the enzymes that catalyze the individual steps in the urea cycle are all preceded by the sequence CCTGCCCTC (or something very close to this). There is simultaneous stimulation of their transcription when large amounts of proteins are ingested, even though the genes are on different chromosomes.

In prokaryotes, where there is no separate nucleus, the translation of an mRNA begins before the process of its production (transcription) is complete. Thus, in these organisms, transcription and translation are tightly coupled. In addition, most bacterial mRNAs have a very short half-life of only a few minutes and, therefore, control of gene expression is generally exerted at the level of transcription initiation. In contrast, in eukaryotes gene transcription takes place in the nucleus and mRNA

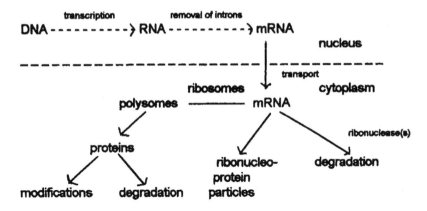

Figure 1. Stages in the synthesis of mRNA and proteins and their fates in the cell.

translation in the cytoplasm. Since these two processes are separated in space and time, there are opportunities for the control of gene expression at different levels. The various steps involved in this process are outlined in Figure 1. DNA is transcribed into a primary RNA transcript that is processed within the nucleus to yield a mature RNA; the mRNA is transported from the nucleus to the cytoplasm where it is translated to produce a protein product.

Since regulation can take place at each of these steps in a eukaryotic cell, the amounts of the different proteins produced can be controlled by regulating:

1. the rate of initiation of transcription (transcriptional control).
2. the way in which a primary RNA is processed (RNA processing control).
3. the choice of which processed RNAs are transported from the nucleus to the cytoplasm (RNA transport control).
4. the rate of degradation of mRNAs in the cytoplasm (mRNA stability control).
5. the rate of translation of mRNAs (translational control).
6. the rate of activation, inactivation or degradation of proteins (posttranslational control).

Regulation of the rate of initiation of gene transcription is the most common form of control of gene expression in eukaryotes. In this chapter we will describe the mechanisms responsible for regulating the rate of initiation of transcription of protein-coding genes which are transcribed by RNA polymerase II (RNA pol II), and then outline various types of posttranscriptional control mechanisms.

CONTROL OF TRANSCRIPTIONAL INITIATION

Evidence for Transcriptional Control

In multicellular organisms each cell expresses only a subset of its genes, the identity of which depends on the tissue and may change according to the developmental stage of the cell. Because these patterns of gene expression are inherited they must be encoded in DNA. Thus, in addition to structural information, each gene must contain the instructions necessary for its correct pattern of expression.

Experimental evidence for the control of eukaryotic gene expression at the transcriptional level has been obtained by a technique known as a nuclear run-on transcription assay. Isolated nuclei are incubated with highly radioactive ribonucleoside tri-phosphates (the precursors of RNA) to label the nascent RNA chains. RNA is then extracted and hybridized to an excess of filter-bound unlabeled complementary DNAs (cDNAs). The fraction of the total radioactivity that binds to a particular cDNA is proportional to the number of molecules of RNA polymerase that were actively transcribing the gene when the nuclei were isolated, and reflects the relative rate of initiation of transcription of that gene. These experiments have demonstrated the importance of transcriptional control in regulating the tissue-specific and developmental stage-specific expression of many proteins, and also in regulating the induction of expression of proteins in response to signals such as steroid hormones.

Core Promoters

A comparison of the 5'-flanking sequences of eukaryotic genes has revealed the presence of several conserved sequence motifs within about 100 base pairs (bp) upstream of the start site of transcription. One of these, TATAA, the so-called TATA box, occurs about 25 bp 5'- of the transcriptional start site, and closely resembles the −10 element of bacterial gene promoters. Other conserved sequences occur at variable positions further upstream, in particular a GC-rich element (usually GGGCGG, often referred to as a GC box) and, rather less frequently, the sequence CCAAT, known as the CAAT box. The GC box appears to be present in all housekeeping genes, i.e., those that are expressed constitutively at low levels in very many cells. In contrast to bacterial promoters, these conserved sequence motifs are not present in the promoters of all eukaryotic genes, suggesting that control of their transcription is more complicated than that of bacterial genes.

It was soon demonstrated that such conserved sequence motifs located just upstream from the start site of transcription of eukaryotic genes did indeed play a functional role in gene transcription. Deletions and point mutations were made in the 5'-flanking regions of cloned eukaryotic genes. The mutant sequences were introduced into a suitable eukaryotic cell and the effect of each mutation determined by measuring the amount of RNA produced from the gene. In a variation of this technique the promoter region was fused to a "reporter" gene encoding an easily measurable bacterial enzyme such as chloramphenicol acetyltransferase (CAT), which acetylates chloramphenicol, and its activity was measured.

The results of such experiments demonstrated that mutations in many regions of the 5'-flanking sequence of a gene had no effect on transcription, but mutations of the TATA, GC, or CCAAT boxes decreased transcriptional activity dramatically. For instance, a single base change in the TATA box can reduce transcriptional efficiency by up to 95%.

Formation of a Stable Transcriptional Initiation Complex

Purified bacterial RNA polymerase can bind directly to a prokaryotic promoter *in vitro* and initiate accurate and efficient transcription. In contrast, purified eukaryotic RNA pol II does not by itself bind specifically to eukaryotic gene promoters, but initiates transcription poorly and at random. However, whole cell or nuclear extracts are capable of transcribing eukaryotic protein-coding genes *in vitro*, indicating that other factors present in these extracts are necessary for the correct binding of RNA pol II to promoters. Considerable progress has been made towards understanding the basic mechanisms responsible for the accurate initiation of transcription of these genes through the use of *in vitro* transcription systems, fractionated nuclear protein extracts and purified proteins.

These experiments have shown that RNA pol II requires at least seven different proteins, known as transcription factors (TFs) to bind correctly to a promoter and initiate transcription. Proteins that bind specifically to isolated DNA sequences can be detected by a technique known as gel-retardation. A double-stranded DNA fragment containing a putative protein-binding site is radiolabeled and incubated in the presence of nonspecific competitor DNA (such as poly dI.dC) with a purified protein or protein extract. If the extract contains a protein that specifically recognizes and binds to a sequence within the DNA fragment, the bound protein will retard the electrophoretic migration of the DNA through a nondenaturing polyacrylamide gel (Figure 2). The use of this technique has revealed that correct initiation of transcription depends on the formation at the promoter of a stable initiation complex involving the TFs and RNA pol II. The stages of assembly of the multi-protein complex are summarized in Figure 3.

A protein termed TFIID, which contains several subunits (one of them the TATA box binding protein, TBP), binds, by itself, specifically to the TATA box motif of the promoter. A second protein, TFIIA, then binds to and stabilizes the TFIID-TATA box complex. Once they are bound, it is difficult to remove these proteins by competition with an excess of a second promoter. Thus, TFIID, TFIIA, and the TATA box represents a stable, "committed" complex. However, before RNA pol II can interact with the promoter, another protein, TFIIB, must bind to the complex,

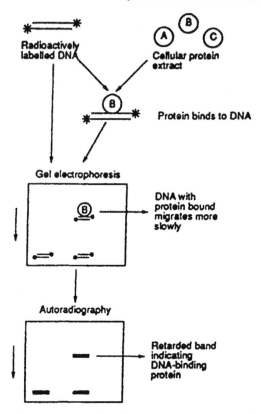

Figure 2. Gel retardation. Binding of a cellular protein (B) to the radioactively labeled DNA causes it to move more slowly on gel electrophoresis resulting in the appearance of a retarded band upon autoradiography to detect the radioactive label. (Reproduced, with permission, from Latchman, D.S. (1991) In: Eukaryotic Transcription Factors, Academic Press, London.

and the polymerase itself is only recruited to the complex when it has bound the protein TFIIF. Finally, three other proteins, TFsIIE, IIH, and IIJ are bound in that order to form a stable initiation complex that is capable of initiating transcription. As the polymerase moves away from the promoter and begins to transcribe the gene, TFIIF remains associated with it, probably acting as a helicase to unwind the double-stranded DNA. TFs IID and IIA remain bound at the TATA box thus allowing the formation of a new initiation complex and further rounds of transcription to take place.

The formation of a stable initiation complex bound to the TATA box results in very low levels of transcription, and genes that are transcribed by RNA pol II may have other sequence motifs within their promoter that contribute to efficient transcription. In addition to the TATA box, two of the most common proximal

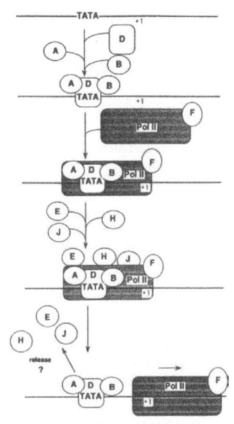

Figure 3. Assembly of the multi-protein transcriptional initiation complex on the TATA box, showing some of the intermediate stages in its formation. The transcription factors are designated by their letters only: A, D, B, F, E, H, J. The site of initiation of transcription is +1. When the Pol II-F complex moves along the gene to transcribe it, the fate of TFs-E, -H, -J is not known.

promoter elements are the GC and CAAT boxes mentioned earlier. Both these elements exert their effect on transcription by binding protein factors that recognize them specifically. The GC box is a common motif in the promoters of housekeeping genes and binds a protein known as Sp1 that is present in essentially all eukaryotic cells. In contrast, the CAAT box can bind several different proteins: for instance, CAAT box transcription factor (CTF), which is ubiquitous; C/EBP, which is found in relatively high concentration in the liver and thus plays a role in liver-specific gene expression; and CDP (CAAT box displacement protein), a negatively acting factor that prevents binding of positively acting proteins to the CAAT box and thus inhibits transcription. CDP is involved in inhibiting the transcription of many genes during the embryonic stages of development.

TATA-Less Promoters

A considerable number of eukaryotic promoters, particularly those of house-keeping genes, do not contain a TATA box. Transcription from these so-called TATA-less promoters requires the participation of all the general transcription factors involved in the assembly of a stable initiation complex at a TATA box, including TFIID. The absence of a TATA box means that, in these cases, TFIID binding cannot be the first stage in the assembly of a transcriptional complex. TATA-less promoters (and many TATA-containing promoters) contain an initiator element (Inr) located at or close to the transcriptional start site. Several families of Inr sequences have been identified, each of which is recognized and bound by different Inr binding proteins (ITFs). The binding of an ITF to an Inr initiates the formation of a transcription complex. The subsequent pathway to complex forma-tion is dependent on the particular ITF bound. Some ITFs may bind to TFIID. Once bound, TFIID serves as a nucleation site for the formation of the complex as described above for TATA-containing promoters. Other ITFs may interact directly with a component of RNA pol II. Whereas the transcription of genes containing a TATA box in their promoters generally initiates precisely at one particular nucleo-tide, the site of initiation of transcription of TATA-less genes is commonly not so rigidly fixed and may start from any one of several different nucleotides within a short sequence of about 20 or 30 nucleotides. Many TATA-less promoters contain multiple upstream GC boxes for binding Sp1. For these promoters there is evidence that a so-called tethering factor binds Sp1 to TF-IID, and thus initiates the formation of a transcription complex.

Regulated Gene Transcription

In the previous section we explained that gene transcription by RNA pol II, even at relatively low levels, requires in addition to the formation of a stable initiation complex near the start site of transcription, the presence of other promoter elements and their cognate binding proteins. Genes that are transcribed at relatively low levels in essentially all the cells of an organism are subject to very little transcrip-tional regulation, and may require only a few additional regulatory elements, such as a GC or CAAT box. However, the transcription of many eukaryotic genes is more complex. For instance, their transcription may be restricted to specific cell-types, such as liver or muscle, and to a particular stage of development. In addition, expression of these genes may depend on the response to a variety of signals, including various hormones, growth factors, and environmental agents. What are the mechanisms for mediating such complex patterns of transcriptional regulation?

The regulatory sequences involved in tissue-specific, developmental stage-specific, and signal-induced systems of gene expression must lie outside the core

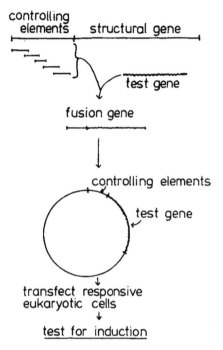

Figure 4. Identification of regulatory elements in DNA. Restriction fragments upstream of the start site of transcription are fused to the basal promoter of a reporter gene. The construct is incorporated into a plasmid and transfected into a suitable eukaryotic cell where the response to inducing agents can be tested.

promoter regions. Although *in vitro* transcription systems were highly successful and essential for unraveling the role of the core promoter elements and general transcription factors in the initiation of transcription, they are not efficient enough to enable the detailed dissection of promoter elements responsible for regulated transcription. A fresh approach for the identification of the responsible regulatory elements was needed. This involved the introduction of cloned genes into established cell lines and has led to the identification of DNA sequences required for the induction of transcription of many genes in response to specific stimuli, such as steroid hormones. These responses are known to be maintained in certain cell lines. A long stretch of the 5'-flanking region of a gene whose transcription is inducible in such a cell line was fused to a reporter gene, coding for an easily identifiable enzyme, such as CAT (see above). The fused gene construct was inserted into a plasmid vector and introduced into the appropriate cell line (Figure 4). If the CAT gene was expressed after treating the cells with the appropriate inducing agent, DNA elements responsible for mediating this response must be present within the 5'-flanking region. It was relatively straightforward to determine the location of the

Table 1. Consensus Sequences of Response Elements to
Some Hormones

Estrogen	A.G.G.T.C.A.N$_3$.T.G.A.C.C.T
Glucocorticoids	A.G.A.A.C.A.N$_3$.T.G.T.T.C.T
Tri-iodothyronine	T.G.G.G.T.C.A.N$_{4-6}$.T.G.A.C.C
Retinoic acid	G.G/T.T.G/C.A.C.C.N$_4$.G.C/T.T.C/G.A.C/G

Note: N signifies any nucleotide.

regulatory DNA element(s) responsible for this induction by constructing a series of deletion mutants of the 5'-flanking region and testing their ability to confer transcriptional inducibility to the reporter gene. The exact boundaries of the regulatory element were defined by fine deletion mapping. Detailed analysis of the effect of point mutations within the element determined which nucleotide residues are of critical importance for its regulatory function.

This approach has been used to analyze the promoters of many genes whose transcription is inducible in response to specific signals. A variety of different DNA promoter elements responsible for controlling the induction of transcription have been identified. The regulation of the expression of different genes by the same signal is mediated by identical or very similar DNA promoter sequences. For example, genes whose transcription is regulated by estrogen each have within their promoters a DNA sequence known as an estrogen response element (ERE), which mediates the induction of transcription of the gene in response to estrogen. Each ERE is about 15 bp in length and contains an imperfect repeat so that it exhibits dyad symmetry. Comparison of these sequences has led to the formulation of a consensus ERE sequence. Other steroid hormones, such as glucocorticoids and progestins, as well as tri-iodothyronine, retinoic acid, and the active form of vitamin D, act via response elements whose sequences are different from, but related to that of the ERE (Table 1).

Whereas these hormones all penetrate into cells and exert their effects by binding to intracellular receptors, some other hormones such as glucagon, the trophic hormones of the anterior pituitary and the catecholamines interact with receptors on the outer face of the cell surface. This leads to the activation of one of a number of enzymes on the inner surface of the cell membrane which catalyze the synthesis of a second messenger. One that is commonly employed is cAMP. This binds to and activates the enzyme protein kinase A, which phosphorylates a protein known as CREB. CREB contains an activation domain whose conformation is altered so that it is now able to stimulate transcription when it has bound to a specific response element (cRE) in the promoter of certain genes. Other second messengers, whose synthesis is stimulated by cytokines and some growth factors, are also involved in a rather similar manner in regulating the transcription of some other genes.

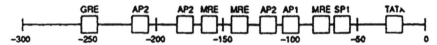

Figure 5. Transcriptional control elements upstream of the start site of transcription of the metallothionein gene. The TATA and GC boxes bind ubiquitous factors (TFIID and Sp1, respectively), while the glucocorticoid response element (GRE), metal response element (MRE) and the AP1 and AP2 sites bind factors involved in the induction of gene expression in response to specific stimuli. (Reproduced, with permission, from Latchman, D.S. (1991). In: Eukaryotic Transcription Factors, Academic Press, London.)

Some additional DNA sequences that respond specifically to other inducing agents or are specific to certain tissues are shown in Table 2. In general, there are distinct, specific DNA regulatory elements for each type of inducing signal. The transcription of some genes (e.g., the metallothionein gene) is regulated by several different signals, and the effect of each inducing agent is mediated independently via a separate and distinct regulatory DNA sequence (Figure 5).

Some promoter sequences that play a role in cell type-specific regulation of gene transcription have been identified through the use of partially differentiated tumor-derived cell lines, but a more complete understanding of the sequences responsible for tissue-specificity and particularly stage-specific gene expression will be possible only through analyzing the regulated genes in the context of the intact animal. This approach involves the introduction of recombinant genes into the germ line of a mouse to produce a transgenic animal. Results indicate that the control of development and differentiation involves a more complex set of regulatory promoter elements than those required for inducing the transcriptional activity of a gene in response to a particular signal.

Table 2. Some DNA Consensus Sequences Involved in Responses to Various Agents, and in Various Tissues

	Sequence
Agent	
Toxic metals	T.G.C.G/A.C/G.N.C
High protein diet	C.C.T.G.C.C.C.T.C
Interferons	G.G.A.A.A.N.T.G.A.A.A.C
Cyclic AMP	T.G.A.C.G.T.C.A
Tissue	
Muscle	C.A.G.C/G.T.G
Pituitary	A.T.T.C.A.T
B lymphocytes	G.G.G.A.N.T.T.C/T.C.C
Hemopoietic tissue	A/T.G.A.T.A.A
Liver	T.A.A.G.G.T.T.A.A.T.T.A.T.T.A.A.C

Enhancers

Two major classes of DNA regulatory elements have been identified. Members of one class, sometimes referred to as upstream promoter elements, are located relatively close to the start site of transcription, usually within 100–200 bp upstream. The position and order of such elements with respect to that site is often crucial for their function. The second class contains elements known as enhancers. The first such sequence to be characterized was discovered in the promoter region of the genome of the animal virus SV40. This promoter contains a TATA box, and 50 to 100 bp upstream from that is an array of six GC boxes organized into three tandem repeats. Mutational analysis revealed the presence, still further upstream, of an additional 144 bp-long region that is important for the transcriptional regulation of the SV40 genome. This comprised two tandemly repeated 72 bp sequences, deletion of which caused a 99% reduction of SV40 transcription (Figure 6).

This so-called enhancer sequence displayed some remarkable properties:

1. it stimulated the transcription of other genes (such as that encoding β-globin) by about 100-fold when attached to their promoters.
2. it acted equally well when placed downstream at various distances from the promoter, even up to several thousand bp away.
3. it worked in either orientation, forward or backward, with respect to the promoter.
4. it functioned when placed upstream or downstream of a gene, or even when placed within an intron.

Enhancer elements have subsequently been found in many cellular genes, and thus are not peculiar to viruses. One of the best understood is that of the immunoglobulin (Ig) heavy-chain genes, which, interestingly, is located within an intron of these genes.

The distinction between upstream promoter elements and enhancers is not always clear-cut. Upstream promoter elements tend to comprise a simple, relatively short sequence motif of up to 20 bp, and are normally found only a fairly short

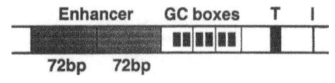

Figure 6. Organization of the SV40 promoter region. Note the TATA box (T), three 21 or 22 bp sequences each containing two GC boxes and the tandem 72 bp repeats in the enhancer. I is the site of initiation of transcription.

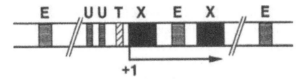

Figure 7. Different classes of eukaryotic transcriptional activating elements. Upstream promoter elements (U) such as GC and CAAT boxes are generally within 200 bp upstream of the transcriptional start site. Enhancers (E) may be very much further away on either side of this site or in an intron, and in either 5'–3' or 3'–5' orientation. T represents the TATA box, X the exons of a hypothetical gene, and I is the site of initiation of transcription.

distance upstream from the transcriptional start site even though they can sometimes function in other locations in artificial constructs. Enhancers are often larger—between 100 and 150 bp—and more complicated (Figure 7). They may be composed of several individual, smaller DNA sequence elements, which are often similar to upstream promoter elements. These elements, many of which are common to several enhancers often occur more than once within an enhancer. Thus, enhancers appear to have a modular structure, consisting of several individual sequence motifs. In some cases, there is functional redundancy within an enhancer, with deletion of single copies of a repeated motif having little effect on overall enhancer function, but in others, the individual elements act in a synergistic manner to magnify the effectiveness of the enhancer. When transcriptional activator proteins are bound by elements that are located at a considerable distance from the promoter, it is believed that these proteins can also interact with promoter sequences or with the transcriptional complex, causing the DNA to bend and form a loop between the two sites of protein binding.

Some enhancers, such as the SV40 enhancer, function in almost all cell types, but most exhibit cell-type specificity, and function best in the cells where the gene with which they are associated is expressed. The Ig enhancer is a good example of the latter type. When this was removed from the Ig gene and linked to an unrelated promoter and introduced into a variety of cell types, it was able to stimulate transcription from the test promoter only in B lymphocytes—the cells to which Ig gene expression is confined. The cell-type specificity of this enhancer appears to be due, in part, to DNA sequence motifs binding proteins and acting positively to stimulate gene transcription in B lymphocytes, but also to negatively acting elements that suppress transcription in other cell types.

Whereas the Ig enhancer is active throughout the life of mature B lymphocytes, other enhancers function only in response to a particular inducing signal. For instance, the heat-shock gene enhancer responds rapidly to elevated temperatures, and the genes whose transcription is activated in response to steroid hormones are regulated by hormone-responsive upstream promoter elements. When such signal-responsive elements are removed from the genes in which they occur and attached to other genes, transcription of the latter becomes inducible in response to the appropriate signal.

Transcriptional Regulatory Proteins

As described above, transcriptional regulation of eukaryotic protein-coding genes is mediated by discrete DNA sequences or regulatory elements. The exact combination of such elements associated with a particular gene defines its regulatory potential, i.e., the stage of development or the cell type in which it is expressed, and the capacity of the gene to respond to specific inducing signals. However, the regulatory elements associated with each gene are present in all cells of an organism and at all times. Thus, although the appropriate DNA regulatory elements must be present for the correct regulation of transcription of a gene, they are clearly not sufficient on their own. Other factors must be required for regulating eukaryotic gene transcription.

As mentioned above, the interaction of RNA pol II with a promoter requires that other protein factors (such as TFIID) first bind to DNA sequence elements, such as the TATA box, located close to the transcription start site. Before efficient transcription can take place, yet more proteins, such as Sp1 and the CAAT box transcription factor, may have to bind to their cognate binding sites. Thus, by analogy, the functioning of DNA elements involved in regulated transcription may also depend on the binding of proteins that specifically interact with them. The presence of these sequence-specific binding proteins was confirmed by the gel-retardation technique (described above). Having detected such a protein, its precise binding site can be determined by a technique known as DNase footprinting.

This involves radiolabeling one end of a DNA restriction fragment, or double-stranded oligonucleotide, containing a putative protein-binding site. The DNA is incubated with a cellular or nuclear protein extract and then treated briefly with DNase I, an endonuclease that makes single-strand cuts, under conditions such that each strand is cut on average only once. In the absence of bound protein, the enzyme will cut throughout the DNA to generate a nest of labeled fragments of different sizes. After electrophoresis through a denaturing polyacrylamide gel and exposure of the gel to an X-ray film, the fragments are visualized as a uniform ladder of bands. If a protein has bound to a sequence within the DNA, it will prevent the DNase from cutting the DNA in that region. This produces a ladder with a gap corresponding to the position of the sequence bound by the protein. This gap is referred to as the footprint of the bound protein (Figures 8 and 9).

This method, together with related techniques for the detection of protein-DNA binding in intact cells, has provided detailed maps of protein-binding sites in promoter and enhancer elements. The positions of the sites within these elements generally correspond to those functionally important regulatory elements that had

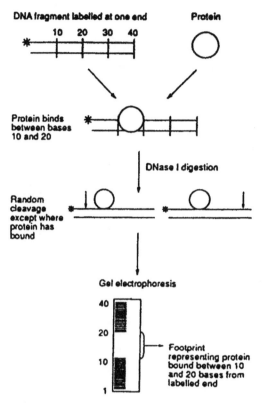

Figure 8. DNase 1 footprinting assay. If a protein binds at a specific site within a DNA fragment labeled at one end, the region of DNA where the protein binds will be protected from digestion with DNase. Hence, on electrophoresis this region will appear as a gap (footprint) in the ladder of bands produced by the DNA being cut at all other points by DNase I. (Reproduced, with permission, from Datchman, D.S. (1991). In: Eukaryotic Transcription Factors. Academic Press, London.)

been identified previously by deletion mutagenesis (see above). These studies confirmed that enhancers are modular structures composed of multiple regulatory elements each of which can interact specifically with a particular regulatory protein.

DNA regulatory elements function only when they have interacted with their cognate binding proteins. Some of these proteins are present in a relatively constant concentration in essentially all cells. Although such proteins are necessary for the active transcription of many genes, they are probably not responsible for regulated expression. In general, DNA regulatory elements that are active in all cells interact with ubiquitous proteins, whereas cell type-specific elements bind proteins that are present only in the cells in which they are active. Elements that regulate gene expression in response to an inducing agent bind proteins only when the agent is

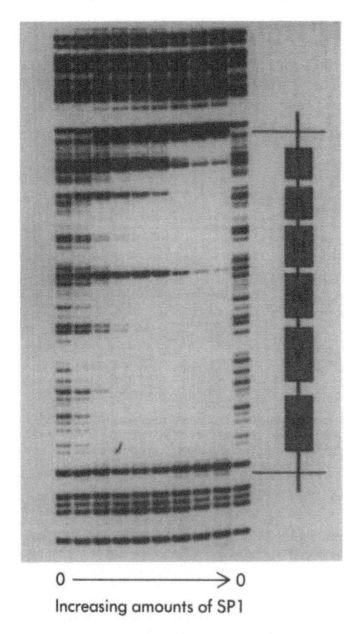

0 ⎯⎯⎯⎯⎯⎯⎯⎯⎯⎯⎯⎯⎯⎯→ 0

Increasing amounts of SP1

Figure 9. Multiple footprints in the promoter of SV40. The six footprints are caused by the binding of Sp1 to the GC boxes (see Figure 6) (Reproduced, with permission, from Tjian, R., et al. (1984). Nature 312, 409.)

present. Most of the eukaryotic regulatory proteins identified so far bind to their cognate DNA element and stimulate transcription from its associated promoter. However, there are some proteins that exert a negative effect by suppressing gene transcription when they are bound.

Although numerous distinct transcriptional regulatory proteins have been identified, many different enhancers contain elements that bind the same protein. Thus, the total number of regulatory proteins is probably not very large—possibly no more than a few hundred.

Positive regulation of eukaryotic gene transcription occurs via a complex set of interactions between various DNA regulatory elements, present upstream of the core promoter, or in enhancers that may be remote from the transcription start site, and their cognate binding proteins. In both situations, the role of stimulatory transcription factors is to expedite access of RNA pol II to the promoter so as to

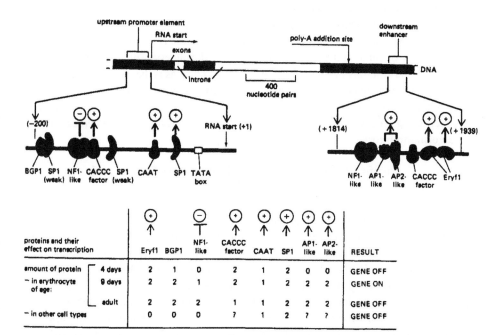

Figure 10. **A.** Binding sites of regulatory proteins in the promoter and enhancer of the chicken β-globin gene that control its expression during erythrocyte development. Those proteins not marked with + (activation) or – (repression) do not appear to have a major effect on transcription. **B.** Relative amounts of each gene regulatory protein present at different times of development. 0, 1, and 2 indicate none, intermediate, and high levels, respectively, on a nonlinear scale. (Reproduced, with permission, from Alberts, B., et al. (1989). In: Molecular Biology of the Cell, 2nd edn., Garland Publishing Inc., New York.)

increase the rate of initiation of transcription. The mechanisms by which this is accomplished will be discussed in subsequent sections.

The transcriptional status of a gene at any particular time will depend on the precise combination of regulatory proteins bound. A good example of this is the regulation of expression of the β-globin gene during chick erythrocyte development. This gene is expressed only in cells of the erythrocyte lineage, where it is inactive up to the four-day stage of development. It becomes switched on between the four- and nine-day stages, but returns to an inactive state in the mature cell. The expression of the gene depends on the presence of various promoter elements located within 200 bp upstream of the transcriptional start site and a 108 bp-long enhancer downstream of the poly-A addition site signal sequence. Deletion analysis of the enhancer attached to a CAT gene under the control of a β-globin gene promoter showed that several distinct DNA elements are necessary for optimal expression of the reporter gene in erythroid cells at the appropriate stage of development (Figure 10). Gel retardation and DNase footprinting studies revealed that the DNA elements corresponded to regulatory protein binding sites. Some of these proteins are ubiquitous, but others are specific to erythroid cells. The relative abundance of some of the latter proteins change during erythroid cell development whereas the amounts of others remain constant. Thus, as is the case for many other genes, cell type and developmental stage-specific expression of the β-globin gene is regulated by a complex pattern of binding of both ubiquitous and cell type-specific regulatory proteins. Some of these stimulate transcription whereas others have an inhibitory effect.

Purification of Transcriptional Regulatory Factors and Cloning of Their Genes

The ability of transcriptional regulatory proteins to bind specifically to particular DNA sequences has been exploited for their purification and for the direct cloning of DNA sequences that encode them. The first step in the purification of a regulatory protein from crude nuclear extracts of cells or tissues that are known to contain it is usually chromatography on a phosphocellulose or heparin-agarose column. As the negative charge density of these matrices is similar to that of DNA, DNA-binding proteins will bind to the column. By washing the column with increasing concentrations of salt the original protein extract can be fractionated. The distribution of particular regulatory proteins among individual column fractions can be monitored by gel-retardation, by DNase footprinting using a DNA fragment containing the requisite protein-binding site, or by assaying the ability of the fraction to stimulate transcription *in vitro* from an appropriate promoter. The protein of interest can be further purified by affinity chromatography on a supporting matrix such as agarose to which double-stranded oligonucleotides containing the binding site for the protein have been coupled. Many of the proteins in the fraction will bind weakly and nonspecifically to the column, as they would to any DNA sequence, and are removed by washing the column with a low concentration of salt. Proteins that bind specifically and therefore more tightly require much higher salt concentrations to be eluted.

Many transcriptional regulatory proteins have been purified by these techniques. Determination of the amino acid sequence of a particular one allows the synthesis of a mixture of oligonucleotides predicted to code for sections of that protein. These are then used to screen a cDNA library for sequences that actually code for the protein. Alternatively, antibodies raised against the purified protein can be used to screen a cDNA expression library for the relevant clones. A more direct approach to the isolation of cDNA clones encoding the transcriptional regulatory protein is to screen a cDNA expression library with synthetic double-stranded oligonucleotides containing the specific binding site for the protein. Once the cDNA for a particular protein has been isolated it can be used to direct the synthesis of large amounts of the protein in a suitable vector.

Many cloned cDNAs encoding different regulatory proteins have now been isolated through the use of these techniques. Analysis of their sequences has identified several novel structural motifs (for details see chapter 5).

Structure/Function Relationships of Transcriptional Regulatory Proteins

One of the first transcriptional regulatory proteins to be purified was Sp1. It was shown, through the use of an *in vitro* transcription system, that Sp1 is able to stimulate transcription only from promoters containing a GC box. DNase footprinting revealed that the protein bound to such boxes, whereas mutations in them that prevented Sp1 binding also abolished its stimulatory effects on transcription. Thus, the activation of transcription by Sp1 is dependent on its ability to bind to this specific sequence element within the promoter.

In addition to binding to DNA and stimulating transcription, some transcriptional regulators such as steroid hormone receptors also bind ligands (in this case, a steroid hormone). Functional analysis of deletion mutants of these receptors has demonstrated that their DNA-binding and steroid-binding abilities are located in separate domains (Figure 11). Point mutations within each of these domains have subsequently identified those amino acid residues that are essential for function. When the glucocorticoid receptor binds a glucocorticoid its conformation is changed so that it is now able to bind to the glucocorticoid response element within the promoter of a glucocorticoid-responsive gene and induce its transcription.

The estrogen receptor, a different but related member of the steroid hormone receptor family, specifically binds estrogens and induces the transcription of estrogen-responsive genes via interaction with an estrogen response element. In a "domain swap" experiment a chimeric receptor was produced in which the DNA-binding domain of an estrogen receptor had been substituted with that of a

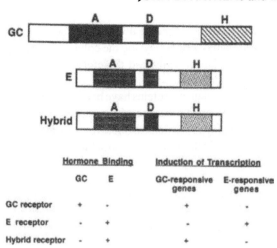

Figure 11. Domain structure of steroid hormone receptors. In each diagram the shaded portions are the activation domain (A), the DNA binding domain (D), and the hormone binding domain (H). The diagrams show the glucocorticoid (GC) receptor, the estrogen (E) receptor, and a hybrid receptor in which the DNA binding domain of the GC receptor is inserted into the E receptor in place of the corresponding domain of the latter. The table at the foot shows the results of a domain swap experiment with the three receptors on hormone binding and gene induction.

glucocorticoid receptor. This bound estrogen normally, but induced the transcription of glucocorticoid-responsive, not estrogen-responsive genes (Figure 11). Thus, the DNA- and steroid-binding domains of these receptors can function independently of each other.

The DNA-binding domain of a transcriptional regulatory protein contains all the structural information necessary for specific interaction with the cognate DNA-binding site, but is unable to activate gene transcription from the associated promoter. Clearly, other regions of the protein are involved in mediating the latter function. The location of such regions was investigated by domain-swapping experiments in which different regions of one transcription factor were fused to the DNA-binding domain of another factor. By determining which of the chimeric proteins were able to stimulate transcription from a promoter containing the appropriate DNA recognition sequences it was possible to identify the regions of the first transcription factor that were responsible for transcriptional activation. Three different types of activation domain (based on amino acid composition) have so far been identified: regions rich in glutamine, proline, or acidic amino acid residues.

Many transcription factors bind DNA as dimers, either with themselves (homodimers), or with other factors (heterodimers). Regions responsible for dimerization have been identified and include helix-loop-helix and leucine zipper structural motifs (see chapter 5).

Thus, structure/function analysis of transcription factors has revealed that they have a modular structure composed of distinct domains for DNA binding, transcriptional activation, and, in some cases, dimerization and ligand binding, each of which can function independently. In most cases the various structural domains of these factors are contained within a single polypeptide, though there are examples where the DNA-binding and activation domains reside in different polypeptides within a multicomponent complex.

Mechanisms of Action of Transcriptional Regulatory Proteins

As described above, in order to exert their stimulatory effect, transcriptional activators must bind to specific recognition sites within a gene promoter or enhancer. In addition to a DNA-binding domain, an activation region is also required. However, sequence-specific binding of an intact activator may not be sufficient to stimulate transcription, suggesting that to exert its effect, the activation domain interacts with a protein component of the transcriptional machinery, often termed the target.

Since the activation domains of many transcription factors are interchangeable, it is likely that they may interact with a common target via similar mechanisms of action. Further support for this hypothesis is provided by the phenomenon of squelching. This arises when two different genes (1 and 2) under the control of two different activators (A and B, respectively) are transcribed together *in vitro*. Normally, each gene is transcribed independently. However, in the presence of a high concentration of one of the activators (for example, A) the gene it controls (gene 1) is transcribed normally, but that controlled by factor B is no longer transcribed. The interpretation of these results is that both factors interact with the same target molecule, which is a component of the basal transcription complex. When one factor is present in high concentration it is able to interact with the target located not only on the promoter of the gene that it regulates, but also in solution, thus sequestering the target with the result that it is unavailable to support the transcription of the second gene.

What is the target of the activation domains? One obvious possibility is the RNA polymerase itself. However there is no direct evidence for this. Alternatively, the target may be a universal transcription factor such as TFIID. Evidence in support of this comes from studies on the mode of action of the mammalian transcription factor ATF, which mediates transcriptional activation of several genes in response to cyclic AMP. DNase footprinting shows that when TFIID binds to the promoter of these genes, it interacts with the TATA box, but not with the transcription start site. When ATF is bound to its cognate binding site the footprint of TFIID is enlarged to cover the transcription start site. This change in the interaction of TFIID with the core promoter facilitates the recruitment of other components of the basal transcription machinery, including RNA pol II, and results in the formation of a stable initiation complex and transcription of the gene.

Coactivators

The purification of general transcription factors and cloning of their genes has revealed the existence of a novel class of regulatory proteins, called coactivators, that are required for the stimulation of basal transcription in response to activator proteins such as Sp1 and CTF.

Almost all of the general transcription factors identified up to now are complexes composed of two or more different polypeptides It has recently been shown that one of the purified polypeptide components of the multimeric TATA box-binding factor, TFIID, is able to bind specifically to the TATA box and interact with other general transcription factors to assemble an initiation complex capable of supporting basal transcription. Owing to its ability to bind to the TATA box, this protein

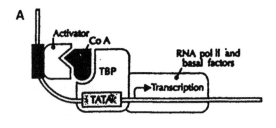

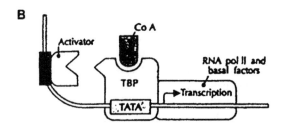

Figure 12. Two models for coactivator function. In **(A)** the coactivator serves as a molecular bridge to connect the regulatory protein with one of the general transcription factors. In **(B)** the coactivator interacts with a general transcription factor causing a change in its conformation so that it can now interact directly with a regulatory protein bound elsewhere. (Reproduced, with permission, from Eukaryotic coactivators associated with the TATA box binding protein. Gill, G. and Tjian, R. (1992). Curr. Opin. Genet. Devel. 2, 236.)

was named the TATA box-binding protein (TBP). However, unlike TFIID isolated from eukaryotic cells, TBP on its own is not able to support the activation of transcription in response to activator proteins such as Sp1 or CTF. Thus, it appears that, in addition to TBP, TFIID contains components (called coactivators) that are required to mediate the activation of transcription. TFIID is a large multiprotein complex containing TBP and at least six different TBP-associated factors (TAFs), some of which have been shown to have coactivator activity.

In contrast to TFIID, purified TBP cannot bind to a TATA-less promoter, though a TBP-TAF complex isolated from TFIID can support transcription from such a promoter. This suggests that one of the TAFs could function as a tethering factor to recruit the TBP to promoters lacking a TATA box.

Two models have been proposed for coactivator function: the coactivator may serve as a molecular bridge between the activator and a general transcription factor; alternatively, it may not interact directly with the activator but, instead, may be involved in propagating the effects of a direct interaction between an activator and a general transcription factor, such as TBP, to the rest of the transcription complex (Figure 12).

THE ROLE OF CHROMATIN STRUCTURE IN REGULATING GENE EXPRESSION

Chromatin Structure and Activation of Gene Transcription

DNA in eukaryotic cells is packaged by combination with proteins into a structure called chromatin. The basic repeating unit of chromatin is the nucleosome. This contains 166 base pairs of DNA wound twice around the outside of a histone octamer comprised of two molecules each of histones H2A, H2B, H3, and H4. Each nucleosome is joined to its neighbor by 15–45 bp of DNA associated with one molecule of histone H1. Within the nucleus, nucleosomes are further compacted by coiling into a solenoid structure with about six nucleosomes per turn to form the 30 nm diameter chromatin fibers. (See Chapter 1 for a detailed description of nucleosomes and chromatin structure.)

Essentially all genes that are transcribed by RNA Pol II remain associated with nucleosomes during transcription. The question arises, therefore, as to how the transcriptional machinery is able to initiate and maintain the transcription of a gene that is packaged into chromatin.

The accessibility of DNA within chromatin to regulatory proteins and general transcription factors has been investigated through the use of endonucleases such as DNase I. After exposure of isolated nuclei to various concentrations of this enzyme, DNA was extracted, digested with a restriction endonuclease, electrophoresed through an agarose gel and analyzed by Southern blotting with cloned DNA probes. It was found that genes in nuclei isolated from cells in which they were

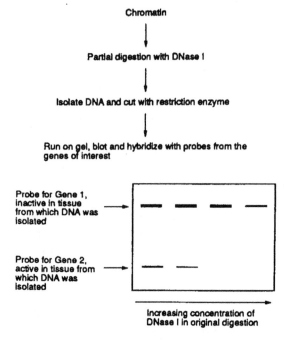

Figure 13. Preferential sensitivity to digestion with DNase I of chromatin containing active genes, as shown by electrophoresis and Southern blotting with cloned DNA probes. (Reproduced, with permission, from Latchman, D.S. (1990). In: Gene Regulation: A Eukaryotic Perspective,Unwin Hyman Ltd., London.)

expressed were more susceptible to endonuclease digestion than those in nuclei from cells that did not express them (Figure 13). This increased accessibility to endonucleases occurs over the entire gene and extends for some distance either side of it because the chromatin structure is more open within and around genes in the cells in which they are transcribed. This alteration in chromatin structure arises before the onset of transcription and persists after transcription has ceased. Thus it is not associated specifically with the act of transcription, but appears, instead, to reflect the ability of a particular gene to be expressed in a given cell type.

The removal from chromatin of two non-histone proteins known as high mobility group 14 and 17 (HMG 14 and 17) results in a loss of the preferential sensitivity to nuclease digestion of "active" genes. When either or both of these proteins is added to the depleted chromatin the pattern of sensitivity characteristic of the original chromatin is restored. Clearly, HMG 14 and 17 play an important role in maintaining the altered chromatin structure associated with transcriptionally competent genes.

Other differences between transcriptionally competent and incompetent genes and their associated proteins have been observed. When cytosine bases occur 5'-

of a guanine base in a DNA sequence they are often methylated at the 5-C position. Such CG sites within a particular gene are often found to be unmethylated in tissues in which the gene is transcribed, but methylated in other tissues. Unmethylated sites correspond to regions of DNase I sensitivity and it has been postulated that differences in patterns of methylation may play a role in regulating alterations in chromatin structure and/or the binding of regulatory proteins to the DNA. It is striking that the methylation pattern of such cytosine residues persists after cell division when the newly synthesized DNA initially contains unmethylated cytosine at these positions. Methylation is discussed in more detail in Chapter 3.

Histones, particularly H3 and H4, can be acetylated *in vivo* on the free amino groups of lysine residues. Hyper-acetylated histones are associated with regions of DNase I hypersensitivity, and there is evidence that such modification may play a role in the generation of a more open structure of chromatin (discussed in more detail in Chapter 4).

Within the regions of enhanced sensitivity to DNase I digestion associated with transcriptionally competent genes, particular sites have been discovered that are even more sensitive to cleavage by the enzyme. These hypersensitive sites were detected by the same type of approach as that used to investigate the general DNase sensitivity of a gene, but with much lower concentrations of the enzyme. As with undermethylation and increased sensitivity of the entire gene to nuclease digestion, DNase I hypersensitive sites appear to be related to the ability of the gene to be transcribed rather than with the act of transcription itself.

These sites seem to be the consequence of localized differences in the chromatin, caused by the absence of nucleosomes or alterations of their structure, resulting in a more open configuration in which the DNA is more accessible to proteins. Such sites are often associated with regions of a gene that are important for transcriptional regulation, and they usually correspond to binding sites for regulatory proteins.

The presence of nucleosomes on promoter sequences has been correlated with repression of transcription. Thus, although transcriptional elongation is not significantly impeded by nucleosomes, initiation of transcription apparently requires the absence of canonical nucleosomes from promoter and enhancer regions. This is reflected by the presence of hypersensitive domains, posing the problem of how such nucleosome-free regions arise. Some evidence comes from transcription of chromatin that has been reconstituted *in vitro*. If basal transcription factors were added to a naked DNA template before the addition of histones, a stable initiation complex formed on the TATA box and the gene was transcribed. But, if histones and basal transcription factors, were added to the DNA simultaneously, the histones formed nucleosomes which bound to the promoter, blocking access of the transcription factors to the TATA box, and prevented initiation of transcription. However, if DNA was simultaneously exposed to histones, basal transcription factors and activator proteins, the latter competed with histones for binding at their cognate binding sites so that the binding of basal factors, but not histones, to the promoter

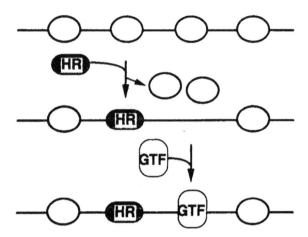

Figure 14. Binding of a steroid hormone receptor complex (HR) to the promoter region of a hormone inducible gene. This displaces nucleosomes (ovals) and allows general transcription factors (GTF) to bind to the promoter so that the gene may be transcribed.

allowed transcriptional initiation to occur. For many promoters, this does not happen when activators are added after histones. It only occurs with DNA that is free, at least briefly, of nucleosomes, as is the case during DNA replication. Hence a "pre-emptive" competition between activator proteins and histones during DNA replication may be involved in defining the transcriptional competency of genes and cell type and developmental stage-specific regulation of gene expression.

There is evidence that some activator proteins are able to disrupt or displace nucleosomes that are already bound to DNA. For instance, on binding to the glucocorticoid response element, the glucocorticoid receptor displaces a nucleosome from the core promoter. This enables an initiation complex to form at the TATA-box and the transcription of the gene to commence (Figure 14). This dynamic competition between activator proteins and nucleosomes may explain how gene expression can be induced in terminally differentiated, nondividing cells.

POSTTRANSCRIPTIONAL MODIFICATIONS OF RNA

RNA Processing

Once RNA has been synthesized by transcription from DNA it is nearly always subject to various kinds of processing before transport out of the nucleus. Nearly all primary transcripts of mRNAs contain nucleotide sequences known as introns that carry no coding information. These are removed by the process of splicing during the maturation of the mRNAs before they are transported into the cytoplasm

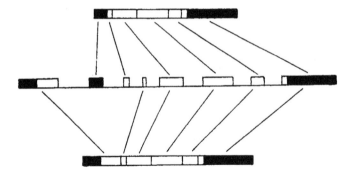

Figure 15. Alternative splicing of the alkali L-chain myosin gene. The middle line shows the genomic structure. The unfilled boxes are coding portions of the exons, while the filled portions are noncoding. The uppermost line shows the structure of the mRNA from embryonic and cardiac muscle. The lowest line shows the structure of the mRNA from adult skeletal muscle. (Reproduced, with permission, from Hawkins, J.D. (1991). In: Gene Structure and Expression 2nd. Ed., Cambridge University Press, Cambridge.)

for translation. The nucleotide sequences that actually contain the coding information are known as exons. In a number of cases two or more mature mRNAs may be formed from the transcription of a single gene by alternative splicing. In this process, specific exons are removed from the primary transcript together with the introns surrounding them. Alternative splicing of the same transcript may occur in different tissues or at different stages of development, giving rise to slightly different forms of a protein, known as isoforms. This is particularly common in transcripts of genes encoding some muscle proteins, such as troponin-T, tropomyosin, and the L and H chains of myosin. Figure 15 shows two ways of splicing the transcript of the gene for the myosin alkali light-chain mRNA.

In this case, different capsites and promoters are used for the two isoforms since the alternatively spliced exons are at the 5'-end of the transcript. The protein tropomyosin exhibits a more complicated set of at least nine different alternative forms that are synthesized in different tissues, such as skeletal and smooth muscles, fibroblasts, and brain.

Alternative splicing is also fairly common in transcripts of genes encoding proteins that modulate transcription. For example, no fewer than seven different isoforms of the retinoic acid receptor have been detected in mouse tissues, though the significance of this finding is not yet apparent. Most of these alternatively spliced transcripts differ in their 5'-untranslated sequences, owing to the presence of several different exons in this region of the gene. Another example is the generation of two forms of the transcription factor *fos-B*. The mRNA for *fos-B2* is formed by using a cryptic splice site that is read through in *fos-B* (Figure 16). *fos-B2* is a shorter protein than *fos-B*, and actually inhibits the actions of the latter. Thus,

GAG GTG AGA . . . . TCT CTC TTT ACA CAC AGT GAA . . .

GAG *GTGAGA . . . . . TCTCTCTTTACACACAG* TGA A . .

Figure 16. Alternative splicing generates either fosB (upper)—nucleotides arranged in triplet codons—or fosB2 (lower) with intron flanked by cryptic splice sites (marked *), leading to termination codon (TGA) immediately 3' to the intron.

control of the relative amounts of the two proteins can modulate the degree of activation of transcription of genes that are under the control of the factor AP-1, which is a heterodimer of *fos* and *jun*.

There are several alternative forms of CREM (cAMP response element modulator). These all bind to the cAMP response element (cRE–TGACGTCA); some forms repress transcription from motifs containing this element, whereas other forms, such as those produced in the testis during spermatogenesis, can stimulate transcription. The alternatively spliced exons encode various functional elements containing cationic amino acid residues responsible for binding to the cRE, as well as exons encoding glutamine-rich sequences that stimulate transcription. In addition, there are domains encoding the leucine-rich zipper motif that can form either homodimers, or heterodimers with CREB—a somewhat similar protein that binds to the cRE and stimulates transcription. Thus, there are numerous possibilities of exerting fine control of transcription according to which form of CREM is produced at a particular time in a particular tissue.

The details of the splicing reaction have not yet been fully worked out, so very little is known about the mechanism(s) of alternative splicing, although a few proteins that can affect the choice of alternative splice sites have been identified.

RNA Editing

In a very few cases it has been shown that some of the bases in a primary RNA transcript may be altered before the RNA is translated. This probably occurs within the nucleus. The best known example is the mRNA for apoB—the protein component of the circulating lipoprotein B. Two forms of the apoprotein are found in the circulating particles. B100 contains 4536 amino acid residues, while B48 contains only the first 2512 of these. The former is secreted from the liver, while the latter comes from the small intestine. Both are the products of the same gene which codes for the larger protein, but in the intestine the mRNA is edited so that a CAA codon (encoding glutamine in the larger protein) is converted to UAA (a termination codon) by deamination of the cytidine residue. Fifty-five nucleotides surrounding the critical cytidine residue are required to allow cell extracts to perform the editing, which is carried out by a protein complex that binds to this site. It consists of several individual proteins, one of which catalyzes the deamination of the cytidine residue to a uridine residue.

Editing of some pre-mRNAs for glutamate receptors in the brain also occurs. Arginine appears at certain positions in the receptors where a glutamine residue would be expected to occur. The details have not been worked out, but it is believed that there may be deamination of an adenine residue in some of the codons for glutamine (CAG) to an inosine residue giving the codon CIG which can code for arginine. There are several almost identical subunits in the glutamate receptors which go to make up calcium channels, and the change from an uncharged glutamine residue to a positively charged arginine one is likely to affect the properties of these channels.

Posttranscriptional editing of RNA occurs widely in the mitochondria of plants and some other organisms and has also been observed in some mitochondrial tRNAs of marsupials, but has not so far been demonstrated in mammals.

Before mRNAs are secreted from the nucleus, they undergo two other reactions. A cap of 7-methyl guanosine is added to their 5'-end. This appears to have two functions: it promotes the binding of the mRNA to the ribosomes and also increases its stability by protecting it against 5'-nucleotidase activity. In addition, the 3'-end of all eukaryotic mRNAs, with the notable exception of those for the histones, have a sequence of adenyl residues (the poly-A tail) added to them.

STABILITY OF mRNA

In general, eukaryotic mRNAs are fairly stable with half-lives measured in hours, or even a few days. However, if the lifetime of an mRNA is altered this leads to changes in the amount of protein whose synthesis it directs. In several cases, particularly following hormonal stimulation or withdrawal, the half-lives of mRNAs are appreciably altered (Table 3). This is generally due to changes in the rates of degradation, though other mechanisms have not been ruled out. mRNA is believed to be degraded from the 3'-end, but the nuclease(s) responsible for this have not been well characterized, though some are probably associated with polyribosomes.

Table 3. The Half-Lives of Some mRNAs Under Various Conditions

mRNA	Tissue	Regulatory Signal	Half-Life	
			+Effector	*−Effector*
Serum albumin	Liver	Estrogen	3	10
Casein	Mammary gland	Estrogen	92	5
Insulin	Pancreas	Glucose	77	29
Histones	HeLa cells	DNA replication	4	0.25

Note: Reprinted, with permission, from Hawkins, J.D. (1991). In: Gene Structure and Expression, 2nd. Ed., Cambridge University Press, Cambridge.

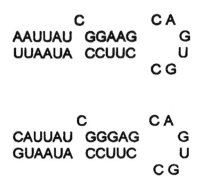

Figure 17. The iron responsive elements in the 3′-region of the human transferrin receptor gene. Note the unpaired C in the stem, which is a constant feature of this element.

A cytoplasmic protein called poly-A binding protein (PABP) binds specifically to the poly-A tails of mature mRNAs and stabilizes them. It binds to a site containing approximately 27 A residues, and mRNAs with a smaller number of these residues are degraded so rapidly as to be almost undetectable. Incubation of purified mRNAs in the absence of this protein or in the presence of antibodies raised against it greatly increases their rate of degradation. It is likely that the rate of dissociation of PABP from mRNAs could play a part in controlling their stability.

The rate of degradation of the mRNA for the transferrin receptor is subject to control under some circumstances. This membrane protein is required for the uptake of transferrin, bearing iron, into cells. It is important to regulate the concentration of this metal ion since, though small amounts are needed for a number of metabolic processes, it is toxic in larger amounts. The 3′-untranslated end of the transferrin mRNA contains two stem and loop structures known as iron-responsive elements (IREs) (Figure 17) that can be bound by a protein called the iron response factor (IRF). This protein has been shown to be the cytosolic form of the enzyme aconitase which itself contains some atoms of iron, bound in a complex with sulfur atoms. When the concentration of iron within the cell is low, the protein is not saturated with iron, is enzymically inactive, and acts as the IRF; when there is an adequate amount of iron in the cell, the protein becomes saturated with iron, is enzymically active, and no longer acts as the IRF. Binding of the IRF to the IREs inhibits the degradation of transferrin mRNA, leading to an increased synthesis of the receptor. Interestingly, the IRE also occurs in multiple copies near the 5′-end of several mRNAs for proteins that either store iron or are concerned with its metabolism in the cell. The binding of the IRF to these elements leads to inhibition of the translation of these mRNAs. Thus, an excellent homeostatic mechanism for the control of the amount of iron in a cell is provided by the interactions of the IREs with the IRF.

UUA<u>UUUA</u>UU<u>AUUUA</u><u>UUUA</u>UU<u>AUUUA</u><u>UUUA</u><u>UUUA</u>

UA<u>UUUA</u><u>UUUA</u>UGUA<u>UUUA</u>UGUA<u>UUUA</u><u>UUUA</u><u>UUUA</u>UU

Figure 18. Sequences in the 3'-untranslated regions of the human tumor necrosis factor mRNA (upper) and mouse Interleukin-3 mRNA (lower) that target them for rapid degradation. The characteristic AUUUA motifs are over- or underlined. (Reproduced, with permission, from Hawkins, J.D. (1991). In: Gene Structure and Expression 2nd. Ed., Cambridge University Press, Cambridge.)

The mRNAs for some cytokines and proto-oncogenes that normally have very short half-lives contain A- and U-rich sequences in their 3'-untranslated regions, where the motif AUUUA is particularly common and generally repeated several times (Figure 18). The transposition of these sequences to the corresponding location in more stable mRNAs greatly increases their rate of degradation. Two cytoplasmic proteins that bind with high affinity to these sequences have been detected, but their role in destabilizing the mRNAs is unknown. mRNAs are sometimes sequestered in the cytoplasm in ribonucleo-protein particles that are transcriptionally inactive. These contain a number of proteins that can bind RNA rather nonspecifically.

GENE AMPLIFICATION

Under some circumstances genes can be amplified, providing the opportunity for the synthesis of greater quantities of the proteins or other products that they encode. This occurs particularly in some cell lines that have been maintained in culture for many generations. Exposure of these cultures to various agents can lead to the overproduction of the proteins whose genes have been amplified. The inducing agents are generally substances or conditions that have some toxic action on the cells. In the case of the former there is generally an increase in the amount of an

Table 4. Some Example of Gene Amplification

Amplifying Agent	Amplified Gene
Methotrexate	Dihydrofolate reductase
Hydroxyurea	Ribonucleotide reductase
Adenine derivatives	Adenosine deaminase
Phosphonoacetyl-aspartate	CAD*
Many drugs	Multi drug resistant protein
Ouabain	Na/K-ATPase (α-subunit)
Nitrogen mustards	Glutathione-S-transferase

Note: *The enzyme complex that contains the first three of the enzymes catalyzing the synthesis of pyrimidines from aspartate and carbamoyl phosphate.

enzyme that the substance inhibits, or that metabolizes the inducing agent (Table 4). It is noteworthy that a number of these amplified genes encode enzymes that are concerned in some way with the metabolism of nucleic acids and their precursors.

The most frequently studied example is the induction of the enzyme dihydrofolate reductase (DHFR), which produces a coenzyme essential for the synthesis of purines. The folic acid antagonist methotrexate inhibits this enzyme and also induces it. Following exposure of Chinese hamster ovary cells to concentrations of methotrexate that severely limit their growth, a small number, typically one in 1,000, become resistant to the action of this antimetabolite. One way in which this resistance arises is by extra replication of the gene for DHFR, possibly with some adjoining DNA, while the rest of the genome is apparently unaffected. In some cases the amplified gene stays in its normal site, giving rise to a well-defined region of homogeneously stained DNA. Frequently the amplified DNA is deleted from the chromosome and forms a minute chromosome with a distinct structure that can be visualized by the usual staining methods. Minute chromosomes are highly unstable self-replicating elements, and may be lost during subsequent cell divisions. Agents that inhibit DNA replication, such as hydroxyurea, can potentiate the induction (up to 10-fold) of DHFR in the presence of methotrexate. In addition to amplification of the gene, more mRNA may be produced by increased transcription, or its rate of degradation may be decreased.

Amplification of oncogenes has also been observed in some tumors, such as the *myc* oncogenes in small cell carcinoma of the lung. Such amplification tends to lead to loss of growth control. Although this phenomenon is obviously of importance in the genesis of cancer, it is doubtful if it is otherwise physiologically relevant since it has been shown that in freshly isolated cell lines from normal tissue there is no detectable gene amplification in response to such agents as methotrexate or hydroxyurea. During oogenesis there is amplification of the genes for ribosomal RNA. The mechanisms for this are, as yet, unknown.

TRANSLATIONAL REGULATION

Although the various mechanisms discussed in connection with transcription are the main ways in which gene expression is controlled, there are also controls at the level of translation. The actual sequence of the mRNA around the initiation codon and 5'- of it may play important roles in determining the rate at which the initiation of the synthesis of a particular protein can occur once the mRNA has reached the ribosomes. The consensus sequence around the initiation codon is GCCA/GCCAUGG and the two purines at −3 and +4 with respect to the translational start site have the strongest effects, so that mRNAs with an A at −3 and a G at +4 are translated most readily. It is believed that protein synthesis is initiated by a scanning mechanism in which the 40S ribosomal subunit, in conjunction with initiation factors, binds to an mRNA at its 5'-end where it bears a 7-methyl guanosine cap, and then moves along the mRNA until it finds an AUG initiation codon. Not infrequently one or more of these may occur upstream of the true initiation site but they are often in a poor nucleotide context and so are not used, and any that are within about 12 nucleotides downstream of the cap site are also not used. It is not uncommon for there to be regions of the mRNA that assume stem-loop structures. If these are in the leader sequence between the cap site and the site of initiation, they tend to inhibit the progress of the ribosomal subunit and hence, the rate of initiation, particularly if they are stabilized by several base pairs in the stem. Similar stem-loop structures just downstream of the initiation codon can actually aid the initiation of translation by causing the ribosome to pause over the initiation codon.

All these factors are inherent in the structure of the mRNA so that they cannot be modified once the mRNA has been made and exported from the nucleus to the ribosomes. Thus, they are not responsible for changes in the rate of synthesis of a particular protein, but they can affect the relative rates at which different mRNAs are translated.

A number of protein factors are required for the initiation and elongation phases of protein synthesis, and some of these play roles in controlling the rate of this process. Eukaryotic initiation factor 2 (eIF-2) is a heterotrimer that is involved in the binding of Met-tRNA$_i$ to the 40S ribosomal subunit at an early stage in protein synthesis. The α-subunit of this complex becomes phosphorylated under various unfavorable physiological conditions so that its action is inhibited. A protein kinase known as protein kinase R (PKR) or double-stranded-RNA activated inhibitor of protein synthesis (DIA) is induced by interferons, produced after viral infection, so that eIF-2 is phosphorylated and protein synthesis is inhibited. This initiation factor is also phosphorylated and its action inhibited in conditions of amino acid or glucose deprivation, after heat shock, or in the presence of heavy metals. The heme controlled repressor (HCR) is a more specific inhibitor of protein synthesis, restricted to reticulocytes and some bone marrow cells. It phosphorylates eIF-2 when the concentration of heme in the cell is low so that there is no wasteful synthesis of globin and other proteins.

The activity of eukaryotic initiation factor-4 (eIF-4) is also regulated by phosphorylation. This factor is a multimeric protein that is required for the binding of mRNA to the complex of the 40S ribosomal subunit and Met-tRNA$_i$. There is good correlation between the level of phosphorylation of the subunit eIF-4E of this factor and the rate of initiation of protein synthesis. Its phosphorylation is inhibited by heat shock, viral infection, and during mitosis, and stimulated by a variety of agents that increase the rate of protein synthesis such as the tumor promoter phorbol myristyl acetate, lipopolysaccharide, insulin, and various growth factors. The kinase(s) responsible for the phosphorylation of eIF-4E has not been well characterized. Overproduction of eIF-4E by the introduction into a cell of its gene under the control of a strong promoter causes the transformation of that cell to a phenotype with unrestricted growth.

The action of eukaryotic elongation factor-2 (eEF-2), which is required for the translocation of the mRNA from the A site to the P site during protein synthesis, can also be inhibited by phosphorylation. The ribosomal protein S6 has several serine residues that can be phosphorylated in the presence of various mitogens and growth factors, and its degree of phosphorylation correlates well with the activation of protein synthesis by these agents. Thus, there are a number of factors whose degree of phosphorylation can affect the rate of protein synthesis. In general they have a fairly indiscriminate effect on this process rather than exerting their effects on the synthesis of specific proteins.

POSTTRANSLATIONAL REGULATION

The control of activity of proteins by phosphorylation is not confined to the factors detailed above that are involved in protein synthesis. Phosphorylation is a very general and widespread mechanism whereby the activity of many enzymes is controlled, particularly by the actions of some hormones and growth factors. The best studied examples involve phosphorylation of serine and threonine residues, but phosphorylation of tyrosine residues, although it occurs to a smaller extent, is also an important control mechanism.

Once proteins have been synthesized, their disposition within the cell may be controlled by various sequences that target them to specific cellular compartments. The best known example of this is the targeting to the endoplasmic reticulum of proteins destined for secretion from the cell. At their N-terminus these proteins have a signal sequence containing a very high proportion of amino acid residues with hydrophobic side chains which directs them into the endoplasmic reticulum while they are still being synthesized. This sequence is cleaved from the proteins before they are secreted. There are also sequences that direct proteins specifically to mitochondria, the nucleus, and other subcellular compartments. The rates at which proteins are degraded within the cell will obviously affect their concentration, and there are wide variations in the rate at which this occurs. The presence of certain amino acid residues at the N-terminus seems to have a large effect on this process, but a detailed discussion of this is beyond the scope of this chapter.

SUMMARY

The major site of regulation of the expression of eukaryotic protein-coding genes is at the level of initiation of transcription. Just upstream from the site of transcription initiation is the promoter, which generally contains a TATA box and frequently the sequences GGGCGG and CCAAT. Proteins, many of which are multimeric, bind to these elements to build up transcription complexes with RNA polymerase II. Additional elements are often present that respond to physiological and pathological changes in the cellular environment. These are mostly fairly short sequences situated closely upstream from the transcription start sites. Other, longer sequences known as enhancers are operative over greater distances on either side of these sites. These sequences are bound by transcriptional regulatory proteins that may be either tissue-specific, or activated after binding ligands such as steroid hormones. Thus, the regulation of eukaryotic gene transcription is mediated by interactions between regulatory proteins and their cognate binding sites generally located within and around gene promoters. Individual domains in transcriptional regulatory proteins function relatively independently for ligand-binding, DNA-binding, and activation of transcription. Most of the factors identified to date stimulate transcription, but there are also some negatively acting factors.

In native chromatin most of the genes are covered by nucleosomes, and this affects the accessibility of the various transcription factors to the genes. There are various mechanisms involved in allowing access of these factors so that transcription may take place. After transcription, RNA is processed by removal of introns, and variations in the way this is carried out can give rise to slightly different isoforms of the protein product.

Further means of regulating the final protein composition of cells are provided by factors that affect the rate of degradation of mRNAs. The quantities of mRNA that are produced can also be controlled by the amplification of certain genes in the DNA, although this seems to occur most frequently in cultured cells. Finally, the actual sequences near the 5'-end of an mRNA may affect its rate of translation. The level of phosphorylation of various translation factors, which affects their activity, is regulated by environmental and hormonal factors.

RECOMMENDED READINGS

General Textbooks

Eukaryotic Gene Transcription and its Regulation

Watson, J.D., Hopkins, N.H., Roberts, J.W., Steitz, J.A., & Weiner, A.M. (1987). In: Molecular Biology of the Gene, 4th edn., Chapter 21, The Benjamin/Cummings Publishing Co. Inc., New York.

Watson, J.D., Witkowski, J., Gillman, M., & Zollner, M. (1992). In: Recombinant DNA, 2nd edn., Chapter 9, Scientific American Book, W.H. Freeman & Co., San Francisco.

Lodish, H., Baltimore, D., Berk, A., Zipursky, S.L., Matsudaira, P., & Darnell, J. (1995). In: Molecular Cell Biology, 3rd edn., Chapters 11 and 12, Scientific American Book, W.H. Freeman & Co., San Francisco.

Alberts, B., Bray, D., Lewis, J., Raff, M., Roberts, K., & Watson, J.D. (1994). In: Molecular Biology of the Cell, 3rd edn., Chapters 8 and 9, Garland Publishing Inc., New York.

Singer, M., & Berg. P. (1991). In: Genes and Genomes, Chapter 8, University Science Books, Sausalito, CA.

More Specialized Books

Hawkins, J.D. (1996). In: Gene Structure and Expression, 3rd edn., Cambridge University Press, Cambridge.

Latchman, D.S. (1995). In: Gene Regulation: A Eukaryotic Perspective, Chapman & Hall, London.

Review Articles

General Transcription Factors and Assembly of Transcriptional Initiation Complexes

Roeder, R.G. (1991). The complexities of eukaryotic transcription initiation: Regulation of preinitiation complex assembly. Trends Biochem. Sci. 16, 402.
Conaway, R.C., & Conaway, J.W. (1993). General initiation factors for RNA polymerase II. Ann. Rev. Biochem. 62, 161–190.
Zawel, L., & Reinberg, D. (1995). Common themes in assembly and functions of eukaryotic transcription complexes. Ann. Rev. Biochem. 64, 533–561.

Transcriptional Regulation of Eukaryotic Genes

Maniatis, T., Goodbourn, S., & Fischer, J.A. (1987). Regulation of inducible and tissue-specific gene expression. Science 236, 1237–1245.
Mitchell, P.J., & Tjian, R. (1989). Transcriptional regulation in mammalian cells by sequence-specific DNA binding proteins. Science 245, 371–378.

Transcriptional Activators and Coactivators

Carey, M. (1991). Mechanistic advances in eukaryotic gene regulation. Curr. Opin. Cell Biol. 3, 425–430.
Gill, G., & Tjian, R. (1992). Eukaryotic coactivators associated with the TATA box binding protein. Curr. Opin. Genet. Dev. 2, 236–240.
Sassone-Corsi, P. (1995). Transcription factors responsible to cAMP. Ann. Rev. Cell Dev. and Biol. 11, 355–377.

Chromatin Structure and Gene Transcription

Felsenfeld, G. (1992). Chromatin as an essential part of the transcriptional mechanism. Nature, 355, 219–224.

Alternative Splicing

McKeown, M. (1992). Alternative mRNA splicing. Ann. Rev. Cell Biol. 8, 133–155.
Beret, S.M. (1995). Exon recognition in vertebrate splicing. J. Biol. Chem. 270, 2411–2414.

RNA Editing

Catteneo, R. (1991). Different types of mRNA editing. Ann. Rev. Genetics 25, 71–88.
Hodges, P., & Scott, J. (1991). Apo lipoprotein B mRNA editing: A new tier for the control of gene expression. Trends Biochem. Sci. 17, 77–81.

mRNA Stability

Atwater, J.A., Wisdom, R., & Verma, I.M. (1990). Regulated mRNA stability. Ann. Rev. Genetics, 24, 519–541.

Sachs, A., & Wahle, E. (1993). Poly(A) tail metabolism and function in eukaryotes. J. Biol. Chem. 268, 22955–22959.

Beelman, C.A., & Parker, A. (1995). Degradation of mRNA in eukaryotes. Cell 81, 179–183.

Gene Amplification

Schimke, R.T. (1988). Gene amplification in cultured cells. J. Biol. Chem. 263, 5989–5992.

Translational Regulation

Kozak, M. (1991). Structural features in eukaryotic mRNAs that modulate the initiation of translation. J. Biol. Chem. 266, 19867–19871.

Rhoads, R.E. (1993). Regulation of eukaryotic protein synthesis by initiation factors. J. Biol. Chem. 268, 3017–3021.

Chapter 10

Signal Transduction to the Cell Nucleus*

ERICH A. NIGG

INTRODUCTION

The life of all eukaryotic organisms depends on tight control of cell proliferation and differentiation. While unicellular organisms such as yeasts must be able to respond to changes in the environment (e.g., nutritional conditions), the task is more formidable for multicellular organisms which need to integrate the behavior of the many distinct cell types forming individual tissues and organs. Of particular

*This article appeared in Advances in Molecular and Cell Biology 4, 103–131 (1992).

Principles of Medical Biology, Volume 5
Molecular and Cellular Genetics, pages 201–228.

medical relevance, a vast majority of (proto-) oncogene and tumor suppressor gene products function in signal transduction, indicating that the deregulation of signaling pathways contributes to the aberrant growth of cancer cells (for reviews see Weinberg, 1985; Kahn and Graf, 1986; Bishop, 1987; Reddy et al., 1989; Cooper, 1990). Thus, it is of great practical and fundamental importance to elucidate how signals regulating DNA template activity are transmitted from the cell surface to the nucleus.

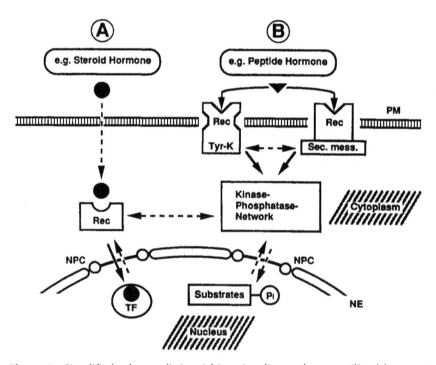

Figure 1. Simplified scheme distinguishing signaling pathways utilized by agents acting upon (**A**) intracellular receptors (e.g., steroid hormones) or (**B**) plasma membrane receptors (e.g., peptide hormones). While it remains uncertain how steroid hormones enter the cell and meet their intracellular receptors, the nuclear action of activated steroid hormone receptor complexes is comparatively well understood. In contrast, while much information is available about the complex biochemical reactions that are triggered in response to plasma membrane receptor occupation by peptide hormones and growth factors, the mechanisms involved in relaying the corresponding signals to the nucleus remain poorly understood. For further discussion see text. Rec, receptor; TF, transcription factor; NPC, nuclear pore complex; NE, nuclear envelope; PM, plasma membrane; sec. mess., second messenger, Tyr-K, tyrosine-specific protein kinase.

Both temporal and spatial patterns of DNA replication and differential gene expression are under the control of a large variety of extracellular stimuli. These include soluble agents such as hormones and growth factors, as well as stimuli acting through direct cell-cell contact, or interactions between cells and extracellular matrix proteins. In broad terms, agents acting upon cells can be divided into two groups, depending on whether they penetrate into cell nuclei or trigger a signaling cascade by binding to receptors at the plasma membrane (Figure 1). The former class of agents includes steroid hormones, thyroid hormones, and certain vitamins such as vitamine D_3 and retinoic acid derivatives (for reviews see Evans, 1988; Green and Chambon, 1986; Yamamoto et al., 1988; Beato, 1989). While it remains to be determined how these agents traverse the plasma membrane, it is clear that they bind to specific *intra*cellular receptors, which then develop a high affinity for gene regulatory sequences, and thus may be considered as ligand-responsive transcription factors (Figure 1A). Agents of the second class, exemplified by peptide hormones and growth factors (Figure 1B), bind to cell surface receptors and trigger the generation and/or activation of molecules that are chemically distinct from the primary regulatory agent (for reviews see Rozengurt, 1986; Carpenter, 1987; Deuel, 1987; Czech et al., 1988; Yarden and Ullrich, 1988). A myriad of biochemical reactions are triggered in response to such stimuli, but there is no doubt that activation of protein kinases plays a predominant role in subsequent propagation of the various signals. An important but largely unresolved question is how such signals are relayed to the cell nucleus.

It is the purpose of this chapter to discuss the evidence pertaining to possible mechanisms of information transfer between the cytoplasm and the nucleus. Rather than providing a complete survey of the literature, an attempt is made to formulate mechanistic models that may apply to several different signaling systems. In the first part, a brief summary will be presented of early events in signal transduction, and basic aspects of nuclear transport will be reviewed. In the second part, possible mechanisms of information transfer to the nucleus will be discussed. Signal transduction to the nucleus is proposed to be based on two major principles. First, a control of the nucleocytoplasmic distribution of regulatory macromolecules (generally proteins), and second, the existence of dynamic equilibria in protein distributions across the nuclear envelope. In the last part, these concepts will be illustrated by briefly considering a few specific examples of intracellular signaling. Although this chapter unavoidably provides an oversimplified, subjective, and partly speculative view of signal transduction, it is my hope that it may stimulate interest in understanding *dynamic* aspects of intracellular organization.

SIGNAL TRANSDUCTION ACROSS THE PLASMA MEMBRANE

It is beyond the scope of this article to review the early events following occupation of plasma membrane receptors by various ligands (some references may be found in Nigg, 1990). Instead, it must suffice to draw attention to a few general principles that have emerged from the study of many different systems (see Figure 1B). First, plasma membrane receptors function in either of two ways. Some are coupled, in a ligand-dependent manner, to effector systems which generate different types of second messengers such as cyclic nucleotides. Most of these second messengers act to stimulate protein kinases that are specific for serine or threonine residues. Other plasma membrane receptors are themselves tyrosine specific protein kinases, or are tightly associated with tyrosine kinases. It is currently thought that activated tyrosine kinases may act predominantly by controlling, directly or indirectly, the activities of serine-threonine specific enzymes (for review see Weiel et al., 1990). Thus, a common consequence of ligand binding to plasma membrane receptors is the activation of protein kinase cascades.

Peptide hormones, growth factors and related agents not only act at the cell surface, but are also internalized. Although this issue remains somewhat controversial (e.g., Miller, 1988), current data indicate that internalization may be more important for down-regulation of the signaling system than for intracellular signal transduction (Goldstein et al., 1985; Johnsson et al., 1985; Carpenter, 1987; Keating and Williams, 1988; Hannink and Donoghue, 1988). Also, numerous reports claim the presence of receptors for peptide hormones and growth factors within nuclei, but due to technical limitations, the significance of many of these studies is difficult to assess (for discussion see Evans and Bergeron, 1988; Burwen and Jones, 1989). There is convincing evidence for a nuclear localization of specific forms of certain growth factors (Bouche et al., 1987; Lee et al., 1987; Maher et al., 1989; Baldin et al., 1990; Imamura et al., 1990). For instance, a nuclear form of platelet derived growth factor (PDGF) was shown to result from differential splicing (Maher et al., 1989), whereas nuclear forms of basic fibroblast growth factor (bFGF) arise from alternative initiation of translation (Bugler et al., 1991). Both mechanisms create proteins containing nuclear localization signals (see below). The existence of nuclear forms of growth factors is intriguing, but their role needs to be studied further. One possibility would be that they contribute primarily to autoregulate the expression of the corresponding factors in the producing cells (as part of feed-back loops) rather than to convey signals to target cells. For the sake of simplicity, the present article is written on the premise that most physiological effects of peptide hormones and growth factors result from the occupation of cell surface receptors, and that propagation of signals is mediated primarily through protein phosphorylation. This should not be taken to imply that no other mechanisms of physiological relevance exist.

Another important aspect to consider is that intracellular signal transduction does not occur in a mere domino-type fashion, but instead involves extensive regulatory networks (for references see Nigg, 1990). As a consequence of "crosstalk" between individual signaling systems, stimulation of any specific receptor will activate not only a unique corresponding signal transduction pathway, but will influence (to variable degrees) most, if not all, other signaling systems present in the same cell. Cross-talk occurs not only between the signaling systems that are operating via second messengers and protein phosphorylation, but also involves those based on ligand-responsive transcription factors (e.g., steroid hormone receptors).

NUCLEAR PORE COMPLEXES
AND NUCLEAR TRANSPORT MECHANISMS

The nuclear envelope separates nuclear activities from those occurring in the cytoplasm and constitutes the most distinctive feature of the eukaryotic cell. It consists of a double membrane system, nuclear pore complexes, and a karyoskeletal structure (the nuclear lamina) underlying the inner nuclear membrane (for reviews see Franke et al., 1981; Gerace and Burke, 1988; Nigg, 1988, 1989; Burke, 1990). The outer nuclear membrane is functionally and structurally continuous with the endoplasmic reticulum, whereas the inner nuclear membrane is in close contact with interphase chromatin. Nuclear pores are believed to mediate most if not all exchange of macromolecules between the nucleus and the cytoplasm.

When viewed by electron microscopy, nuclear pores appear as large, highly symmetrical structures displaying eightfold radial symmetry (Franke, 1974; Unwin and Milligan, 1982; Akey, 1989; Reichelt et al., 1990). A highly simplified model of nuclear pore structure is shown in Figure 2 (for detailed discussions of morphological aspects of pore function see Franke, 1974; Scheer et al., 1988; Akey, 1990; Reichelt et al., 1990). Nuclear pores have a mass of approximately 125×10^6 daltons (Reichelt et al., 1990) and may be composed of as many as 100–200 different proteins and glycoproteins, only few of which have so far been characterized at a molecular level (Wozniak et al., 1989; Davis and Fink, 1990; Greber et al., 1990; Nehrbass et al., 1990; Starr et al., 1990).

Electron microscopic evidence unequivocally shows that nuclear pore complexes are the sites of transport of both RNA and protein (Stevens and Swift, 1966; Skoglund et al., 1983; Feldherr et al., 1984). Moreover, these studies show that individual pores can function in both import and export processes (Dworetzky and Feldherr, 1988). On the one hand, nuclear pores may be considered as molecular sieves, allowing free diffusion of ions, metabolites and small macromolecules; on the other hand, they mediate active, signal-dependent translocation of large macromolecules (for reviews see Dingwall and Laskey, 1986; Peters, 1986; Goldfarb, 1989; Silver and Goodson, 1989; Garcia-Bustos et al., 1991a). Microinjection studies using various types of labeled tracers have defined a functional open diameter of nuclear pores of about 9–10 nm (reviewed in Paine and Horowitz, 1980; Peters, 1986). Thus, the pores are expected to provide an aqueous channel for diffusion of small proteins (up to about 40–50 kDa assuming a globular shape), a notion supported by microinjection experiments using exogenous tracer proteins (e.g., trypsin inhibitor, or lysozyme; Breeuwer and Goldfarb, 1990). One might

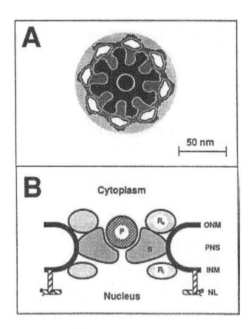

Figure 2. A schematic view of the nuclear pore complex (adapted from Reichelt et al., 1990). (**A**) An illustration of the striking eightfold symmetry of the pore as it emerges from electron microscopic studies of spread nuclear envelopes. (**B**) Schematic side view of the nuclear pore complex. According to this particular model (Reichelt et al., 1990), major constituents of the pore are outer (R_o) and inner (R_i) rings, spokes (S) and a central plug (P). (For a somewhat divergent model and nomenclature see Akey 1989, 1990; Akey and Goldfarb, 1989). Note the continuity of the lipid bilayer of the outer (ONM) and inner (INM) nuclear membranes at the edges of the pore; the two membranes enclose the so-called perinuclear space (PNS), a compartment which is in continuity with the lumen of the endoplasmic reticulum. The inner nuclear membrane is lined, at the nucleoplasmic surface, by a karyoskeletal structure, the nuclear lamina (NL).

have envisaged, therefore, that nuclear accumulation of physiologically relevant small proteins (e.g., histones) might result from diffusion into nuclei, followed by binding to non-diffusible intranuclear receptors (e.g., DNA; Bonner, 1978). However, there is virtually no evidence at present to support such a simple model (Zimmer et al., 1988). To the contrary, recent studies indicate that nuclear transport of histone H1 is based on a receptor-facilitated process (Breeuwer and Goldfarb, 1990). This interesting observation suggests that perhaps all nuclear proteins, irrespective of their sizes, may be imported into nuclei by active mechanisms.

Information about nuclear export of tRNA (Zasloff, 1983; Tobian et al., 1985), ribonucleoprotein particles (Guddat et al., 1990; Hamm and Mattaj, 1990) and pre-ribosomes (Khanna-Gupta and Ware, 1989; Bataillé et al., 1990) is only beginning to emerge. In contrast, much has been learned over the last few years about signals specifying nuclear import of proteins (for reviews see Dingwall and Laskey, 1986; Goldfarb, 1989; Roberts, 1989; Garcia-Bustos et al., 1991a). In particular, nuclear proteins were shown to contain sequence motifs, so-called nuclear localization signals (NLS), that are both necessary and sufficient to confer nuclear localization. Importantly, these sequence motifs are not generally located at the termini of proteins, nor are they removed during translocation into the nucleus. Although no single consensus sequence has emerged, most NLSS are comparatively short, generally containing multiple basic residues and frequently prolines. Whereas early studies on a viral nuclear protein (simian virus 40 (SV40) large T antigen) identified a single and short (less than 10 residue) NLS (Kalderon et al., 1984a,b; Lanford and Butel, 1984; Colledge et al., 1986), more recent studies suggest that many proteins may actually contain two signal sequences (e.g., Richardson et al., 1986; Picard and Yamamoto, 1987; Greenspan et al., 1988; Kleinschmidt and Seiter, 1988; Morin et al., 1989; Hall et al., 1990; Nath and Nayak, 1990; Underwood and Fried, 1990). Moreover, nuclear transport was found to be influenced by both the number and the positioning of particular NLS sequences within a given protein (Lanford et al., 1986; Welsh et al., 1986; Roberts et al., 1987; Dingwall et al., 1988; Dworetzky et al., 1988; Fischer-Fantuzzi and Vesco, 1988).

Progress has also been made toward understanding nuclear import of a class of small ribonucleoprotein particles, the so-called U snRNPs. These snRNPs are composed of a U snRNA (Reddy and Busch, 1988) and several (probably more than 10) different U snRNP proteins (Lührmann, 1988). Particles are assembled in the cytoplasm before they are imported into the nucleus (DeRobertis et al., 1982; Zeller et al., 1983; Zieve et al., 1988), where they function in splicing and other RNA processing events (for review see Steitz et al., 1988). Early studies had indicated that both the RNA and the protein components of U snRNPs were somehow required for nuclear uptake (Mattaj and DeRobertis, 1985), and results now point to a role of both the snRNP core proteins (the so-called Sm proteins) as well as the trimethyl-G cap structure characteristic of most U snRNAs (Fischer and Lührmann, 1990; Hamm et al., 1990). These findings strongly suggest that distinct pathways may exist for transport of different classes of macromolecules to the nucleus. Also, the discovery of multiple signals in both proteins and ribonucleoprotein particles raises the question of their precise functional significance. While it is conceivable that some of these dual signals may merely act additively, in other cases they might fulfill distinct roles. For instance, it is possible that some signals may function in transport to the nuclear envelope whereas others may be important for actual translocation through the nuclear pores (see below).

One major challenge in the nuclear transport field is to identify the receptors for NLS sequences (and for the trimethyl-G cap structure of snRNPs). A number of laboratories have applied various affinity approaches to search for NLS-binding proteins, mostly using the prototypic SV40 large T antigen NLS as a model (Adam et al., 1989; Benditt et al., 1989; Lee and Melese, 1989; Li and Thomas, 1989; Silver et al., 1989; Yamasaki et al., 1989; Meier and Blobel, 1990). According to a very simple model of nuclear transport one might have predicted NLS-binding proteins to be located specifically at the nuclear pores (see Figure 3). It is remarkable, therefore, that the majority of the candidate NLS receptors identified so far were found to reside in either the cytoplasm and/or the nucleus. One possible interpretation of these findings is that some of the proteins identified by virtue of their binding to (highly charged) NLS peptides may not actually function in nuclear transport, and that the true receptors yet await discovery. However, other, more interesting explanations are equally possible. It is attractive to think that some of the proteins identified as potential NLS receptors may function as either docking proteins (Figure 4A) or shuttling carrier proteins (Figure 4B). In order to evaluate these possibilities it will be important to further characterize the candidate NLS receptor proteins. One NLS-binding protein of 140 kDa was recently purified and shown by immunochemical techniques to reside predominantly in the nucleolus (Meier and Blobel, 1990). This result is somewhat puzzling because proteins carrying the corresponding NLSs are generally excluded from the nucleolus. Thus, one might conclude that this 140 kDa protein plays no role in nuclear transport. Alternatively, it is an intriguing possibility that the nucleolus might function in the

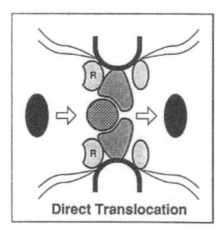

Direct Translocation

Figure 3. A simple model of nuclear transport, based on the assumption that NLS-receptors are integral parts of the nuclear pore itself. Tentatively, and rather arbitrarily, NLS-receptors (R) are assumed here to be associated with the outer ring components of the pore (see Newmeyer and Forbes, 1988; Akey and Goldfarb, 1989). Also depicted are filamentous structures emanating from the nuclear pores into both the cytoplasmic and the nucleoplasmic compartments; conceivably, these structures might play a role in recognition and/or targeting of transport substrates to the pores (Frank, 1974; Richardson et al., 1988; Scheer et al., 1988).

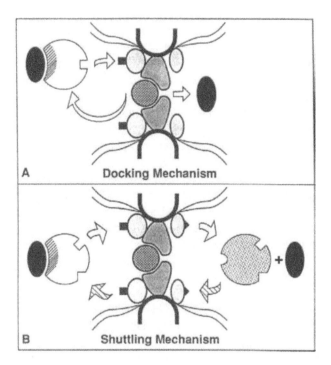

Figure 4. Multistep models of nuclear transport; these are based on the assumption that the primary NLS-receptors are not stably associated with the nuclear pore. (**A**) According to a "docking mechanism," cytoplasmic NLS-receptor would deliver the transport substrate to the pore, but would not itself be translocated through the pore. In contrast, a "shuttling mechanism" would involve translocation of a complex between NLS-receptor and transport substrate through the pore (**B**). In theory, shuttling carrier proteins might function in either uni- or bidirectional transport. The shuttle model would predict that NLS-receptors are endowed with appropriate signals for both nuclear import and export, and that mechanisms exist to trigger release of transport substrates in the appropriate compartment. For further explanation see text and the legend to Figure 3.

nuclear accumulation of certain classes of proteins. In this context it is worth considering that the nucleolus is not only the site of ribosomal RNA transcription, but also functions in assembly of precursors to ribosomal subunits (for reviews see Scheer and Benavente, 1990; Warner, 1990). As a consequence, there is a constant flow of ribosomal proteins from the cytoplasm to the nucleolus and back to the cytoplasm.

In the next few years, the study of nuclear import may be expected to benefit greatly from the availability of cell free nuclear transport systems (Newmeyer et al., 1986a,b; Finlay et al., 1987; Kalinich and Douglas, 1989; Adam et al., 1990;

Garcia-Bustos et al., 1991b). Interestingly, these systems were generally found to depend on the addition of cytosolic components, indicating a need for soluble factors in nuclear transport (see Adam et al., 1990). Fractionation of cytosolic extracts may thus provide a valuable alternative route for identifying NLS-receptor proteins, as well as other factors involved in transport (Newmeyer and Forbes, 1990). Moreover, *in vitro* systems lend themselves to depletion and reconstitution experiments (Finlay and Forbes, 1990), and in the case of yeast systems, there is the potential to combine biochemical studies with genetic approaches (e.g., Sadler et al., 1989; for review see Garcia-Bustos et al., 1991a).

The mechanics of nucleocytoplasmic transport remain largely unknown, but two distinct steps in nuclear import have been resolved (Richardson et al., 1988; Newmeyer and Forbes, 1988; Akey and Goldfarb, 1989; Stewart et al., 1990). One appears to be recognition and binding of NLS motifs by appropriate receptors, the other is energy-dependent translocation across the nuclear pores. How the pore "opens up" in response to signals remains to be determined. According to one recent model, opening of the pore might be compared to the opening of an iris diaphragm of a photographic camera (for discussion see Akey, 1990).

SIGNALING TO THE NUCLEUS: CONCEPTUAL CONSIDERATIONS

Dynamic Equilibria Between Nucleus and Cytoplasm

Frequently, physiological transport processes are assumed to occur unidirectionally, i.e., from cytoplasm to nucleus for most proteins, and from nucleus to cytoplasm for most RNAs. Yet, early studies with amoebae had long indicated the existence of proteins shuttling back and forth between nucleus and cytoplasm (Goldstein, 1958; for review see Goldstein and Ko, 1981). The concept of shuttling proteins has been revitalized by the demonstration that two major nucleolar proteins, called nucleolin and No38, shuttle constantly between nucleus and cytoplasm (Borer et al., 1989). While it remains unclear how many proteins shuttle between the two compartments, available evidence indicates that the phenomenon may be more frequent than generally appreciated (e.g., Madsen et al., 1986; Bachmann et al., 1989; Mandell and Feldherr, 1990). Shuttling of small proteins might conceivably occur by diffusion (see, however, Breeuwer and Goldfarb, 1990), but bidirectional transport of large macromolecules through nuclear pores would be expected to require ATP hydrolysis, and structurally distinct signals specifying import and export, respectively.

With respect to the question of generality of protein shuttling, it is important to emphasize that most techniques commonly used to monitor protein distributions (i.e., immunocytochemistry and subcellular fractionation) merely report the equilibrium distributions of proteins. Although these techniques are useful to detect shifts in equilibria, they do not provide information about the *dynamics* of protein distributions under steady-state conditions. For instance, both immunofluorescence microscopy and subcellular fractionation indicated an almost completely nuclear localization of the two shuttling nucleolar proteins nucleolin and No38. The existence of dynamic equilibria could be detected only by experimental approaches that were designed specifically to detect protein mobility (Borer et al., 1989). Thus, shuttling of proteins between nucleus and cytoplasm may easily escape detection by common steady-state analyses of protein distributions.

There is no definitive information, at present, on the functional significance of protein shuttling across the nuclear envelope. However, it is possible to make two specific proposals. One obvious possibility is that shuttling proteins may play a major role as carriers in nucleocytoplasmic transport processes (see Figure 4B, above). Another intriguing possibility is that they may contribute to signal transduction (Figure 5A). For instance, if nuclear transcription and replication factors were to shuttle between nucleus and cytoplasm, their activities might be controlled while transiently exposed to cytoplasmic enzymes (e.g., kinases and phosphatases). It is interesting to consider the kinetic properties of (hypothetical) signaling systems based on cytoplasmic modification of shuttling nuclear proteins (Figure 5B). If one assumes that even a few modified proteins entering the nucleus were sufficient to elicit a physiological response, then such a mechanism might conceivably be very rapid (minutes). However, if a critical threshold level of modified proteins were required to produce nuclear responses, then a signaling mechanism based on shuttling proteins would be expected to exhibit a comparatively slow response time (hours). Thus, signals would be transmitted only if cytoplasmic stimuli persisted for a sufficient length of time.

Regulated Compartmentalization of Proteins

Studies on developing embryos of both vertebrates and invertebrates have revealed dramatic shifts in the nucleocytoplasmic distribution of individual proteins at precise stages of development (Dreyer et al., 1982; Dreyer and Hausen, 1983; Dequin et al., 1984; Dreyer, 1987; Servetnick and Wilt, 1987). Similarly, redistributions of proteins between cytoplasm and nucleus were reported in response to hormonal stimulation (Nigg et al., 1985a; Picard and Yamamoto, 1987), passage of particular stages of the cell cycle (e.g., McMorrow et al., 1990; Nasmyth et al., 1990; Pines and Hunter, 1991), and heat shock (Velasquez and Linquist, 1984; Welch and Feramisco, 1984; Lewis and Pelham, 1985; Collier and Schlesinger, 1986). That alterations in steady-state distributions of proteins between the two compartments can have profound physiological consequences is illustrated also by studies on the products of the (proto-) oncogenes *abl* (Van Etten et al., 1989) and *fos* (Roux et al., 1990). In both cases, differences in the nucleocytoplasmic distri-

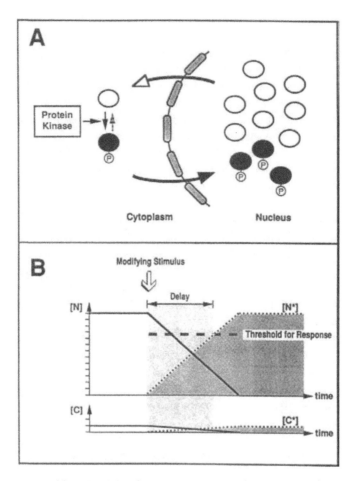

Figure 5. A possible role of shuttling proteins in signal transduction between cytoplasm and nucleus. Cytoplasmic modification (e.g., by phosphorylation) of shuttling regulatory proteins would be expected to result in nuclear accumulation of proteins with altered properties (**A**). The kinetic properties of such a hypothetical signaling system are illustrated in (**B**). Shown are, as a function of time, the cytoplasmic and nuclear concentrations of a regulatory protein in its unmodified state ([C] and [N]), and those of the same protein in its modified state ([C*] and [N*]). In the example shown, the regulatory protein is presumed to exist in a dynamic equilibrium between nucleus and cytoplasm; in the absence of a modifying stimulus, its steady-state distribution is presumed to be predominantly nuclear. For the sake of simplicity, the modifying stimulus is assumed to alter only the functional properties of the protein, but not its transport properties. As a consequence, neither the rates of nuclear import and export, nor the equilibrium distribution would be expected to change. Nevertheless, cytoplasmic modification of the regulatory protein would result in progressive nuclear accumulation of a protein with altered functional properties. (Continued)

bution between the cellular and viral versions of the corresponding proteins were correlated with differences in transforming potential.

All of the above observations clearly demonstrate that the nucleocytoplasmic distribution of particular proteins can be subject to regulation, and that changes in subcellular distributions can have profound physiological consequences. Controls of nucleocytoplasmic protein distributions may be expected to operate at several levels. On the one hand, there is evidence for changes in both passive and NLS-dependent permeability properties of nuclear pores, depending on the physiological state of the cell (Schindler and Jiang, 1987; Jiang and Schindler, 1988; Feldherr and Akin, 1990). However, changes in pore properties are expected to influence the rates of transport of entire classes of macromolecules, and cannot account for controlled redistributions of individual proteins. Thus, one has to postulate the existence of more specific mechanisms. In support of this notion, increasing evidence suggests that the efficiency of NLS sequences can be modulated in at least two ways: first, through posttranslational modification (particularly phosphorylation), and second, through protein-protein interactions.

Posttranslational Control of NLS Function

Several studies strongly suggest that the efficiency of NLS sequences can be altered by phosphorylation of residues located in close proximity to (or within) an NLS. Systematic studies on the SV40 NLS indicate that the rate of nuclear transport of NLSS containing proteins may be controlled by phosphorylation of flanking residues (Rihs and Peters, 1989; Rihs et al., 1991). Based on the observation that many proteins contain potential casein kinase II (CKII) phosphoacceptor sites in proximity to NLS sequences, Peters and co-workers have proposed that CKII might play a general role in controlling the rate of protein transport, and thereby contribute to transmitting growth regulatory signals to the nucleus (Rihs et al., 1991). While this proposal is very intriguing, it will be important to explore the generality of the phenomenon by directly demonstrating that the potential CKII phosphoacceptor sites are actually used *in vivo* in proteins other than SV40 large T antigen. Another example for a role of phosphorylation in controlling nuclear transport is provided by recent studies on the transcriptional activator SWI5 in *Saccharomyces cerevisiae*. This gene product is translocated to the cell nucleus at a specific stage of the

---

Figure 5. (Continued) Depending on the threshold of modified protein required for triggering a physiological response within the nucleus, signal transduction might be either rapid or slow. While not shown in this figure, a modifying stimulus might change not only the functional properties of a regulatory molecule, but also its transport properties. As a consequence, it would also induce alterations in the steady-state distribution of the protein across the nuclear envelope. (For further discussion see text and Figure 6.)

cell cycle, G1 phase, when it activates the HO endonuclease gene (Nasmyth et al., 1990). Strong evidence indicates that the cytoplasmic location of the SWI5 gene product during both S and M phase is due to phosphorylation, by the CDC28 kinase, of a residue located within the predicted NLS of the protein (Moll et al., personal

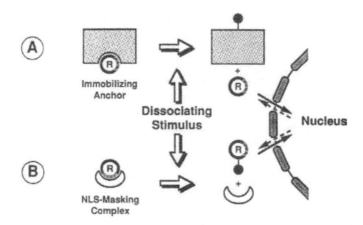

Figure 6. Possible mechanism controlling the nucleocytoplasmic distribution of regulatory macromolecules. Proteins (R) with a regulatory intranuclear function may be sequestered in the cytoplasm due to (**A**) interactions with insoluble cytoplasmic components (e.g., the cytoskeleton or endomembrane systems), or, alternatively (**B**), masking of NLS-sequences (stippled box) by formation of complexes with cytoplasmic proteins. In the former case, dissociating stimuli would induce release of the regulatory protein from the immobilizing anchor (the anchorage-release model); in the latter case, the dissociating stimulus would lead to exposure of the NLS on the regulatory protein (the signal-unmasking model). It is important to note that, in both models, the dissociating modification (depicted by a filled circle) may target either the regulatory protein or the complex partner; for the sake of simplicity, only two of the four possible models are drawn here. Comparison of several physiologically distinct signal transduction pathways illustrates that they are all conceptually similar, differences relating predominantly to the precise way in which the subcellular distribution of a regulatory macromolecule is controlled: the best example for the anchorage-release model (**A**) may be provided by the cAMP-dependent signaling pathway. The compartmentalization of the glucocorticoid receptor might be considered as an illustration of the signal-unmasking model (**B**), although contributions by anchoring proteins (hsp90?) are not excluded. In this case, the modifying stimulus is provided by steroid hormone and targets the regulatory protein (i.e., the steroid receptor), as depicted in panel B. While the subcellular distribution of the NF*kB/dorsal/rel* transcription factor appears to be controlled by a similar mechanism, the modifying stimulus most probably affects the complex partner rather than NF*k*B itself; also, the relative contributions of anchoring versus NLS-masking complex partners of NF*k*B remain to be determined. (For further explanation see text.)

communication). It appears to be a safe prediction that future studies will reveal many other examples for regulation of NLS function by phosphorylation.

Anchorage-Release and NLS-Masking/Unmasking Mechanism

The nucleocytoplasmic distribution of proteins can also be controlled by mechanisms involving complex formation with other proteins (Figure 6). On the one hand, NLS sequences may be prevented from functioning by tight binding of a protein to an insoluble structure: an "immobilizing anchor" may be provided, for instance, by a protein that itself is attached to the cytoskeleton or to endomembranes (Figure 6A). On the other hand, NLS motifs may be masked within protein-protein complexes (Figure 6B). Even if such complexes were soluble, they would not be transported to the nucleus, due to lack of recognition of the masked NLS sequence by corresponding receptors. Irrespective of the precise mechanism inactivating an NLS, the important aspect of the above models is that appropriate stimuli can trigger dissociation of protein complexes and restore NLS function (Figure 6). Such stimuli may target either the protein containing the NLS sequence or its complex partner. Below, a few specific examples of signal transduction pathways will be briefly discussed. These were chosen to illustrate that several apparently distinct physiological situations can readily be interpreted in terms of the models shown in Figure 6.

SIGNALING TO THE NUCLEUS: EXAMPLES

Protein Kinase Translocations to the Nucleus

The cAMP-dependent signal transduction pathway is comparatively well understood (Figure 7): cAMP is generated by adenylyl cyclase in response to binding of various ligands to appropriate (e.g., β-adrenergic) receptors. In eukaryotes, elevation of this second messenger then elicits a large variety of physiological responses, all of which are believed to result from activation of cAMP-dependent protein kinases (PKAs) (reviewed in Lohmann and Walter, 1984). These enzymes are composed of two regulatory (R) and two catalytic (C) subunits (Flockhart and Corbin, 1982; Edelman et al., 1987; Taylor et al., 1990). On the basis of biochemical fractionation, two distinct types of PKAs differing in their R subunits (RI and RII) can be distinguished. Additional isoforms of both R and C subunits have been identified by cDNA cloning, but structural differences appear to be minor and their physiological significance remains to be explored.

Upon binding of cAMP to the R subunits, the inactive holoenzyme dissociates to release the C subunits in their active form. Regulation of cellular processes is then effected through phosphorylation of specific target proteins. Many genes are known to be regulated at the transcriptional level by cAMP (Roesler et al., 1988;

Karin, 1989). To explain gene-regulatory effects of cAMP, two fundamentally different mechanisms were proposed (for references see Jans and Hemmings, 1988; Nigg, 1990). According to the R-signaling hypothesis, RI or RII subunits were envisaged to translocate to the nucleus in response to cAMP, and to activate gene transcription by directly binding to regulatory sequences in the promoter regions of target genes. In contrast, the alternative C-signaling hypothesis proposed that the C subunit activates transcription by phosphorylating specific transcription factors. This latter hypothesis is now strongly supported by multiple lines of evidence (e.g., Grove et al., 1987; Handler et al., 1988; Jans and Hemmings, 1988; Nakagawa et al., 1988; Riabowol et al., 1988; Day et al., 1989; Maurer, 1989; Mellon et al., 1989; Pearson et al., 1991). In particular, the activities of at least two nuclear transcription factors (CREB in vertebrates and ADR1 in yeast) were shown to be regulated via

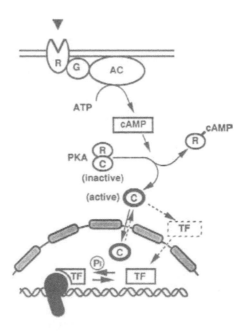

Figure 7. Highly simplified scheme illustrating likely mechanisms of gene regulation by cAMP-dependent protein kinases. In response to stimulation of appropriate plasma membrane receptors by agonists, adenylyl cyclase (AC) is activated (in a G-protein dependent process) and produces cAMP. This second messenger binds to the regulatory subunits (R) of cAMP-dependent protein kinases (PKA) and triggers release of active catalytic (C) subunits. Following translocation to the nucleus, active kinase subunits may phosphorylate (and thereby regulate the activity) of intranuclear transcription factors (TF). In other cases, they may regulate, through cytoplasmic phosphorylation, the subcellular distribution of transcription factors (see Figure 6). Note that these two mechanisms are not mutually exclusive.

phosphorylation by PKA *in vivo* (K.K. Yamamoto et al., 1988; Cherry et al., 1989; Gonzalez and Montminy, 1989).

Most interesting in the context of the present discussion is that immunocyto-chemistry provided strong evidence for a reversible cAMP dependent nuclear translocation of C (but not RII) subunits (Nigg et al., 1985a, 1988; Pearson et al., 1991). In unstimulated cells, C and RII subunits were shown to be codistributed in the area of the Golgi apparatus and to be associated with microtubule organizing centers (Nigg et al., 1985b; De Camilli et al., 1986). Recent immunoelectron microscopic studies have provided more precise information on the location of RII in the Golgi area and demonstrate that this protein is associated with the cytoplasmic surface of various membrane systems, specifically the prelysosomal compartment and the *trans*-Golgi network (Griffiths et al., 1990). In response to cAMP, C but not RII subunits were found to redistribute throughout the cell, and particularly, to the nucleus. Nuclear translocation of C subunits was rapid and reversible (Nigg et al., 1985a: see also Nigg, 1990). One straightforward way to interpret these results is to postulate the existence of a dynamic equilibrium in the distribution of the C subunit across the nuclear envelope. Cytoplasmic versus nuclear localization of this kinase would then be determined by the (cAMP-dependent) binding affinity of the cytoplasmic R subunits. According to this interpretation, the endomembrane asso-ciated RII subunits would serve as cytoplasmic immobilizing anchors for C subunits (see Figure 6A).

Thus, the fate of the C subunit of PKA may be considered as a prototypic example for an anchorage-release mechanism controlling the nucleocytoplasmic distribution of a protein kinase. Other kinases may undergo similar nuclear translo-cations in response to appropriate physiological stimuli. One prime candidate for such a dynamic behavior is protein kinase C (PKC). This kinase is activated by numerous growth factors, as well as tumor promoting phorbol esters, and it probably acts predominantly on cytoplasmic targets in close proximity to the plasma membrane (for reviews see Kikkawa and Nishizuka, 1986; Nishizuka, 1988; Jaken, 1990). In fact, the bulk of PKC is well known to undergo a "translo-cation" to the plasma membrane in response to physiological stimulation (e.g., Kraft and Anderson, 1983; Ito et al., 1988). This type of translocation presumably reflects a change in the affinity of PKC for the plasma membrane; it should not be confused with the nuclear translocation discussed above for PKA. However, there are many isoforms of PKC, and these may well differ from each other with respect to their subcellular distribution (Nishizuka, 1988). Nuclear translocations of PKC isoforms were reported in several studies (Cambier et al., 1987; Halsey et al., 1987; Kiss et al., 1988; Thomas et al., 1988; Leach et al., 1989; Masmoudi et al., 1989; Fields et al., 1990; Rogue et al., 1990), and there is strong evidence for phospho-rylation of nuclear proteins (e.g., lamins) by a kinase with substrate specificity closely resembling that of PKC (Fields et al., 1988, 1990; Friedman and Ken, 1988; Hornbeck et al., 1988, Peter et al., 1990; Tsuda and Alexander, 1990). It is tempting to think, therefore, that particular isoforms of PKC might play a role in signal transduction from the cytoplasm to the nucleus, as suggested previously for PKA.

Other kinases potentially involved in signaling to the nucleus are casein kinase II (CKII), and, conceivably, members of the MAP kinase family. CKII has been reported to be stimulated (although to a generally rather modest extent) by various growth factors, suggesting that it may play a role in controlling cell proliferation (for reviews see Carroll et al., 1988; Krebs et al., 1988; Pinna, 1990). In support of this view, CKII activities were described to be increased in neoplastically transformed cell lines and in tumors (Münstermann et al., 1990), and several nuclear proteins implicated in cell proliferation control were shown to be *in vivo* substrates of CKII (for review see Pinna, 1990). Of particular interest, putative physiological substrates for CKII include transcription factors as well as nuclear products of oncogenes and tumor suppressor genes (Grässer et al., 1988; Firzlaff et al., 1989; Lüscher et al., 1989, 1990; Meek et al., 1990; Manak et al., 1990). Clearly, it will be very important, therefore, to learn more about the subcellular distribution and the mechanisms of regulation of CKII.

MAP kinase was named originally according to its ability to phosphorylate the microtubule associated protein MAP2; more recently, however, it was recognized that this kinase is likely to function downstream of receptors for various mitogens (Hoshi et al., 1988; Ray and Sturgill, 1988; Sturgill et al., 1988), and, accordingly, the abbreviation MAP is now taken to stand for mitogen-activated protein kinase. Recent sequencing information indicates that there may be a family of structurally related MAP kinases (Boulton et al., 1990) and that these enzymes are related to two kinases (KSS1 and FUS3) involved in the pheromone response pathways in *S. cerevisiae* (Courchesne et al., 1989; Elion et al., 1990). What is particularly intriguing about MAP kinases is that these serine/threonine-specific enzymes appear to require phosphorylation on both tyrosine and threonine residues for activity (Cooper, 1989; Rossomando et al., 1989; Anderson et al., 1990; Ferrell and Martin, 1990). Thus, MAP kinases are likely to act downstream of both tyrosine and serine/threonine kinases, and may therefore play a pivotal role in integrating signals derived from independently activated pathways. Whether or not members of the MAP kinase family function to transmit signals across the nuclear envelope remains to be determined, but this clearly is an attractive possibility.

NF*k*B *(dorsal/rel)*

The transcription factor NF*k*B was first identified as a protein binding to the enhancer of the κ light chain gene in B lymphocytes (Sen and Baltimore, 1986a). Because NF*k*B was constitutively present only in cells expressing light chains, it was originally considered to be a tissue-restricted transcription factor (Atchison and Perry, 1987; Lenardo and Baltimore, 1989). Subsequently, however, NF*k*B activity was found to be latent in the cytoplasm of most (if not all) mammalian cells, and its activation was shown to influence the activities of many different target genes (for reviews see Lenardo and Baltimore, 1989; Baeuerle, 1991; Baeuerle and Baltimore, 1991). NF*k*B is of great interest in the context of the signal transduction mechanisms discussed here, because it re-emphasizes the importance of subcellular compartmentalization of regulatory proteins (see Figure 6).

NF*k*B is inducible by several agents, notably phorbol esters and lipopolysaccharides (Sen and Baltimore, 1986b). Activation of NF*k*B does not require protein synthesis, but instead results from dissociation of the DNA binding factor from a cytosolic inhibitor protein called I*k*B (Baeuerle and Baltimore, 1988a,b). The available data indicate that I*k*B is a substrate for activated protein kinases (particularly PKC) and that phosphorylation of I*k*B triggers release and nuclear translocation of NF*k*B (see Baeuerle, 1991 and Figure 6).

Recent cDNA cloning revealed the existence of a whole family of NF*k*B-like proteins; these form distinct multiprotein complexes which may differ with respect to their specificities for target genes, as well as their sensitivities to regulatory stimuli (for discussion see Ballard et al., 1990; Baeuerle, 1991). In this respect, the NF*k*B family would appear to resemble the AP1/*fos*/*jun* family of transcription factors (for reviews see Curran and Franza, 1988; Vogt and Bos, 1989; Ransone and Verma, 1990). Most intriguingly, the DNA binding subunit of NF*k*B is structurally related to the products of two genes, *dorsal* and *rel*, that had previously been studied for their roles in development and neoplastic transformation, respectively (Ballard et al., 1990; Bours et al., 1990; Ghosh et al., 1990; Kieran et al., 1990; for review see Gilmore, 1990). The *dorsal* gene in *Drosophila* encodes a transcription factor required for the development of ventral structures in the embryo (for references see Roth et al., 1989). Remarkably, the spatial regulation of *dorsal* activity in the fly embryo depends on the subcellular distribution of the *dorsal* gene product rather than its abundance; the *dorsal* protein is in fact located in the nuclei of ventral cells but in the cytoplasm of dorsal cells, with a continuous gradient of distribution inbetween (Roth et al., 1989; Rushlow et al., 1989; Steward, 1989; for review see Hunt, 1989). It clearly is an interesting question, therefore, what determines the nucleocytoplasmic distribution of *dorsal*. While wild-type *dorsal* protein expressed in tissue culture cells was cytoplasmic, deletion of eight amino acids from the carboxy-terminus resulted in nuclear localization (Rushlow et al., 1989). This evidence strongly suggests that the carboxy-terminus of *dorsal* somehow interacts with a cytoplasmic anchoring or NLS-masking protein (see Figure 6). The identity of the putative cytoplasmic complex partner(s) of *dorsal* is not presently known, but one candidate is the product of the *cactus* gene (Roth et al., 1989). Given the relationship between *dorsal* and NF*k*B, *cactus* might correspond to I*k*B, and it may be anticipated that a molecular analysis of *dorsal* function by genetic approaches will shed light on other components interacting with NF*k*B.

NF*k*B and *dorsal* are also related to the product of the avian (proto-) oncogene *rel* (for review see Gilmore, 1990). v-*rel* was identified originally as the oncogene encoded by the avian retrovirus Rev-T; it is presumed to function as a *trans*-activating gene regulatory protein (Gelinas and Temin, 1988), but the molecular basis of its transforming potential is unclear. Interestingly, the product of the cellular homolog (*c-rel*) is located predominantly in the cytoplasm (at least when overexpressed), while the viral gene product (*v-rel*) is cytoplasmic in lymphoid cells but nuclear in fibroblasts (Gilmore and Temin, 1986; Capobianco et al., 1990). In line with the results obtained for *dorsal* (Rushlow et al., 1989), a carboxy-terminal sequence was found to be required for cytoplasmic localization of *rel* (Hannink and Temin, 1989; Capobianco et al., 1990). Deletion of this sequence in the virally transduced version of the *rel* oncogene (*v-rel*) may thus account for the nuclear localization of the corresponding gene product.

Somewhat surprisingly, no correlation was observed between the subcellular distribution and the transforming potential of various *rel* proteins. Thus, nuclear as well as cytoplasmic forms of *rel* proteins were found to be equally transforming (Gilmore and Temin, 1988). One way to rationalize this observation is to propose that *rel* proteins might shuttle between nucleus and cytoplasm. A dynamic equilibrium across the nuclear envelope might then provide sufficient amounts of *rel* protein within the relevant compartment to allow cell transformation, irrespective of any particular steady-state distribution.

The Steroid Hormone Receptor Family

Steroid hormone receptors may be considered as prototypes for a large family of gene regulatory proteins, comprising receptors also for a considerable variety of nonsteroid ligands (for reviews see Green and Chambon, 1986; Evans, 1988; K.R. Yamamoto et al., 1988; Beato, 1989). When occupied, these receptors directly regulate the rate of transcription of hormone-responsive genes, and they thus function as ligand-dependent transcription factors. According to a paradigm formulated more than 20 years ago, signal transduction by steroid hormones was thought to depend on hormone-induced translocation of receptors from the cytoplasm to the nucleus (Gorski et al., 1968; Jensen et al., 1968). More recently, however, this two-step model of hormone action has been challenged (King and Green, 1984; Welshons et al., 1984) and the bulk of the available evidence now seems to indicate that the receptors for most steroid hormones reside predominantly in the nucleus even in their unoccupied state (for discussion see Fuxe et al., 1985; Perrot-Applanat et al., 1985; Wikström et al., 1987; Gasc et al., 1989; Picard and Yamamoto, 1987). Among the receptors that have been studied carefully so far, only the one for glucocorticoid hormones appears to redistribute from the cytoplasm to the nucleus, as proposed in the original translocation model (Picard and Yamamoto, 1987).

It is interesting to consider these observations in light of the models drawn in Figure 6. Analyses of various mutant receptor constructs in fact suggest that the accessibility (and hence function) of the NLS sequences of different receptors is differentially regulated. In the case of the glucocorticoid receptor, a major NLS was found to be masked by the receptor's own hormone binding domain; accordingly, one of the functions of glucocorticoid hormone consists in unmasking an NLS within its receptor (Picard and Yamamoto, 1987). In contrast, the NLS of the progesterone receptor was found to be constitutively exposed and active even in the absence of hormone (Picard et al., 1990a). Thus, differential exposure of NLS sequences may account, at least in part, for the different steady-state distributions of different unoccupied hormone receptors.

Another factor contributing to receptor inactivation is complex formation between unoccupied steroid hormone receptors and other proteins. Prominent among the complex partners of steroid receptors is hsp90, an abundant member of the heat shock protein family. In line with the scheme shown in Figure 6, activation of receptors by hormones appears to require dissociation from the hsp90 complexes (e.g., Catelli et al., 1985; Sanchez et al., 1985; Howard and Distelhorst, 1988; Chambraud et al., 1990; Picard et al., 1990b). To what extent hsp90 (or other heat shock proteins; Sanchez, 1990) play a role in determining the nucleocytoplasmic distribution of steroid hormone receptors remains to be elucidated.

Finally, recent evidence is consistent with the idea that steroid hormone receptors might shuttle between nucleus and cytoplasm (Guiochon-Mantel et al., 1989). The existence of a dynamic equilibrium between nuclear and cytoplasmic pools of receptors might facilitate the transport of steroid hormones into nuclei. The kinetics of such a process might be more directly relevant to signal transduction than the distribution of any particular receptor in the steady-state.

SUMMARY

This chapter has emphasized dynamic aspects of intracellular organization. Signal transduction from cytoplasm to nucleus was proposed to depend on the existence of equilibria in protein distributions across the nuclear envelope, and on various mechanisms controlling the subcellular distribution of regulatory macromolecules. Prominent roles in the control of nucleocytoplasmic protein distributions were attributed to cytoplasmic anchorage-release or NLS-masking/unmasking mechanisms, and to posttranslational modifications, notably phosphorylation.

Although the notions presented here may contribute to the rationalization of experimental findings in several different systems, it should be emphasized that other fundamentally different mechanisms of signal transduction to the nucleus are not excluded. For instance, it is obvious that nuclear activities depend on the availability of cytoplasmically produced metabolites (e.g., building blocks required for nucleic acid synthesis). The regulation of the cytoplasmic production of such components clearly provides a means to modulate nuclear activities. Also, it is possible to entertain models of signal transduction based on purely mechanical aspects of cellular organization. For instance, there is no doubt that the extracellular matrix is linked across the plasma membrane to the cytoskeleton (e.g., Burridge, 1986), and interactions between the cytoskeleton and karyoskeletal proteins have also been proposed (Fey et al., 1984; Georgatos and Blobel, 1987a,b; Fey and Penman, 1988). Alterations in cell morphology are well known to influence patterns of gene expression (e.g., Benecke et al., 1978; Farmer et al., 1983), and it is an interesting question, therefore, to what extent changes in nuclear activities can be attributed directly to changes in the architecture of the cytoskeleton and the nucleoskeleton.

ACKNOWLEDGMENTS

I am most grateful to Ms. C. Ravussin and Mr. P. Dubied for help with the preparation of this manuscript, and to Drs. V. Simanis and H. Hennekes for helpful comments. Work in the author's laboratory was supported by the Swiss National Science Foundation (31.8782.86 and 31.26413.89) and the Swiss Cancer League (424.90.1).

REFERENCES

Adam, S.A., Lobl, T.J., Mitchell, M.A., & Gerace, L. (1989). Nature 337, 276–279.
Adam, S.A., Marr, R.S., & Gerace, L. (1990). J. Cell Biol. 111, 807–816.
Akey, C.W. (1989). J. Cell Biol. 109, 955–970.
Akey, C.W. (1990). Biophys. J. 58, 341–355.
Akey, C.W., & Goldfarb, D.S. (1989). J. Cell Biol. 109, 971–982.
Anderson, N.G., Maller, J.L., Tonks, N.K., & Sturgill, T.W. (1990). Nature 343, 651–653.
Atchison, M.L., & Perry, R.P. (1987). Cell 48, 121–128.
Bachmann, M., Pfeifer, K., Schroder, H.C., & Muller, W.E. (1989). Mol. Cell Biochem. 85, 103–114.
Baeuerle, P.A. (1991). Biochim. Biophys. Acta, in press.
Baeuerle, P.A., & Baltimore. D. (1988a). Cell 53, 211–217.
Baeuerle, P.A., & Baltimore, D. (1988b). Science 242, 540–546.
Baeuerle, P.A., & Baltimore, D. (1991). In: The Hormonal Control Regulation of Gene Transcription (Cohen, P., & Foulkes, J.G., eds.), pp. 409–432. Elsevier Science Publishers BV (Biomedical Division, Amsterdam.
Baldin, V., Roman, A.M., Bosc Bierne, I., Amalric, F., & Bouche, G. (1990). EMBO J. 9, 1511–1517.
Ballard, D.W., Walker, W.H., Doerre, S., Sista, P., Molitor, J. A., Dixon, E.P., Peffer, N.J., Hannink, M., & Greene, W.C. (1990). Cell 63, 803–814.
Bataillé, N., Helser, T., & Fried, H.M. (1990). J. Cell Biol. 111, 1571–1582.
Beato, M. (1989). Cell 56, 335–344.
Benditt, J.O., Meyer, C., Fasold, H., Barnard, F.C., & Riedel, N. (1989). Proc. Natl. Acad. Sci. USA 86, 9327–9331.
Benecke, B.-J., Ben Ze'ev, A., & Penman, S. (1978). Cell 14, 931–939.
Bishop, J.M. (1987). Science 235, 305–311.
Bonner, W.M. (1978). In: The Cell Nucleus, Vol. 6, Busch, H., ed.), pp. 97–148, Academic Press, New York.
Borer, R.A., Lehner, C.F., Eppenberger, H.M., & Nigg, E.A. (1989). Cell 56, 379–390.
Bouche, G., Gas, N., Prats, H., Baldin, V., Tauber, J.P., Teissie, J., & Amalric, F. (1987). Proc. Natl. Acad. Sci. USA 84, 6770–6774.
Boulton, T.G., Yancopoulos, G.D., Gregory, J.S., Slaughter, C., Moomaw, C., Hsu, J., & Cobb, M.H. (1990). Science 249, 64–67.
Bours, V., Villalobos, J., Burd, P.R., Kelly, K., & Siebenlist, U. (1990). Nature 348, 76–80.

Breeuwer, M., & Goldfarb, D.S. (1990). Cell 60, 999–1008.

Bugler, B., Amalric, F., & Prats, H. (1991). Mol. Cell. Biol. 11, 573–577.

Burke, B. (1990). Curr. Opin. Cell Biol. 2, 514–520.

Burridge, K. (1986). Cancer Rev. 4, 18–78.

Burwen, S.J., & Jones, A.L. (1989). Trends Biochem. Sci. 12, 159–162.

Cambier, J.C., Newell, M.K., Justement, L.B., McGuire, J.C., & Leach, K.L. (1987). Nature 327, 629–632.

Capobianco, A.J., Simmons, D.L., & Gilmore, T.D. (1990). Oncogene 5, 257–265.

Carpenter, G. (1987). Ann. Rev. Biochem. 56, 881–914.

Carroll, D., Santoro, N., & Marshak, D.R. (1988). Cold Spring Harb. Symp. Quant. Biol. 53, 91–95.

Catelli, M.G., Binart, N., Jung Testas, I., Renoir, J.M., Baulieu, E.E., Feramisco, J.R., & Welch, W.J. (1985). EMBO J. 4, 3131–3135.

Chambraud, B., Berry, M., Redeuilh, G., Chambon, P., & Baulieu, E.E. (1990). J. Biol. Chem. 265, 20686–20691.

Cherry, J.R., Johnson, T.R., Dollard, C., Shuster, J.R., & Denis, C.L. (1989). Cell 56, 409–419.

Colledge, W.H., Richardson, W.D., Edge, M.D., & Smith, A.E. (1986). Mol. Cell Biol. 6, 4136–4139.

Collier, N.C., & Schlesinger, M.J. (1986). J. Cell Biol. 103, 1495–1507.

Cooper, J.A. (1989). Mol. Cell Biol. 9, 3143–3147.

Cooper, J.A. (1990). Curr. Opin. Cell Biol. 2, 285–295.

Courchesne, W.E., Kunisawa, R., & Thorner, J. (1989). Cell 58, 1107–1119.

Curran, T., & Franza, B.R., Jr. (1988). Cell 55, 395–397.

Czech, M.P., Klarlund, J.K., Yagaloff, K.A., Bradford, A.P., & Lewis, R.E. (1988). J. Biol. Chem. 263, 11017–11020.

Davis, L.I., & Fink, G.R. (1990). Cell 61, 965–978.

Day, R.N., Walder, J.A., & Maurer, R.A. (1989). J. Biol. Chem. 264, 431–436.

De Camilli, P., Moretti, M., Donini, S.D., Walter, U., & Lohmann. S.M. (1986). J. Cell Biol. 103, 189–203.

De Robertis, E.M., Lienhard, S., & Parisot, R.F. (1982). Nature 295, 572–577.

Dequin, R., Saumweber, H., & Sedat, J.W. (1984). Dev. Biol. 104, 37–48.

Deuel, T.F. (1987). Ann. Rev. Cell Biol. 3, 443–492.

Dingwall, C., & Laskey, R.A. (1986). Ann. Rev. Cell Biol. 2, 367–390.

Dingwall, C., Robbins, J., Dilworth, S.M., Roberts, B., & Richardson, W.D. (1988). J. Cell Biol. 107, 841–849.

Dreyer, C. (1987). Development. 101, 829–846.

Dreyer, C., & Hausen, P. (1983). Dev. Biol. 100, 412–425.

Dreyer, C., Scholz, E., & Hausen, P. (1982). Wilhelm Roux's Archives 191, 228–233.

Dworetzky, S.I., & Feldherr, C.M. (1988). J. Cell Biol. 106, 575–584.

Dworetzky, S.I., Lanford, R.E., & Feldherr, C.M. (1988). J. Cell Biol. 107, 1279–1287.

Edelman, A.M., Blumenthal, D.K., & Krebs, E.G. (1987). Ann. Rev. Biochem. 56, 567–613.

Elion, E.A., Grisafi, P.L., & Fink, G.R. (1990). Cell 60, 649–664.

Evans, R.M. (1988). Science 240, 889–895.

Evans, W.H., & Bergeron, J.J.M. (1988). Trends Biochem. Sci. 13, 7–8.

Farmer, S., Wan, K., Ben-Ze'ev, A., & Penman, S. (1983). Mol. Cell. Biol. 3, 182–189.

Feldherr, C.M., & Akin, D. (1990). J. Cell Biol. 111, 1–8.

Feldherr, C.M., Kallenbach, E., & Schultz, N. (1984). J. Cell Biol. 99, 2216–2222.

Ferrell, J.E., & Martin, G.S. (1990). Mol. Cell. Biol. 10, 3020–3026.

Fey, E.G., & Penman, S. (1988). Proc. Natl. Acad. Sci. USA 85, 121–125.

Fey, E.G., Wan, K.M., & Penman, S. (1984). J. Cell Biol. 98, 1973–1984.

Fields, A.P., Pettit, G.R., & May, W.S. (1988). J. Biol. Chem. 263, 8253–8260.

Fields, A.P., Tyler, G., Kraft, A.S., & May, W.S. (1990). J. Cell Sci. 96, 107–114.

Finlay, D.R., Newmeyer, D.D., Price, T.M., & Forbes, D.J. (1987). J. Cell Biol. 104, 189–200.

Finlay, D.R., & Forbes, D.J. (1990). Cell 60, 17–29.

Firzlaff, J.M., Galloway, D.A., Eisenman, R.N., & Lüscher, B. (1989). New Biologist 1, 44–53.

Fischer, U., & Lührmann, R. (1990). Science 249, 786–790.

Fischer-Fantuzzi, L., & Vesco, C. (1988). Mol. Cell Biol. 8, 5495–5503.

Flockhart, D.A., & Corbin, J.D. (1982). CRC Crit. Rev. Biochem. 12, 133–186.

Franke, W.W. (1974). Int. Rev. Cytol. Suppl. 4, 71–236.

Franke, W.W., Scheer, U., Krohne, G., & Jarasch, E. (1981). J. Cell Biol. 91, 39s–50s.

Friedman, D.L., & Ken, R. (1988). J. Biol. Chem. 263, 1103–1106.

Fuxe, K., Wikstrom, A.C., Okret, S., Agnati, L.F., Harfstrand, A., Yu, Z.Y., Granholm, L., Zoli, M., Vale, W., & Gustafsson, J.A. (1985). Endocrinology 117, 1803–1812.

Garcia-Bustos, J., Heitman, J., & Hall, M.N. (1991a). Biochim. Biophys. Acta, in press.

Garcia-Bustos, J.F., Wagner, P., & Hall, M.N. (1991b). Exp. Cell Res. 192, 213–219.

Gasc, J.M., Delahaye, F., & Baulieu, E.E. (1989). Exp. Cell Res. 181, 492–504.

Gelinas, C., & Temin, H.M. (1988). Oncogene 3, 349–355.

Georgatos, S.D., & Blobel, G. (1987a). J. Cell Biol. 105, 105–115.

Georgatos, S.D., & Blobel, G. (1987b). J. Cell Biol. 105, 117–125.

Gerace, L., & Burke, B. (1988). Ann. Rev. Cell Biol. 4, 335–374.

Ghosh, S., Gifford, A.M., Riviere, L.R., Tempst, P., Nolan, G.P., & Baltimore, D. (1990). Cell 62, 1019–1029.

Gilmore, T.D., & Temin, H.M. (1986). Cell 44, 791–800.

Gilmore, T.D., & Temin, H.M. (1988). J. Virol. 62, 703–714.

Gilmore, T.D. (1990). Cell 62, 841–843.

Goldfarb, D.S. (1989). Curr. Opin. Cell Biol. 1, 441–446.

Goldstein, J.L., Brown, M.S., Anderson, R.G.W., Russell, D.W., & Schneider, W.J. (1985). Ann. Rev. Cell Biol. 1, 1–39.

Goldstein, L. (1958). Exp. Cell Res. 15, 635–637.

Goldstein, L., & Ko, C. (1981). J. Cell Biol. 88, 516–525.

Gonzalez, G.A., & Montminy, M.R. (1989). Cell 59, 675–680.

Gorski, J., Toft, D.O., Shyamala, G., Smith, D., & Notides, A. (1968). Rec. Prog. Horm. Res. 24, 45–80.

Grässer, F.A., Scheidtmann, K.H., Tuazon, P.T., Traugh, J.A., & Walter, G. (1988). Virology 165, 13–22.

Greber, U.F., Senior, A., & Gerace, L. (1990). EMBO J. 9, 1495–1502.

Green, S., & Chambon, P. (1986). Nature 324, 615–617.

Greenspan, D., Palese, P., & Krystal, M. (1988). J. Virol. 62, 3020–3026.

Griffiths, G., Hollinshead, R., Hemmings, B.A., & Nigg, E.A. (1990). J. Cell Sci. 96, 691–703.

Grove, J.R., Price, D.J., Goodman, H.M., & Avruch, J. (1987). Science 238, 530–533.

Guddat, U., Bakken, A.H., & Pieler, T. (1990). Cell 60, 619–628.

Guiochon-Mantel, A., Loosfelt, H., Lescop, P., Sar, S., Atger, M., Perrot Applanat, M., & Milgrom, E. (1989). Cell 57, 1147–1154.

Hall, M.N., Craik, C., & Hiraoka, Y. (1990). Proc. Natl. Acad. Sci. USA 87, 6954–6958.

Halsey, D.L., Girard, P.R., Kuo, J.F., & Blackshear, P.J. (1987). J. Biol. Chem. 262, 2234–2243.

Hamm, J., Darzynkiewicz, E., Tahara, S.M., & Mattaj, I.W. (1990). Cell 62, 569–577.

Hamm, J., & Mattaj, I.W. (1990). Cell 63, 109–118.

Handler, J.D., Schimmer, B.P., Flynn, T.R., Szyf, M., & Seidman, J. G. (1988). J. Biol. Chem. 263, 13068–13073.

Hannink, M., & Donoghue, D.J. (1988). J. Cell Biol. 107, 287–298.

Hannink, M., & Temin, H.M. (1989). Mol. Cell. Biol. 9, 4323–4336.

Hornbeck, P., Huang, K.P., & Paul, W.E. (1988). Proc. Natl. Acad. Sci. USA 85, 2279–2283.

Hoshi, M., Nishida, E., & Sakai, H. (1988). J. Biol. Chem. 263, 5396–5401.

Howard, K.J., & Distelhorst, C.W. (1988). J. Biol. Chem. 263, 3474–3481.

Hunt, T. (1989). Cell 59, 949–951.

Imamura, T., Engleka, K., Zhan, X., Tokita, Y., Forough, R., Roeder, D., Jackson, A., Maier, J.A., Hla, T., & Maciag, T. (1990). Science 249, 1567–1570.

Ito, T., Tanaka, T., Yoshida, T., Onoda, K., Ohta, H., Hagiwara, M., Itoh, Y., Ogura, M., Saito, H., & Hidaka, H. (1988). J. Cell Biol. 107, 929–937.

Jaken, S. (1990). Curr. Opin. Cell Biol. 2, 192–197.

Jans, D.A., & Hemmings, B.A. (1988). Adv. Second Messenger Phosphoprotein Res. 21, 109–161.

Jensen, E.V., Suzuki, T., Kawashima, T., Stumpf, W.E., Jungblut, P.W., & DeSombre, E.R. (1968). Proc. Natl. Acad. Sci. USA 59, 632–638.

Jiang, L.W., & Schindler, M. (1988). J. Cell Biol. 106, 13–19.

Johnsson, A., Betsholtz, C., Heldin, C.H., & Westermark. B. (1985). Nature 317, 438–440.

Kahn, P., & Graf, T. (eds.) (1986). In: Oncogenes and Growth Control, Springer Verlag, New York.

Kalderon, D., Richardson, W.D., Markham, A.F., & Smith, A.E. (1984a). Nature 311, 33–38.

Kalderon, D., Roberts, B.L., Richardson, W.D., & Smith, A.E. (1984b). Cell 39, 499–509.

Kalinich, J.F., & Douglas, M.G. (1989). J. Biol. Chem. 264, 17979–17989.

Karin, M. (1989). Trends Biochem. Sci. 5, 65–67.

Keating, M.T., & Williams, L.T. (1988). Science 239, 914–916.

Khanna-Gupta, A., & Ware, V.C. (1989). Proc. Natl. Acad. Sci. USA 86, 1791–1795.

Kieran, M., Blank, V., Logeat, F., Vandekerckhove, J., Lottspeich, F., Le Bail, O., Urban, M.B., Kourilsky, P., Baeuerle, P.A., & Israel, A. (1990). Cell 62, 1007–1018.

Kikkawa, U., & Nishizuka, Y. (1986). Ann. Rev. Cell Biol. 2, 149–178.

King, W.J., & Greene, G.L. (1984). Nature 307, 745–747.

Kiss, Z., Deli, E., & Kuo, J.F. (1988). FEBS. Lett. 231, 41–46.

Kleinschmidt, J.A., & Seiter, A. (1988). EMBO J. 7, 1605–1614.

Kraft, A., & Anderson, W.B. (1983). Nature 301, 621–623.

Krebs, E.G., Eisenman, R.N., Kuenzel, E.A., Litchfield, D.W., Lozeman, F.J., Luscher, B., & Sommercorn, J. (1988). Cold Spring Harb. Symp. Quant. Biol. 53, 77–84.

Lanford, R.E., & Butel, J.S. (1984). Cell 37, 801–813.

Lanford, R.E., Kanda, P., & Kennedy, R.C. (1986). Cell 46, 575–582.

Leach, K.L., Powers, E.A., Ruff, V.A., Jaken, S., & Kaufmann, S. (1989). J. Cell Biol. 109, 685–695.

Lee, B.A., Maher, D.W., Hannink, M., & Donoghue, D.J. (1987). Mol. Cell Biol. 7, 3527–3537.

Lee, W.C., & Melese, T. (1989). Proc. Natl. Acad. Sci. USA 86, 8808–8812.

Lenardo, M.J., & Baltimore, D. (1989). Cell 58, 227–229.

Lewis, M.J., & Pelham, H.R. (1985). EMBO J. 4, 3137–3143.

Li, R.H., & Thomas, J.O. (1989). J. Cell Biol. 109, 2623–2632.

Lohmann, S.M., & Walter, U. (1984). Adv. Cyclic Nucleotide Protein Phosphorylation Res. 18, 63–117.

Lührmann, R. (1988). In: Structure and Function of Major and Minor Small Nuclear Ribonucleoprotein Particles (Birnstiel, M., ed.). pp. 71–99, Springer Verlag, Heidelberg.

Lüscher, B., Kuenzel, E.A., Krebs, E.G., & Eisenman, R.N. (1989). EMBO J. 8, 1111–1119.

Lüscher, B., Christenson, E., Litchfield, D.W., Krebs, E.G., & Eisenman, R.N. (1990). Nature 344, 517–522.

Madsen, P., Nielsen, S., & Celis, J.E. (1986). J. Cell Biol. 103, 2083–2089.

Maher, D.W., Lee, B.A., & Donoghue, D.J. (1989). Mol. Cell Biol. 9, 2251–2253.

Manak, J.R., de Bisschop, N., Kris, R.M., & Prywes, R. (1990). Genes. Dev. 4, 955–967.

Mandell, R.B., & Feldherr, C.M. (1990). J. Cell Biol. 111, 1775–1783.

Masmoudi, A., Labourdette, G., Mersel, M., Huang, F.L., Huang, K.-P., Vincendon, G., & Malviya, A.N. (1989). J. Biol. Chem. 264, 1172–1179.

Mattaj, I.W., & De Robertis, E.M. (1985). Cell 40, 111–118.

Maurer, R.A. (1989). J. Biol. Chem. 264, 6870–6873.

McMorrow, I., Souter, W.E., Plopper, G., & Burke, B. (1990). J. Cell Biol. 110, 1513–1523.

Meek, D.W., Simon, S., Kikkawa, U., & Eckhart, W. (1990). EMBO J. 9, 3253–3260.

Meier, U.T., & Blobel, G. (1990). J. Cell Biol. 111, 2235–2245.

Mellon, P.L., Clegg, C.H., Correll, L.A., & McKnight, G.S. (1989). Proc. Natl. Acad. Sci. USA 86, 4887–4891.

Miller, D.S. (1988). Science 240, 506–509.

Morin, N., Delsert, C., & Klessig, D.F. (1989). Mol. Cell Biol. 9, 4372–4380.

Münstermann, U., Fritz, G., Seitz, G., Yiping, L., Schneider, H.R., & Issinger, O.-G. (1990). Eur. J. Biochem. 189, 251–257.

Nakagawa, J., von der, Ahe, D., Pearson, D., Hemmings, B.A., Shibahara, S., & Nagamine, Y. (1988). J. Biol. Chem. 263, 2460–2468.

Nasmyth, K., Adolf, G., Lydall, D., & Seddon, A. (1990). Cell 62, 631–647.

Nath, S.T., & Nayak, D.P. (1990). Mol. Cell Biol. 10, 4139–4145.

Nehrbass, U., Kern, H., Mutvei, A., Horstmann, H., Marshallsay, B., & Hurt, E.C. (1990). Cell 61, 979–989.

Newmeyer, D.D., Finlay, D.R., & Forbes, D.J. (1986a). J. Cell Biol. 103, 2091–2102.

Newmeyer, D.D., Lucocq, J.M., Bürglin, T.R., & De Robertis, E.M. (1986b). EMBO J. 5, 501–510.

Newmeyer, D.D., & Forbes, D.J. (1988). Cell 52, 641–653.

Newmeyer, D.D., & Forbes, D.J. (1990). J. Cell Biol. 110, 547–557.

Nigg, E.A. (1988). Int. Rev. Cytol. 110, 27–92.

Nigg, E.A. (1989). Curr. Opinions Cell. Biol. 1, 435–440.

Nigg, E.A. (1990). Adv. Cancer Res. 55, 271–310.

Nigg, E.A., Eppenberger, H.M., Jans, D.A., Hemmings, B.A., & Hilz, H. (1988). In: Mechanisms of Control of Gene Expression, UCLA Symposia on Molecular and Cellular Biology, New Series, (Cullen, B., Gage, L.P., Siddiqui, M.A.Q., Stalka, A.M., & Weissbach, H., eds.). pp. 169–178, Alan R. Liss, New York.

Nigg, E.A., Hilz, H., Eppenberger, H.M., & Dutly, F. (1985a). EMBO J. 4, 2801–2806.

Nigg, E.A., Schaefer, G., Hilz, H., & Eppenberger, H.M. (1985b). Cell 41, 1039–1051.

Nishizuka, Y. (1988). Nature 334, 661–665.

Paine, P.L., & Horowitz, S.B. (1980). In: Cell Biology: A Comprehensive Treatise (Prescott, D.M., & Goldstein, L., eds.), pp. 299–338, Academic Press, New York.

Pearson, D., Nigg, E.A., Nagamine, Y., Jans, D.A., & Hemmings, B.A. (1991). Exp. Cell Res. 192, 315–318.

Perro-Applanat, M., Logeat, F., Groyer Picard, M.T., & Milgrom, E. (1985). Endocrinology. 116, 1473–1484.

Peter, M., Nakagawa, J., Dorée, M., Labbé, J.C., & Nigg, E.A. (1990). Cell 61, 591–602.

Peters, R. (1986). Biochim. Biophys. Acta 864, 305–359.

Picard, D., & Yamamoto, K.R. (1987). EMBO J. 6, 3333–3340.

Picard, D., Kumart, V., Chambon, P., & Yamamoto, K.R. (1990a). Cell Regulation 1, 291–299.

Picard, D., Khursheed, B., Garabedian, M.J., Fortin, M.G., Lindquist, S., & Yamamoto, K.R. (1990b). Nature 348, 166–168.

Pines, J., & Hunter, T. (1991). Cell, submitted.

Pinna, L.A. (1990). Biochim. Biophys. Acta 1054, 267–284.

Ransone, L.J., & Verma, I.M. (1990). Ann. Rev. Cell Biol. 6, 539–557.

Ray, L.B., & Sturgill, T.W. (1988). Proc. Natl. Acad. Sci. USA 85, 3753–3757.

Reddy, E.P., Skalka, A.M., & Curran, T. (1989). In: The Oncogene Handbook, Elsevier, Amsterdam.

Reddy, R., & Busch, H. (1988). In: Structure and Function of Major and Minor Small Nuclear Ribonucleoprotein Particles (Birnstiel, M., ed.). pp. 1–37, Springer Verlag, Heidelberg.

Reichelt, R., Holzenburg, A., Buhle, E.L. Jr., Jarnik, M., Engel, A., & Aebi, U. (1990). J. Cell Biol. 110, 883–894.

Riabowol, K.T., Fink, J.S., Gilman, M.Z., Walsh, D.A., Goodman, R.H., & Feramisco, J.R. (1988). Nature 336, 83–86.

Richardson, W.D., Roberts, B.L., & Smith, A.E. (1986). Cell 44, 77–85.

Richardson, W.D., Mills, A.D., Dilworth, S.M., Laskey, R.A., & Dingwall, C. (1988). Cell 52, 655–664.

Rihs, H.-P., & Peters, R. (1989). EMBO J. 8, 1479–1484.

Rihs, H.-P., Jans, D.A., Fan, H., & Peters, R. (1991). EMBO J., in press.

Roberts, B. (1989). Biochim. Biophys. Acta 1008, 263–280.

Roberts, B.L., Richardson, W.D., & Smith, A.E. (1987). Cell 50, 465–475.

Roesler, W.J., Vandenbark, G.R., & Hanson, R.W. (1988). J. Biol. Chem. 263, 9063–9066.

Rogue, P., Labourdette, G., Masmoudi, A., Yoshida, Y., Huang, F.L., Huang, K.P., Zwiller, J., Vincendon, G., & Malviya, A.N. (1990). J. Biol. Chem. 265, 4161–4165.

Rossomando, A.J., Payne, D.M., Weber, M.J., & Sturgill, T.W. (1989). Proc. Natl. Acad. Sci. USA 86, 6940–6943.

Roth, S., Stein, D., & Nüsslein-Volhard, C. (1989). Cell 59, 1189–1202.

Roux, P., Blanchard, J.M., Fernandez, A., Lamb, N., Jeanteur, P., & Piechaczyk, M. (1990). Cell 63, 341–351.

Rozengurt, E. (1986). Science 234, 161–166.

Rushlow, C.A., Han, K., Manley, J.L., & Levine, M. (1989). Cell 59, 1165–1177.

Sadler, I., Chiang, A., Kurihara, T., Rothblatt, J., Way, J., & Silver, P. (1989). J. Cell Biol. 109, 2665–2675.

Sanchez, E.R. (1990). J. Biol. Chem. 265, 22067–22070.

Sanchez, E.R., Toft, D.O., Schlesinger, M.J., & Pratt, W.B. (1985). J. Biol. Chem. 260, 12398–12401.

Scheer, U., & Benavente, R. (1990). Bioessays 12, 14–21.

Scheer, U., Dabauvalle, M.C., Merkert, H., & Benavente, R. (1988). Cell Biol. Int. Rep. 12, 669–689.

Schindler, M., & Jiang, L.W. (1987). J. Cell Biol. 104, 849–853.

Sen, R., & Baltimore, D. (1986a). Cell 46, 705–716.

Sen, R., & Baltimore, D. (1986b). Cell 47, 921–928.

Servetnick, M.D., & Wilt, F.H. (1987). Dev. Biol. 123, 231–244.

Silver, P., & Goodson, H. (1989). Crit. Rev. Biochem. Mol. Biol. 24, 419–435.

Silver, P., Sadler, I., & Osborne, M.A. (1989). J. Cell Biol. 109, 983–989.

Skoglund, U., Andersson, K., Bjorkroth, B., Lamb, M.M., & Daneholt, B. (1983). Cell 34, 847–855.

Starr, C.M., D'Onofrio, M., Park, M.K., & Hanover, J.A. (1990). J. Cell Biol. 110, 1861–1871.

Steitz, J.A., Black, D.L., Gerke, V., Parker, K.A., Krämer, A., Freudewey, D., & Keller, W. (1988). In: Structure and Function of Major and Minor Small Nuclear Ribonucleoprotein Particles (Birnstiel, M., ed.), pp. 115–154, Springer Verlag, Heidelberg.

Stevens, B.J., & Swift, H. (1966). J. Cell Biol. 31, 55–77.

Steward, R. (1989). Cell 59, 1179–1188.

Stewart, M., Whytock, S., & Mills, A.D. (1990). J. Mol. Biol. 213, 575–582.

Sturgill, T.W., Ray, L.B., Erikson, E., & Maller, J.L. (1988). Nature 334, 715–718.

Taylor, S.S., Buechler, J.A., & Yonemoto, W. (1990). Ann. Rev. Biochem. 59, 971–1005.

Thomas, T.P., Talwar, H.S., & Anderson, W.B. (1988). Cancer Res. 48, 1910–1919.

Tobian, J.A., Drinkard, L., & Zasloff, M. (1985). Cell 43, 915–922.

Tsuda, T., & Alexander, R.W. (1990). J. Biol. Chem. 265, 1165–1170.

Underwood, M.R., & Fried, H.M. (1990). EMBO J. 9, 91–99.

Unwin, P.N.T., & Milligan, R.A. (1982). J. Cell Biol. 93, 63–75.

Van Etten, R.A., Jackson, P., & Baltimore, D. (1989). Cell 58, 669–678.

Velazquez, J.M., & Lindquist, S. (1984). Cell 36, 655–662.

Vogt, P.K., & Bos, T.J. (1989). Trends Biochem. Sci. 14, 172–175.

Warner, J.R. (1990). Curr. Opin. Cell Biol. 2, 521–527.

Weiel, J.E., Ahn, N.G., Seger, R., & Krebs, E.G. (1990). Adv. Second Messenger Phosphoprotein Res. 24, 182–195.

Weinberg, R.A. (1985). Science 230, 770–776.

Welch, W.J., & Feramisco, J.R. (1984). J. Biol. Chem. 259, 4501–4513.

Welsh, J.D., Swimmer, C., Cocke, T., & Shenk, T. (1986). Mol. Cell Biol. 6, 2207–2212.

Welshons, W.V., Lieberman, M.E., & Gorski, J. (1984). Nature 307, 747–749.

Wikström, A.C., Bakke, O., Okret, S., Bronnegard, M., & Gustafsson, J.A. (1987). Endocrinology 120, 1232–1242.

Wozniak, R.W., Bartnik, E., & Blobel, G. (1989). J. Cell Biol. 108, 2083–2092.

Yamamoto, K.K., Gonzalez, G.A., Biggs, W.H., & Montminy, M.R. (1988). Nature 334, 494–498.

Yamamoto, K.R., Godowski, P.J., & Picard, D. (1988). Cold Spring Harb. Symp. Quant. Biol. 53, 803–811.

Yamasaki, L., Kanda, P., & Lanford, R.E. (1989). Mol. Cell. Biol. 9, 3028–3036.

Yarden, Y., & Ullrich, A. (1988). Ann. Rev. Biochem. 57, 443–478.

Zasloff, M. (1983). Proc. Natl. Acad. Sci. USA 80, 6436–6440.

Zeller, R., Nyffenegger, T., & De Robertis, E.M. (1983). Cell 32, 425–434.

Zieve, G.W., Sauterer, R.A., & Feeny, R.J. (1988). J. Mol. Biol. 199, 259–267.

Zimmer, F.J., Dreyer, C., & Hausen, P. (1988). J. Cell Biol. 106, 1435–1444.

Chapter 11

DNA Damage and Its Repair

ANDREW R. COLLINS

Principles of Medical Biology, Volume 5
Molecular and Cellular Genetics, pages 229–254.
Copyright © 1996 by JAI Press Inc.
All rights of reproduction in any form reserved.
ISBN: 1-55938-809-9

INTRODUCTION:
THE SIGNIFICANCE OF DNA DAMAGE

The stability of the genome is of crucial importance for the proper functioning of the cell, for the well-being of the organism, and for its successful reproduction (Figure 1). This has been understood ever since the discovery of the structure of DNA and the nature of the genetic code. Even a minor change in one of the four bases in DNA can alter its coding properties, resulting in a substituted amino acid in the peptide sequence specified by the DNA.

Unless DNA damage is corrected before the DNA is replicated, it may be fixed as permanent DNA alterations or mutations. Most of these simply contribute to the harmless variability or polymorphism of the genetic material; only if they are in coding regions of DNA are they expressed as an alteration in a protein. Even these changes are not necessarily harmful; in fact, they constitute the natural variations that are the basis for selection and evolution. However, the minority of mutations that affect the nature or amount of a particular protein in a deleterious way can result in heritable disorders, birth defects, or abortion if they occur in germ cells or embryo. In the case of somatic cells, mutations in genes associated with regulation of cell growth can cause tumorigenesis and cancer, and mutations in other genes probably contribute to the general loss of cellular function with age. It therefore comes as no surprise that the cell has efficient repair mechanisms to deal with most DNA damage.

Mutagenesis is fully dealt with in chapter 12. Here we will concentrate on the kinds of DNA lesions, the agents that cause them, and on the various cellular responses to DNA damage.

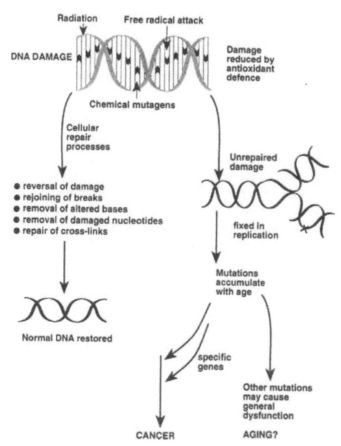

Radiation   Free radical attack

DNA DAMAGE

Damage
reduced by
antioxidant
defence

Chemical mutagens

Cellular
repair
processes

Unrepaired
damage

● reversal of damage
● rejoining of breaks
● removal of altered bases
● removal of damaged nucleotides
● repair of cross-links

fixed in
replication

Mutations
accumulate
with age

Normal DNA restored

specific
genes

Other mutations
may cause
general
dysfunction

CANCER   AGING?

Figure 1. The biological consequences of DNA damage. In this simplified diagram, the consequences of germ cell mutations are not considered.

HOW DOES DNA DAMAGE ARISE?

Errors in Replication

The accuracy with which the DNA sequence is copied during replication depends on the fidelity of the DNA polymerases in selecting the correct deoxyribonucleotide to insert against its complementary partner in the parental DNA template strand. Rates of misincorporation in *in vitro* systems are low but significant; DNA polymerase α, for instance, makes one error in 30,000 correct incorporations. However, the error rate estimated from the number of mutations arising spontaneously in proliferating cultured cells is far below the level expected from the misincorporation rate, and so it seems that newly replicated DNA is scanned for mispairings which are excised and replaced. Replication errors are increased if the balance of the four deoxyribonucleotides in the cell's DNA precursor pool is disturbed; if one precursor is present in great excess it can overwhelm the normally fastidious polymerase and be incorporated in place of the correct nucleotide.

Spontaneous Damage

DNA is actually surprisingly unstable. At 37° C, thousands of purine bases (and hundreds of pyrimidines) are lost spontaneously in every human cell each day, leaving apurinic/apyrimidinic (or AP) sites incapable of determining correct base insertion on the nascent complementary strand during replication.

Deamination of cytosine to uracil also occurs spontaneously. Uracil, like thymine, pairs with adenine, and so a change from a CG base-pair to TA will be seen following the next round of replication unless the uracil is replaced with

Radiolysis of water

$$H_2O \xrightarrow[\text{radiation}]{\text{ionising}} e^-_{aq}, \cdot OH, \cdot H, H_2, H_2O_2$$

$$e^-_{aq} + O_2 \rightarrow O^-_2$$

$$\cdot H + O_2 \rightarrow HO_2^- \rightleftharpoons H^+ + O_2^-$$

Successive addition of electrons to oxygen during normal metabolism

$$O_2 \xrightarrow{e^-} O^-_2 \xrightarrow{e^-} H_2O_2$$

Fenton reaction involving transition metal catalysis

$$Fe^{2+} + H_2O_2 \rightarrow Fe^{3+} + \cdot OH + OH^-$$

$\cdot OH$ hydroxyl radical	e^-_{aq} hydrated electron
O^-_2 superoxide radical	HO_2^- hydroperoxyl radical

Figure 2. Free radicals; the origin of reactive oxygen species.

cytosine. Deamination of 5-methylcytosine (which occurs in certain CpG sequences as a signal preventing transcription) produces mispaired thymine.

Endogenous Damage

A major cause of DNA damage is thought to be the occurrence of free radicals as by-products of normal aerobic respiration (Figure 2). Free radicals are molecules, or molecular fragments, with unpaired electrons, which renders them reactive. The hydroxyl radical, ·OH, is capable of reacting with DNA to produce breaks as well as many oxidized forms of the bases, such as 8-OH-guanine (Figure 3).

Inflammatory leukocytes (neutrophils and macrophages), present at sites of infection and tissue damage, respond to the presence of foreign organisms with a burst of active oxygen released into the extracellular environment. As well as destroying the target, the active oxygen may initiate chain reactions in neighboring host cells that result in damage to their DNA. A further factor is cell proliferation—part of the normal restorative response to tissue damage; during the DNA replication induced in these proliferating cells, damage can become fixed as mutations. This promotional effect of cell proliferation probably contributes to the well-established association between chronic inflammation and the subsequent occurrence of tumors at the same site. Another twist to this story of free radicals comes with the observation that chemicals that promote tumorigenesis, such as phorbol myristate acetate (TPA), are also capable of inducing a respiratory burst in macrophages or other cells, with a concomitant release of active oxygen.

Exogenous Agents

Ionizing Radiation

X and γ rays cause some DNA damage directly, by ionization within the DNA molecule itself; additional damage arises from attack by free radicals, produced by ionization in the water surrounding the double helix (Figure 2). Ionizing radiation damage takes several forms:

1. Single-strand breaks (SSBs) result from breakage of phosphodiester linkages or from degradation of deoxyribose via radical formation. Most SSBs are in fact gaps, produced by destruction of at least one nucleotide.
2. Double-strand breaks (DSBs) are induced at a frequency about one-tenth that of SSBs. They are most likely to occur as a result of a single energy deposition event, causing breaks by direct ionization within one or both strands, and/or indirect damage via free radicals in solution. A DSB can be regarded as two SSBs in close proximity; if exactly opposed, a blunt-ended

DSB will result, but staggered breaks (i.e., where the two strand breaks are several nucleotides apart) are also effectively DSBs.

3. Damaged bases are induced by direct ionization or by free radical attack, in similar amounts to SSBs. The variety of oxidized bases is enormous; over 20 derivatives are known for thymine alone.

Ultraviolet Radiation

UV light is the most commonly studied DNA-damaging agent, being easy to administer in well-defined doses, producing a limited range of lesions, and, as the damaging component of sunlight, having universal biological relevance. UV is an intrastrand cross-linking agent. Adjacent pyrimidines are covalently linked to form cyclobutane pyrimidine dimers: T̂T, ĈT, and ĈC, of which T̂T is the most abundant (Figure 3). At a lower frequency than the cyclobutane dimers, pyrimidine (6–4) pyrimidone lesions, or (6–4) photoproducts are formed; the TC species is more common than CC or TT (the CT product is apparently not formed). Dimers cause loss of base pairing and local denaturation and distortion of the double helix.

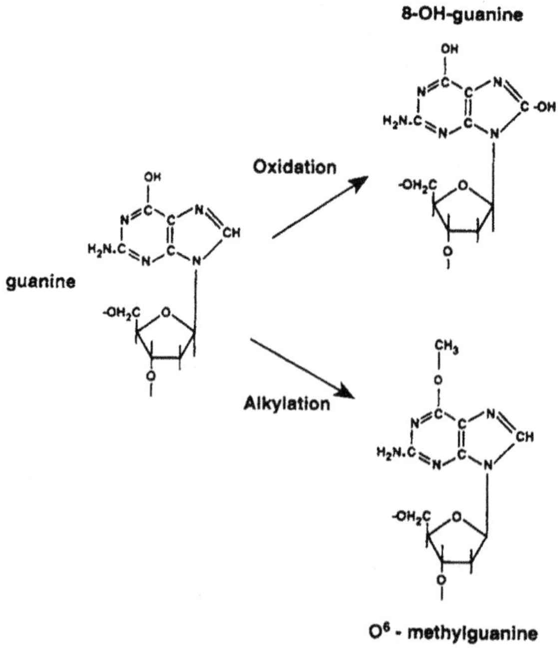

Figure 3. Some examples of DNA damage.

Alkylating Agents

There are many electrophilic organic compounds with alkyl groups that can react with nucleophilic centers in DNA. Alkyl groups are small but the changes that they make to the DNA can be very significant, since they can affect hydrogen bonding and alter base pairing. The sites in DNA most susceptible to alkylation are the ring nitrogens of the bases, in particular the N^7 of guanine and the N^3 of adenine; O^6 of guanine is also commonly alkylated (Figure 3), as are oxygens in the phosphodi-esters of the backbone (converted into alkylphosphotriesters).

Sulphur mustard and nitrogen mustards are probably the most notorious alkyl-ating agents, having been developed for chemical warfare, but others occur in everyday life (e.g., epoxides, in adhesives) and several have been used in cancer chemotherapy (e.g., the mustards, cyclophosphamide, and *N*-methyl-*N*-nitrosourea).

Bulky Adducts

Most of the chemicals described so far make small modifications to the DNA, such as the addition of an ethyl or methyl group. There are others, though, that form large adducts with DNA and cause disruption/distortion of the normal double helical conformation. The formation of a bulky lesion is the only common feature

psoralen cross-link

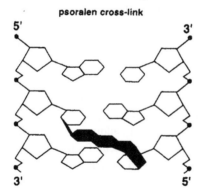

Cyclobutane pyrimidine dimer

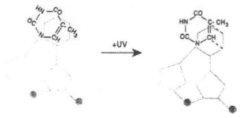

Figure 3. (Continued)

of these otherwise disparate agents. The cyclobutane pyrimidine dimer—or the distortion of the helix that it causes—can be regarded as a bulky lesion. Other agents in this group include the following, which most commonly attack guanine:

1. *N*-2-acetyl-2-aminofluorene (AAF), an aromatic amine, activated metabolically in the liver.
2. Mitomycin C, an antibiotic from *Streptomyces caespitosus*.
3. Aflatoxin B_1, a potent mycotoxin activated in the liver.
4. Benzo(a)pyrene, a product of partially burned hydrocarbons (smoke, soot, etc.), converted to an active form in the liver.
5. 4-nitroquinoline-1-oxide, reduced to 4-hydroxyaminoquinoline-1-oxide.
6. Furocoumarins, such as psoralens and angelicin (common in umbelliferous plants) have three aromatic rings, and can intercalate in DNA; on absorption of long-wavelength UV light, reactive sites form cyclobutane-type links with pyrimidines.

Cross-Linking Agents

The nitrogen and sulphur mustards are, in fact, bifunctional alkylating agents. If both the reactive groups of one of these molecules are involved in alkylation reactions within the same DNA molecule, interstrand (if the reactions occur in opposite strands) or intrastrand (within the same strand) cross-links result. Various other agents form cross-links, generally in a reaction that follows the initial production of a monoadduct. They include mitomycin C, *cis*-platinum(II)diammine dichloride (*cis*-Pt), and certain of the furocoumarins (Figure 3). Interstrand cross-links are of particular interest because they prevent DNA strand separation, profoundly disrupting DNA replication. In addition to cross-links within the DNA, covalent bonding of DNA to protein can occur. Protein-DNA cross-links occur after ultraviolet irradiation or treatment with various other agents.

Free Radicals

Already described under the section entitled "Endogenous Damage," free radicals are also implicated in damage by exogenous chemicals. Bleomycin, an antibiotic (from *Streptomyces verticillus*), is a DNA-damaging agent with enzymelike properties, with chelated iron at its active center. Fe^{2+}-bleomycin reduces molecular oxygen to form activated bleomycin with iron in the Fe^{2+} state plus $\cdot OH$, O_2^-, and H_2O_2; cellular reduction restores Fe^{2+} and further cycles of ferrous oxidase activity can occur. Many chemicals go through a free radical-mediated activation before they are able to form adducts with DNA.

Metabolic Activation of Carcinogens

The liver is responsible for detoxifying foreign chemicals in the body. It does this by means of the microsomal cytochrome P450 metalloenzyme system, which carries out a range of oxidative reactions with free radicals as intermediates. The products are typically phenols, diols, epoxides, and hydroxylamines, which are more soluble than the precursor (and therefore more readily excreted), but at the same time more reactive. Unfortunately, certain carcinogens, typically those in the classes of polycyclic aromatic hydrocarbons, *N*-nitrosamines, and aromatic amines, are converted to the active form via P450 oxidation. Variations in the spectrum of P450 activities occur among individuals in a population, and may account for differences in susceptibility to these carcinogens.

Protection Against DNA Damage

Obviously, the amount of damage suffered by DNA will be affected by physical factors; UV irradiation, for example, causes damage in the skin, but very little UV penetrates to deeper tissues. In the case of chemicals, cell membranes may act as a barrier, and the tight packing of DNA in the nucleus limits its accessibility to some chemical damaging agents.

Chemical protection against free radical attack is provided by antioxidants. Dietary constituents, such as vitamin C, vitamin E, carotenoids, and flavonoids, scavenge free radicals and break free radical chain reactions. There are also cellular antioxidant defenses. Glutathione is a tripeptide with a cysteine sulfhydryl group, present at high concentration in the nucleus, which readily reacts with H_2O_2 and organic peroxides. Finally, enzymes such as catalase and superoxide dismutase break down intracellular active oxygen species (peroxides and superoxide, $O_2^{\cdot}$, respectively).

DNA REPAIR PATHWAYS

Cells, whether bacterial, plant, or animal, show an impressive vigilance in dealing with DNA damage. In a few cases, damage is simply reversed. Photoreactivation of UV-induced pyrimidine dimers is carried out by an enzyme, photolyase, using the energy of longer wavelength light to break the bonds of the cyclobutane ring. Another enzyme, the O^6-alkyltransferase, accepts methyl or ethyl groups (from the O^6 position of guanine) onto a cysteine sulfhydryl group in its active site; this reaction is irreversible, and the alkyltransferase is described as a suicide enzyme. True repair pathways involve the excision of damaged nucleotides and their replacement; they therefore involve a small but significant amount of DNA synthesis.

Base Excision Repair

Typically, this pathway (Figure 4) deals with relatively small changes to bases, such as alkylation or oxidation. The altered base is removed by a glycosylase; specific forms of this enzyme occur for different kinds of damage, for example uracil-DNA glycosylase, which removes the product of cytosine deamination, and 3-methyladenine-DNA glycosylase, active against 3-methyladenine, 7-methyl-guanine, and 3-methylguanine. Loss of the base leaves an AP site, and a second enzyme, AP endonuclease, is responsible for cutting the phosphodiester backbone. Further enzyme action trims off the residual baseless sugar and a small number of

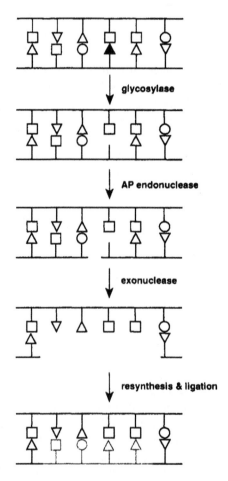

Figure 4. Base excision repair.

adjacent nucleotides, creating a gap which is then filled in by DNA polymerase β, which inserts the appropriate nucleotides to complement the undamaged sequence on the other strand. The final step is ligation of the repair patch into the preexisting DNA. In the case of oxidative damage, enzymes with rather broad specificity have been identified. They are known as redoxyendonucleases, and combine glycosylase and AP endonuclease activities on the same protein molecule.

Nucleotide Excision Repair

The bulky adducts, and UV-induced pyrimidine dimers, are repaired by an alternative excision pathway (Figure 5). A distortion of the double helix seems to be involved in recognition by an excinuclease which cuts the DNA on both sides of the adduct and removes an oligonucleotide. The gap to be filled is larger than is the case with base excision, and a different enzyme, polymerase δ, is involved.

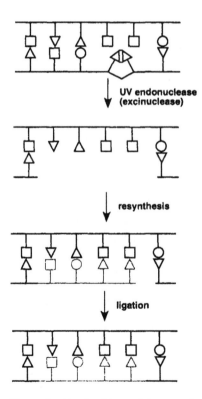

Figure 5. Nucleotide excision repair.

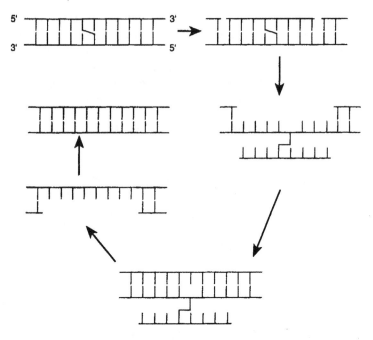

Figure 6. A model of cross-link repair (based on van Houten et al., 1986).

Cross-Link Repair

In bacteria, cross-links are apparently repaired by the excinuclease responsible for bulky adduct repair (Figure 6). Once the enzyme has nicked on each side of one half of the cross-link, the damage can be regarded as a rather large adduct attached to the other strand. If this is then excised, and a repair patch inserted, the lesion is removed, but with the potential loss of correct base-pairing at the site of the cross-link. A way around this, in bacteria, is for the DNA to undergo some form of recombination with a homologous duplex, so that a complementary template DNA is provided for patch synthesis. Details of the process in eukaryotes are not yet known, though it has some features in common with nucleotide excision repair.

Single-Strand Break Repair

SSBs, induced by ionization or free radicals, are repaired extremely quickly: they have a half-life of a few minutes, compared with nucleotide and base excision repair which takes hours or even days for the cell to process the damage. Not all SSBs are clean breaks requiring merely religation, and so the speed is unlikely to be because the ligase reaction is a very simple one. Breaks in the continuity of DNA seem to signal the need for a rapid response, perhaps because they are potentially capable of disrupting nuclear structure and function via a release of supercoiling.

Double-Strand Break Repair

DSBs could easily lead to the degradation of DNA, by nuclease digestion from the exposed ends. Loss of sequence, as in the case of cross-link repair, would be a serious consequence. Surprisingly, perhaps, DSBs are repaired slowly, with a half-life measured in hours. The mechanism is not well understood.

Mismatch Correction

Mismatches include mispairings, where opposed bases in the double helix are not complementary, as well as deletions or additions of a small number of nucleotides, resulting from faulty replication. Since the bases themselves are normal, there is no indication within the mismatch as to which strand should be repaired to restore the correct sequence. However, for a short time after DNA replication, the daughter strand is distinguishable because it lacks the methylation signals (related to transcription status) that are present, at specific nucleotide sequences, on the parental strand. In bacteria, and probably also in eukaryotes, mismatch correction depends on this differentiation of the two strands. The process is initiated by specific repair proteins binding to the mismatched bases, thereby activating an endonuclease which is able to nick DNA at a nearby methylation signal sequence, on the unmethylated, daughter strand. Exonucleolytic digestion of DNA then proceeds up to the mismatch itself.

Repair of Active Genes

It is found that after UV irradiation of mammalian cells, genes that are actively transcribing are repaired more quickly than the bulk of DNA. The phenomenon is most marked in rodent cells, in which repair of the genome overall is inefficient, but it also applies to human cells. Furthermore, the selective repair is seen only in the transcribed strand of the active gene, which suggests that blocked transcription—rather than, say, a more accessible chromatin conformation in active DNA—is the signal for repair to take place.

Unrepaired DNA Damage Detected *In Vivo*

Since repair is not instantaneous, the effect of exposure to environmental or endogenous damaging agents should be apparent as a steady-state level of lesions, reflecting both the rate of input and the rate of removal of damage. Strand breaks are present at barely detectable levels in, for example, normal human lymphocytes, whereas oxidized bases are several times more frequent. Pyrimidine dimers can be detected in sun-exposed skin. Alkylation adducts have been demonstrated, perhaps resulting from such agents as nitrosamines in food. The detection of bulky adducts correlates with known exposure, occupational or environmental, to specific agents; adducts from tobacco smoke are found in the DNA of smokers. Aflatoxin B1 adducts occur at measurable levels in the DNA of people living in regions where fungally contaminated peanuts are consumed. Only recently has it become feasible to screen for these lesions at the level of individuals or populations; information is likely to accumulate rapidly over the next few years.

MEASURING DNA DAMAGE AND REPAIR

Lesion Detection: Direct Assays

In principle, the ideal assay for DNA damage and repair is one in which specific lesions can be detected and measured directly, and their removal by a repair process followed with time. In the case of agents that introduce a new chemical group to the DNA, appropriate labeling of the agent with radioisotope will produce labeled adducts in the DNA. After enzymic or chemical hydrolysis of the DNA, and chromatographic separation, the modified bases are identified and relative yields of different products can be calculated. This approach has been used extensively with alkylating agents, and agents forming bulky adducts with DNA can also be applied in labeled form. Because radioactive groups are introduced to otherwise unlabeled DNA, these assays can be very sensitive. A different approach is required to measure modifications of the DNA which do not involve introduction of a new group, such as the damaged bases produced by ionizing radiation and active oxygen, or the dimerized pyrimidines that are the predominant damage following ultraviolet irradiation. It is possible to prelabel the DNA (normally by incubating cells with ^{3}H-thymidine) and subsequently to separate the altered constituent from the DNA digestion products by chromatography. In a few cases, high performance liquid chromatography (HPLC), with its high resolution and sensitivity, has allowed the detection of altered bases without the need for prelabeling.

Lesion Detection: Indirect Assays

To circumvent the sensitivity problem, indirect assays have been developed, making use of antibodies or enzymes specific to particular lesions. Antibodies that recognize ultraviolet-induced damage, thymine glycol, alkylation products, and various bulky adducts are available. Measurement of binding of antibody to DNA can be done by radioimmunoassay or immunoprecipitation. Antibodies tagged with radioactive, fluorescent, or enzyme-linked probes can also be used to study the distribution of lesions by microscopic examination.

Enzymes have been isolated from bacteria or phage-infected bacteria that have specific endonucleolytic activity against one kind of lesion. UV endonuclease (from *Micrococcus luteus* or T4 phage) recognizes pyrimidine dimers in isolated DNA and introduces strand breaks at dimer sites. Measuring the decline in UV endonuclease-sensitive sites in the DNA of cells incubated for increasing times after irradiation allows an estimate of cellular repair activity. Endonucleases specific for other lesions can be used in this way, too.

^{32}P-Postlabeling

This is a sensitive and general method for detecting DNA adducts. The DNA is totally digested to deoxyribonucleoside-3′-monophosphates (including some that are damaged), and these are then phosphorylated in the 5′ position using ^{32}P-ATP. Nucleotides can be separated by thin layer chromatography, and detected by autoradiography of the ^{32}P; the presence of an adduct affects mobility, and so DNA damage is reflected in the appearance of extra spots on the chromatogram. Because of the high sensitivity of detection of ^{32}P-labeled material, very low levels of damage can be seen, down to one adduct in 10^7–10^8 normal nucleotides. However, it is generally not easy to identify specific adducts without prior knowledge of the type of damage present, and not all adducts give good discrimination from the normal nucleotide.

Measuring Single-Strand Breaks

SSBs are common as lesions; they also occur as intermediates in the cellular repair process, and as the product of the enzymic assays described in the section on Indirect Assays above. They are probably the easiest forms of DNA damage to measure, and methods based on different principles are available. Generally, DSBs are detected too, but as a relatively infrequent lesion their contribution can be safely ignored.

Alkaline elution depends on the ability of DNA to pass through a filter; the longer the strands of DNA are, i.e., the less frequent the breaks, the more slowly the strands escape from the filter.

Alkaline unwinding occurs as a consequence of gentle lysis of cells at moderately high pH. The separation of the two strands depends on the presence of ends to act as unwinding points, and the fraction of DNA single-stranded after a given time measured by hydroxyapatite chromatography or differential fluorescence accurately reflects the frequency of breaks.

Nucleoids are the residual tightly supercoiled DNA structures left after detergent and hypertonic treatment of cells. SSBs cause relaxation of supercoiling and DNA expands as a halo around the nucleoid. The expanded nucleoids sediment at a reduced rate in a sucrose gradient, and this feature is indicative of DNA breaks.

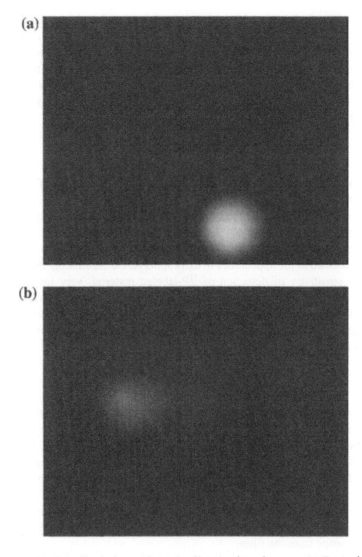

Figure 7. Single cell gel electrophoresis of human lymphocytes. (**a**) Typical image of undamaged cell. (**b**) Cells were treated with H_2O_2 to introduce strand breaks; these allowed DNA loops to move in the electrophoretic field to form a "comet tail." (**c**) Cells, not treated with any DNA damaging agent *in vitro*, were incubated (after lysis) with endonuclease III which makes breaks at sites of oxidized pyrimidines. The significant release of DNA into the tail reflects the presence of endogenous oxidative DNA damage.

(c)

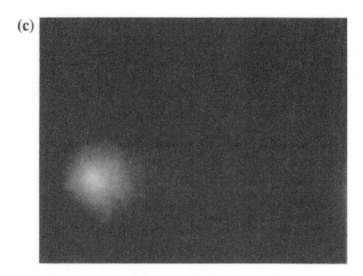

Figure 7. (Continued)

Single cell gel electrophoresis is a close relative of nucleoid sedimentation. Cells embedded in agarose and lysed in detergent and 2M NaCl are electrophoresed at high pH. Nucleoids appear as compact round bodies if DNA is undamaged, but as "comets" if damage is present, with the fraction of DNA in the comet tail representing the extent of breakage. The nucleoid halo, made up of relaxed DNA loops, is pulled towards the anode to form the tail (Figure 7).

Measuring Double-Strand Breaks

Whereas both SSBs and DSBs allow separation of strands in alkali, only DSBs actually change the length of the double helix. Assays employing neutral pH permit the specific measurement of DSBs. Neutral elution is the counterpart of alkaline elution. On a neutral sucrose gradient, DNA sediments at a rate that depends on fragment length, i.e., the length between two DSBs. Pulsed field gel electrophoresis allows migration of very large DNA molecules, such as are produced by a few DSBs; intact chromosomal DNA remains at the origin. Repair of DSBs can be detected as a restoration of DNA to the non-migrating position.

Repair DNA Synthesis

The small amount of DNA synthesis associated with repair can be detected only if normal replicative DNA synthesis is somehow excluded. In cells treated with DNA-damaging agent and then incubated with the DNA precursor ^{3}H-thymidine,

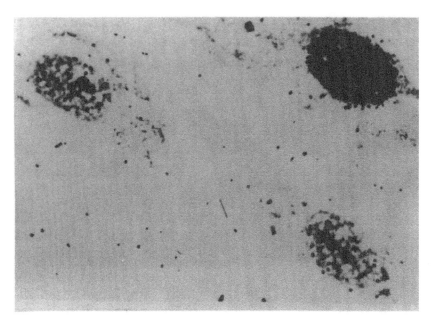

Figure 8. Unscheduled DNA synthesis. Normal human fibroblasts were UV-irradiated and incubated with ³H-thymidine. Autoradiographs show one heavily labeled nucleus (of a cell in S-phase) and two nuclei only lightly covered in silver grains, indicating a low level of repair synthesis. (From Mullinger et al., 1983, with permission.)

and processed for autoradiography, S-phase nuclei are heavily labeled, but repair activity, or unscheduled DNA synthesis (UDS), is seen as a low density of silver grains over the nuclei of cells not engaged in replication (Figure 8). Repair synthesis is an indirect measure; the amount of it reflects not only the number of lesions, but the size of repair patches, and the specific activity of labeled ³H-dTTP in the cellular pool.

OTHER CELLULAR RESPONSES TO DNA DAMAGE

Interactions With Replication

DNA damage causes a reduction in the rate of replication, perhaps simply because the replication complex is unable to proceed past the lesion. Nascent DNA fragments made after treatment of cells with sublethal doses of UV light are smaller than normal. However, a recovery process quickly takes effect, with elongation of the nascent fragments and an increase in the overall DNA synthetic rate to control levels within a few hours. Recovery may involve translesion synthesis, i.e., the stalled replication machinery somehow by-passes the damage, resuming proper

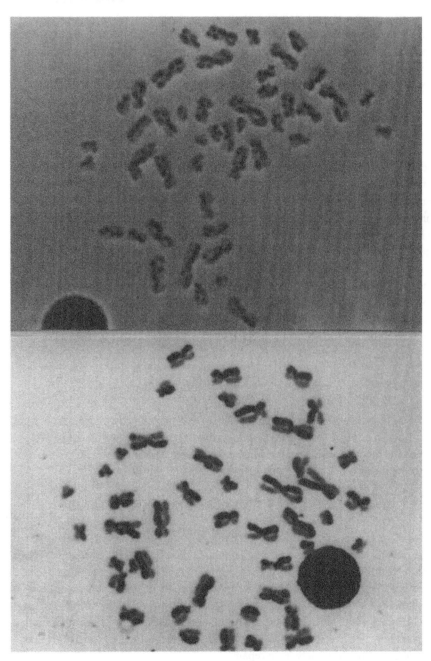

Figure 9. Sister chromatic exchange. In untreated human lymphocytes (lower photograph), few harlequin chromosomes are present. The upper photograph, showing many such chromosomes, is of lymphocytes treated with mitocyin C. (Kindly provided by Dr. James Mackay, Zeneca.)

247

complementary strand synthesis beyond it; or it may depend on extended synthesis from the adjacent replication unit to fill the gap. The so-called postreplication recovery is error-prone, since normal base pairing is impossible at the damage site. RNA synthesis is also sensitive to DNA damage.

Recombination

Genetic recombination of the kind that occurs in meiosis, i.e., the exchange of sections of homologous chromosomes, is also seen as a consequence of DNA damage. Sister chromatid exchange (SCE) gives rise to harlequin patterns when chromosomes are stained in a particular way that distinguishes the individual chromatids (Figure 9). SCE is elevated in various mutant cell lines that are defective in DNA repair, and also occurs spontaneously at a high rate in cells from patients with Bloom's syndrome (a cancer-prone condition associated with sunlight sensitivity, caused by a defect in DNA helicase, an enzyme involved in cellular manipulation of DNA. SCE can be seen as a response to the presence of broken DNA owing to disrupted replication or unrepaired damage, the breaks allowing strands from one double helix to hybridize, illicitly, with a complementary strand from the other double helix in the replicated chromosome, thus initiating a crossing-over process.

Induced RNA and Protein Synthesis

DNA damage, like other kinds of cellular stress, induces transcription from specific genes. These include the *GADD* (growth arrest and DNA damage) and *fos* genes, and the gene for the enzyme heme oxygenase. It is assumed that the proteins coded by these genes contribute to some kind of cellular protection.

VARIATIONS IN RESPONSE TO DNA DAMAGE

Whether repair varies significantly between individuals, as a result of aging, or in different tissues, are questions that have not been satisfactorily answered. Repair is more rapidly completed in proliferating than in quiescent cells, probably because there is a richer supply of DNA precursors (needed for the repair patch), while tumor cells are relatively deficient in repair compared with their normal counterparts (Figure 10). There are some important special cases; first, a group of human diseases in which a repair function is affected, and, second, the artificially produced mutant cells lines with defects in repair.

Table 1 lists the hereditary human disorders with abnormalities in the cellular response to DNA damage. Most of these are also associated with an elevated risk of cancer. Some are illustrated in Figure 10.

Table 1. Human Hereditary Disorders Associated With Photosensitivity and Defective Processing of DNA Lesions*

Syndrome	Clinical Features	Likelihood of Cancer	Cellular Characteristics
Xeroderma pigmento-sum (classical XP)	Sunlight hypersensitivity Often neurological abnormalities Ocular defects Immune deficiency (some patients)	+++ (tumors on exposed skin)	Hypersensitive to and hypermutable by UV and some chemical carcinogens Defective in early step of excision repair Seven complementation groups
Xeroderma pigmento-sum (XP variant)	Similar to classical XP but no neurological problems	+	Hypermutable by UV Deficient in DNA replication on UV damaged templates
Fanconi's anemia	Growth retardation Pancytopenia Bone marrow deficiency Anatomical defects	+ (leukemia)	Hypersensitive to some cross linking agents High frequency of spontaneous chromosome aberrations
Cockayne's syndrome	Cachectic dwarfism Mental retardation (microcephaly) Premature aging Sunlight hypersensitivity	—	Hypersensitive to UV and chemical carcinogens RNA and DNA synthesis abnormally UV-sensitive
Trichothiodystrophy	Brittle hair Mental retardation Growth retardation Ichthyosis Photosensitivity (some patients)	—	Altered regulation of keratin synthesis UV sensitivity and reduced excision repair in some patients
Ataxia-telangiectasia	Cerebellar ataxia Cutaneous and ocular telangiectasia Hypersensitive to ionizing radiation Immune deficiency (variable)	+++ (lymphomas)	Hypersensitive but hypomutable to ionizing radiation Radioresistant DNA synthesis High frequency of chromosome aberrations

Note: *Based on Hanawalt and Sarasin, 1986, with additions and revision.

Xeroderma Pigmentosum

It was the discovery that cells from xeroderma pigmentosum (XP) patients (known to be at high risk of skin cancer) are unable to carry out nucleotide excision repair after UV irradiation that confirmed DNA repair as a crucial element in the defense against carcinogenesis.

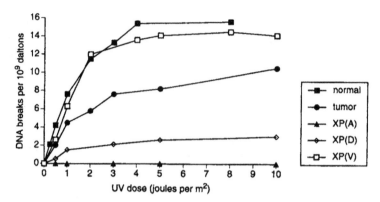

Figure 10. Cellular repair activity indicated by incision. Various human cell types were incubated for 30 minutes after UV irradiation in the presence of DNA synthesis inhibitors to accumulate incomplete repair sites (reflecting enzymic breakage at dimer sites without subsequent repair synthesis or ligation). (Redrawn from Squires et al., 1982.)

Cell lines derived from a large number of XP patients have been characterized genetically by creating hybrids or heterokaryons between cell lines and measuring DNA repair activity, usually as UV-induced UDS. A heterokaryon from cells carrying defects in different genes shows substantially restored repair activity, because the repair product missing in one line is provided by the partner in the heterokaryon. The two lines are said to complement each other. However, two cells that are defective in the same gene, when fused together, remain unable to make the functional product of that gene, and are said to be in the same complementation group. Seven complementation groups of excision repair-defective XP are known: XP(A) to XP(G). Since all these groups are deficient in the earliest detectable step of repair, i.e., incision, the implication is that at least seven distinct gene products (enzymes or other proteins) are involved at or before incision—an indication of the complexity of the repair process. As well as the repair defect, XP cells show greatly reduced catalase activity. This enzyme is responsible for decreasing H_2O_2 within cells, and so XP cells would be expected to suffer excessive oxidative damage. However, oxidative damage is not repaired by the same pathway as UV damage, and the co-occurrence of diminished excision repair and low catalase in several complementation groups is not readily explained genetically. The condition known as XP (variant) does not show any excision repair defect, but replication seems to be abnormally susceptible to blockage by DNA damage.

Much progress has recently been made in the isolation of the genes affected in XP, and the characterization of their products. XPA protein binds to UV-damaged DNA. XPB and XPD are helicases (DNA unwinding enzymes) that are components—along with 6 other proteins—of the human transcription factor TFIIH. An association between repair and transcription is not unexpected, in view of the evidence for preferential repair of transcribing genes. (Another link with transcription is seen in XP(C) cells, which show significant incision and repair synthesis after UV irradiation, but only in DNA which is actively transcribing.) (XPG is an endonuclease, and a second endonuclease comprises a complex between XPF and ERCC1 (see below). Reconstitution of repair *in vitro* has been successfully attempted; a combination of TFIIH, XPA, XPC (which binds single-stranded DNA), XPG, the ERCC1/XPF complex, and RPA (a protein essential for replication) is capable of carrying out the dual incision required to remove a bulky adduct or dimer. XPE, a protein that binds to damaged DNA, stimulates the process but is not essential.

Cockayne's Syndrome

There are two distinct complementation groups for Cockayne's syndrome (CS), CS(A) and the more common CS(B). In addition, patients in the rare XP complementation group B invariably show clinical symptoms of both XP and CS. Isolated CS cases also fall into the XP(D) and XP(G) groups. Excision repair in (non-XP) CS cells appears normal, but they do not recover from the depression of RNA and DNA synthesis caused by UV irradiation. It has recently been shown that CS cells have a subtle excision repair defect; they lack the preferential repair of actively transcribing genes. A protein with a role in both transcription and repair is the likely product of the gene *ERCC-6* implicated in CS(B).

Trichothiodystrophy

Some patients with trichothiodystrophy (TTD) show photosensitivity and a deficiency in excision repair that is genetically identical with that of XP(D). Curiously, they do not show an increased incidence of cancer. It is possible that extra features present in XP (such as the low catalase activity, or poor immune surveillance) contribute to carcinogenesis. It seems likely that the mutations responsible for TTD and XP(D) are at closely linked loci or even within the same gene.

Ataxia-Telangiectasia

The main cellular characteristic of ataxia-telangiectasia (A-T), a disease associated with extreme X-ray sensitivity, is lack of the normal ionizing radiation-induced depression of DNA synthesis. In normal cells, the level of the tumor suppressor nuclear protein p53 increases at the same time as the decrease in replicative synthesis following irradiation. A-T cells fail to show the radiation-induced increase in p53. p53 is implicated in controlling cell cycle checkpoints, and seems to be directly responsible for inducing transcription of the *GADD45* gene: also involved in cell cycle arrest, and known not to be expressed in A-T cells. The actual reason for A-T cell killing by radiation remains obscure.

Fanconi's Anemia

Cells from Fanconi's anemia (FA) patients are typically sensitive to killing by bifunctional agents such as psoralens. However, it may be the degree of helical perturbation rather than the cross-linking that is important, since even monoadduct repair is impaired in FA cells. In cells from the two FA complementation groups (A and B) two different, specific endonucleases show decreased affinity for psoralen adducts.

Mutant Mammalian Cell Lines With DNA Repair Defects

Mutant rodent cell lines selected for sensitivity to DNA damaging agents have been classified biochemically, in terms of defects in repair (or other) response, as well as genetically, into complementation groups. There are at least 10 complementation groups for UV-sensitivity, and human genes that are able to restore the normal UV-resistance to these cells have been isolated in several cases. They form the excision repair cross complementing series of human genes, some of which identify with the genes defective in known human disorders—*ERCC*-2 with XP(D), *ERCC3* with XP(B), *ERCC6* with CS(B). Analogous series of mutants sensitive to ionizing radiation or to cross-linking agents are less well characterized; but one of the latter is homologous to FA(A). These mutants have been useful for basic investigations of repair pathways; for example, UV-sensitive mutants tend to be cross-sensitive to bulky adducts and often to cross-linking agents, which points to common steps in the repair of all these lesions.

SUMMARY AND PERSPECTIVE

DNA damage is indisputably a crucial factor in carcinogenesis. Therefore, DNA repair is essential for protection from carcinogenesis. There are likely to be strong links between kinds of damage—or sources of damaging agents—and particular cancers; for example, between UV exposure and skin cancer, and between diet and gastrointestinal cancers. However, complications immediately become apparent; the influence of diet is not simply a matter of whether food contains carcinogens, but also whether it lacks protective agents. There are other potentially important variables, such as individual differences in carcinogen metabolism, in cellular protection, and in repair capacity. Such variations may be genetic in origin; the subtle effect of being an A-T heterozygote, i.e., defective in one of the two copies of the A-T gene, has been claimed to be responsible for a significant fraction of breast cancer incidence. Screening for these genetic differences at the population level may eventually be feasible. In addition, the possible influences of environmental, dietary, or physiological factors on the induction or modulation of repair and protective mechanisms represent virtually unexplored territory, with the potential for controlling susceptibility to cancer by appropriate intervention.

ACKNOWLEDGMENTS

I acknowledge the financial support of the Scottish Office Agriculture Environment and Fisheries Department. I would like to thank my colleagues Dr. Susan Duthie and Mrs. Catherine Gedik for their helpful comments during the writing of this chapter.

REFERENCES

Hanawalt, P.C., & Sarasin, A. (1986). Cancer-prone hereditary diseases with DNA processing abnormalities. Trends Genet. 2, 124–129.

Mullinger, A.M., Collins, A.R.S., & Johnson, R.T. (1983). Cell growth state determines susceptibility of repair DNA synthesis to inhibition by hydroxyurea and 1-β-D-arabinofuranosylcytosine. Carcinogenesis 4, 1039–1043.

Squires, S., Johnson, R.T., & Collins, A.R.S. (1982). Initial rates of DNA incision in UV-irradiated human cells. Differences between normal, xeroderma pigmentosum and tumour cells. Mutation Res. 95, 389–404.

Van Houten, B., Gamper, H., Holbrook, S.R., Hearst, J.E., & Sancar, A. (1986). Action mechanism of ABC excision nuclease on a DNA substrate containing a psoralen crosslink at a defined position. Proc. Natl. Acad. Sci. USA 83, 8077–8081.

RECOMMENDED READINGS

Ames, B.N. (1983). Dietary carcinogens and anticarcinogens. Science 221, 1256–1264.

Culotta, E., & Koshland, D.E. (1994). DNA works its way to the top. (Editorial.) Science 266, 1926–1929.

Friedberg, E.C. (ed.) (1990). The enzymology of DNA repair. Mutation Res. 236, 145–311.

Friedberg, E.C., Walker, G.C., & Siede, W. (1995). DNA Repair and Mutagenesis. pp. 698, ASM Press, Washington.

Garner, C. (1992). Molecular potential [comments on molecular epidemiology]. Nature (London) 360, 207–208.

Halliwell, B., & Aruoma, O. (1991). DNA damage by oxygen-derived species. FEBS Lett. 281, 9–19.

Hanawalt, P.C. (1994). Transcription-coupled repair and human disease. Science 266, 1957–1958.

Huang, J.-C., Svoboda, D.L., Reardon, J.T., & Sancar, A. (1992). Human nucleotide excision nuclease removes thymine dimers from DNA by incising the 22nd phosphodiester bond 5' and the 6th phosphodiester bond 3' to the photodimer. Proc. Natl. Acad. Sci. USA 89, 3664–3668.

Kunkel, T.A., & Loeb, L.A. (1981). Fidelity of mammalian DNA polymerases. Science 213, 765–767.

Lindahl, T. (1993). Instability and decay of the primary structure of DNA. Nature (London) 362, 709–715.

Modrich, P. (1987). DNA mismatch correction. Ann. Rev. Biochem. 56, 435–466.

Modrich, P. (1994). Mismatch repair, genetic stability, and cancer. Science 266, 1959–1960.

Mu, D., Park, C-H., Matsunaga, T., Hsu, D.S., Reardon, J.T., & Sancar, A. (1995). Reconstitution of human DNA repair excision nuclease in a highly defined system. J. Biol. Chem. 270, 2415–2418.

Sancar, A., & Sancar, G.B. (1988). DNA repair enzymes. Ann. Rev. Biochem. 57, 29–67.

Sancar, A. (1994). Mechanisms of DNA excision repair. Science 266, 1954–1956.

Santella, R.M. (1988). Application of new techniques for the detection of carcinogen adducts to human population monitoring. Mutation Res. 205, 271–282.

Smith, C.A., & Mellon, I. (1990). Clues to the organization of DNA repair systems gained from studies of intragenomic repair heterogeneity. In: Advances in Mutagenesis Research (Obe, G., ed.), pp. 153–194, Springer-Verlag, Berlin.

Chapter 12

The Nature and Significance of Mutations

ANDREW R. COLLINS

Principles of Medical Biology, Volume 5
Molecular and Cellular Genetics, pages 255–269.
Copyright © 1996 by JAI Press Inc.
All rights of reproduction in any form reserved.
ISBN: 1-55938-809-9

WHAT ARE MUTATIONS?

A very serviceable definition was formulated by Louis Siminovitch in 1976: a mutational event is one which involves "any hereditable nucleotide base change, deletion, or rearrangement in the primary structure of DNA...Mutation is taken as a global term to include not only point and deletion mutations, but also chromosomal rearrangements and chromosome loss, as long as hereditable changes result." Mutations can occur spontaneously, or through exposure to a mutagenic, DNA-damaging agent.

The organization of eukaryotic DNA into chromosomes results in complexities that are unknown in bacterial systems. The genome (in somatic cells) is diploid, i.e., there are two copies of most genes, which means that inactivation of one allele does not necessarily affect the cell's functions, since the other copy can provide enough transcripts for protein synthesis. Individuals who are heterozygous for a mutation in an essential gene (i.e., only one copy is defective) do not, generally, display symptoms of disease associated with the homozygous condition (both copies defective). In the case of genes on the X chromosome, there is only one copy in males, and in females most of the genes on one of the two X chromosomes are permanently inactive. An example is the *hprt* gene, which codes for hypoxanthine phosphoribosyltransferase, a purine salvage enzyme. Because only one normal copy needs to be inactivated to express the ostensibly recessive phenotype, mutant cell lines arise at relatively high frequency. However, mutations occur in various genes on the autosomal chromosomes of the standard laboratory Chinese hamster ovary (CHO) cell line with similar high probability, even though two copies must be inactivated. To explain this observation, it was proposed that many genes in CHO cells may be in the hemizygous state; i.e., only one of the two alleles is functioning in transcription. Non-transcribing genes are marked by methylation of specific cytosines in CpG dinucleotides; the methylation signal is passed from one cell generation to the next. The extent to which hemizygosity operates in normal cells is not clear.

TYPES OF MUTATION

Base Substitution: Point Mutations

Exchange of one base for another in a sequence may result in substitution of an amino acid in the protein coded by the gene. If the base substitution creates a "stop" codon, the protein will be prematurely terminated. A base substitution is called a transition if the change is simply from one purine to the other purine or one pyrimidine to the other pyrimidine, and a transversion if the change is from pyrimidine to purine or vice versa. Simple base changes are generally reversible; a "back-mutation" will restore the original sequence, and the reversion frequency, whether spontaneous or induced, is of the same order as the original forward mutation rate.

Deletions and Additions

Deletions can be as small as a single nucleotide, or as large as a visible portion of a chromosome or even a whole chromosome. Deletion of one or two nucleotides constitutes a frameshift mutation; i.e., the translation reading frame in the coding sequence is shifted, and incorrect amino acids are specified from that point on. Frameshifts situated towards the end of the coding sequence will obviously have less impact on structure and function of the protein than mutations near the beginning of the sequence.

More extensive deletions may lead to a missing part of the protein product (truncation), to fusion of products of adjacent genes, or to loss of entire gene(s). Reversion of deletion mutations is virtually impossible, except for the single base deletions which revert at a very low rate. Addition of one or a few bases to a sequence is also possible. Larger scale additions are described under "Amplification" below.

Rearrangements

DNA rearrangement is an essential process in the development of immunoglobulin genes. In other cases, inappropriate movement of a piece of DNA from one chromosomal site to another may alter genetic function. Rearrangements may result in one gene coming under the control of a different regulatory system at the new site, leading to inappropriate transcription which can have serious consequences. On a large scale, rearrangements can be seen as translocations between chromosomes.

Amplification

A rather unexpected way of upregulating synthesis of a particular protein is by an increase in the number of copies of the relevant gene. Tumors that are initially kept under control by chemotherapy often develop a resistance at the cellular level, which extends to drugs other than those originally applied. This is referred to as the multidrug resistance phenotype; its genetic basis is the presence of tandem arrays of many identical copies of a gene coding for one of the efflux proteins present in cell membranes. The cells containing the amplified gene synthesize large quantities of this protein which keeps the intracellular drug concentration below toxic levels. Gene amplification can also be induced in established cell lines in the laboratory. Continued culture of CHO cells in gradually increasing concentrations of methotrexate, an inhibitor of an essential enzyme of folic acid metabolism, dihydrofolate reductase, leads to the emergence of methotrexate-resistant cells which carry hundreds of copies of the *dhfr* gene rather than the two copies present in a normal diploid cell.

Mutations that survive to form a significant fraction of cells in a population, whether in the laboratory or in the organism, do so because they have a selective advantage over the wild-type cells. The selection may be imposed by the experimenter (or the clinician) as a regime of drug treatment, or the advantage may consist of an ability to overcome normal controls on cell growth and division, as in the case of tumor cells expressing oncogenes. We will return to the biological significance of mutations after dealing with the ways in which mutations arise.

MECHANISMS OF MUTAGENESIS

Replication Errors

The intrinsic inaccuracy of the DNA replication process results in misincorporation at a low but significant level. Most of the mispaired nucleotides are, however, removed almost immediately by a proofreading activity associated with the replication machinery.

Another kind of error that occurs during replication is slippage (Figure 1). When a sequence containing a run of identical bases, or a series of short repeats, is copied, the polymerase has a (slight) tendency to slip back to an apparently identical region

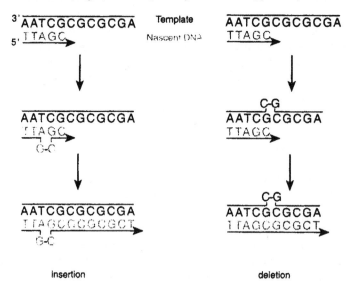

Figure 1. Replication errors resulting in insertion or deletion. The schematic diagram shows the result of looping out of the nascent DNA strand (left) or the template strand (right) during replication.

of the sequence and to rereplicate from that point. Thus an extra loop of DNA is generated. Alternatively, if the polymerase slips forward, a length of sequence will be omitted from the copy. When the daughter strand with the deletion or addition itself acts as a template strand in subsequent replication, the mutation is fixed. Frameshift mutations, larger deletions, and insertions can result from slippage. Repeated replication of an entire gene may be a cause of gene amplification.

Effects of DNA Damage

Small changes to a base, such as deamination, alkylation, or oxidation, can change its base pairing. For example, the deaminated form of cytosine is uracil, which pairs with adenine, not guanine (causing a transition); 8-OH-guanine pairs with adenine rather than cytosine, generating a potential transversion. Such base pairing changes affect the sequence of RNA transcribed from the damaged DNA. However, this is a transient problem, overcome by DNA repair. If DNA is replicated before the base sequence is restored, the altered base pairing results in a change in the sequence of the daughter strand which is perpetuated as a mutation.

If a base is lost completely, during replication the base to be inserted on the daughter strand cannot be determined from the template sequence; the site is noninstructional and potentially mutagenic. There is a strong tendency for the polymerase to incorporate adenosine monophosphate opposite the abasic site. If the original base was a guanine, $G \rightarrow T$ transversion will occur; if cytosine, the result will be a $C \rightarrow T$ transition (Figure 2). This phenomenon is referred to as the A-rule.

Some lesions, for example the pyrimidine dimer induced by UV light, prevent base pairing and thus block the processes of transcription and replication. Postreplication recovery, or translesion synthesis allows the daughter strand to extend across the damaged section, but the bases inserted are not defined by the sequence and this "repair" process is essentially error-prone or mutagenic. The A-rule applies here, too. The presence of a replication-blocking lesion in a repetitive sequence can cause frameshifts, apparently by the damaged nucleotide looping out from the template strand to allow the replication complex to pass (a process analogous to slippage; see above).

Cross-links between strands in the DNA duplex prevent movement of the replication fork and have a profoundly disruptive effect on DNA synthesis with consequent chromosomal aberrations. Agents that induce strand breaks at high frequency, such as ionizing radiation, typically cause deletion mutations, including gross rearrangements and deletions visible at the chromosomal level. These mutations may result directly from the damage (a double-strand break (DSB), if not stabilized by chromatin structure, will obviously cause a chromosome break); or they follow replication on the damaged template. Repair itself may be inaccurate; rejoining of DSBs *in vitro* sometimes leads to loss of sequence around the break point.

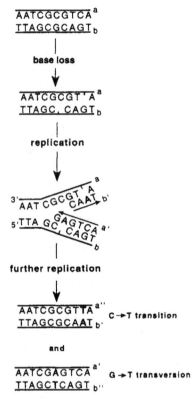

Figure 2. The A-rule. Noninstructional abasic sites result from loss of a C (from strand a) or a G (from strand b). During replication, application of the A-rule produces a transition or transversion mutation.

Gene Amplification

There are two models for the origin of amplified genes. Agents that induce amplification typically block replication either by damaging the DNA or by inhibiting polymerization. During the period of blocked replication, initiation factors are thought to accumulate, so that on release from inhibition, a burst of replication occurs even in DNA that has already replicated in that round. Thus extra copies are made and subsequently arranged into a tandem series (Figure 3).

An alternative suggestion is that amplification is a consequence of fragmentation of DNA, for example as a result of DNA damage or disrupted replication. If a fragment containing a particular gene segregates asymmetrically in mitosis, one daughter cell will contain an extra copy at the expense of the other daughter cell. Asymmetric segregation might occur through unequal sister chromatid exchange. Under appropriate selection (which is essential for amplification to occur), cells

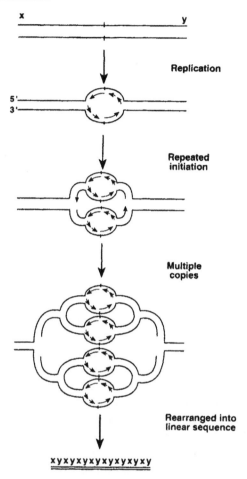

x y x y x y x y x y x y x y x y

Figure 3. One possible mechanism of gene amplification involving repeated initiation of replication (the "onion-skin" model). (Based on a diagram in Stark and Wahl, 1984.)

with extra copies will be favored and, with repeated cycles of asymmetric segregation, they will dominate the population.

THE BIOLOGICAL SIGNIFICANCE OF MUTATIONS

Mutations are by no means all bad. Most are probably neutral, with no discernible effect on the functioning of the cell. Even those that alter an amino acid sequence, or rearrange genes by translocating within or between chromosomes, may be beneficial in the long run, since they provide the background of variability that is the raw material for evolution. Mutations in exons, the segments of a gene that are the raw material for evolution. Mutations in exons, the segments of a gene that are ultimately represented in an amino acid sequence, clearly would be expected to alter that sequence unless, by chance, the new codon specifies the same amino acid because of the degeneracy of the code. Mutations in the intervening sequences (introns), though not directly affecting the amino acid sequence, may interfere with normal splicing of the RNA and have quite drastic effects on protein structure. Regulatory sequences upstream from the coding region determine the level of transcription of a particular gene, in response to signals that reflect the cell's needs. Mutations in the regulatory elements may disrupt this control mechanism.

Mutations as a Source of Biochemical Insight

Mutant cell lines, defective in specific genes, are extremely useful in the dissection of biochemical pathways, since the step catalyzed by the product of a gene defective in a particular mutant can be identified by the accumulation of intermediates prior to that step. However, this approach works only if the cells bearing the mutation can be kept alive. Some mutations are nonlethal; for example, the HPRT salvage system is supplementary to *de novo* purine biosynthesis, an optional extra for the cell. (However, for the organism as a whole the lack of this enzyme is very damaging; its absence causes the serious human disorder known as the Lesch-Nyhan syndrome, characterized by self-destructive behavior.)

Some mutations are only noticed if the cell receives an appropriate challenge; mutants in pathways for DNA repair grow and divide just as normal cells until they receive DNA damage. It is sometimes possible to bypass the cell's requirement for a particular enzyme; mutants (known as auxotrophs) defective in synthesis of some essential metabolite depend on addition of the metabolite to the growth medium. As an example, a series of purine auxotrophs have been isolated with defects in the enzymes of purine biosynthesis; they grow normally if hypoxanthine is provided in the medium as a purine source. Mutations causing conditions such as these are described as conditionally lethal.

Temperature-sensitive mutants are another variation on this theme. Some mutations are only expressed at relatively high temperature, say 40° C, which renders the altered protein unstable. The mutant cells can be cultured routinely at a low temperature (about 33° C) and placed at the elevated temperature in order to study the effect of the mutation. These mutants are particularly important for investigating vital processes such as DNA replication and other events in the cell cycle.

Mutations in Germ Cells and Somatic Cells

Mutations occurring in germ cells have a very different significance from those in somatic cells. In the former case, mutations can be passed to succeeding generations as heritable disorders, material for future evolutionary change, or, at least, as restriction fragment length polymorphisms; or RFLPs (see the next section). Severe mutations in the germ line (or in the early embryo) may prevent normal development and lead to abortion or birth defects. On the other hand, a mutation in a somatic cell may be lethal to that cell, but the death of an individual cell is of no consequence to the organism. The important somatic cell mutations are those which bestow on the cell one of the characteristics of tumorigenic transformation, such as the escape from normal control of proliferation. Two classes of genes are involved. Oncogenes are mutated or abnormally activated forms of normal cellular genes, proto-oncogenes. These code for components of a signaling pathway, originating with a growth factor acting at the cell membrane and ending in the nucleus with the switching on of genes responsible for DNA replication and cell division. The oncogene expresses this normal function but in an uncontrolled way. Oncogenes tend to operate in a genetically dominant mode, with mutation of only one of the two alleles sufficient for the tumorigenic phenotype to be expressed. The second class comprises tumor suppressor genes which, when normally expressed, provide cell cycle checkpoints that switch off the progression of the cycle if proliferation is inappropriate. Tumor suppressor genes are expected to show a recessive mode of inheritance; defective expression of one allele is not enough to transform cells as the normal product is synthesized from the other allele.

Restriction Fragment Length Polymorphism

Harmless mutations accumulated in noncoding DNA are passed from one generation to the next together with any recent mutations in the germ line. Because of the segregation that occurs in meiosis and the subsequent combination of maternal and paternal genetic material, each individual has a unique set of mutations. An ingenious method for detecting these mutations is based on the use of restriction endonucleases, which cleave DNA at specific short recognition sites of between four and eight base pairs. Mutations that change the endonuclease recognition sites in a particular sequence of DNA alter the pattern of bands that is seen, after electrophoretic separation on an agarose gel, by probing with radiolabeled DNA complementary to the sequence being examined. Thus, for example, abolition of a site by a base change might mean that two short fragments (relatively fast-moving bands) are replaced by one longer (slower-moving) fragment. Restriction endonucleases with different recognition sites are used to reveal different, characteristic patterns. These genetic variations are known as restriction fragment length polymorphisms.

Another kind of DNA polymorphism is the variable number tandem repeat (VNTR), which reflects the existence of different numbers of repeats of the same sequence (sometimes as small as a dinucleotide) between two restriction sites, altering restriction fragment length, but not the number of fragments, since the sites themselves are not changed. The ability to detect these polymorphisms has revolutionized the study of inherited human disease, since they allow investigation of the distribution of defective genes (to which RFLPs may be closely linked) within families.

TUMORIGENESIS

Cancer Depends on Multiple Hits

Cancer incidence rises very sharply with age (Figure 4). It is known that cancer has a clonal origin, i.e., all the cells in a tumor derive from a common progenitor; a curve such as this is what would be expected if cancer depends on several genetic changes taking place in one cell lineage over a period of time. Mutations occur with a more or less constant frequency through life, and the probability that a particular mutation has happened is therefore proportional to age. If N mutations are required to convert a normal cell to a tumor cell, the probability that all have occurred is proportional to age^N. N is somewhere between two and seven for different types of cancer.

Hereditary and Environmental Factors: Retinoblastoma

The underlying assumption is that mutations are the result of exposure to mutagens. In addition there is a tendency for certain cancers to be inherited. This predisposition may indicate that one of the N mutations is already present in all cells of the body, so that only N-1 mutations are needed for a tumor to result. Retinoblastoma, a tumor of the eye affecting children, occurs in two forms: hereditary, and sporadic. The hereditary form is associated with a high incidence

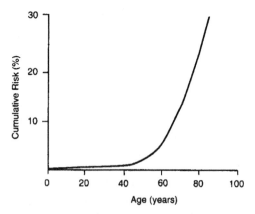

Figure 4. The age-dependence of cancer deaths in humans. The cumulative risk of mortality from all cancers increases with approximately the fifth power of age. (Redrawn from Ames et al., 1990.)

of other tumors. Deletions at a band on the long arm of chromosome 13 are commonly seen in retinoblastoma tumor cells, and a gene *RB-1* that hybridizes to this point has been cloned. It behaves in a recessive fashion and is probably a tumor suppressor gene. The inherited condition—most likely a point mutation or small deletion at one allele—is without effect until the other allele is inactivated or lost, and there is a high probability of such a change occurring in at least one cell through, for example, abnormal mitotic segregation. Ninety-five percent of carriers of the recessive mutation develop retinoblastoma, and it often occurs independently in both eyes. In the sporadic variety of the disease, the homozygous condition arises through two chance inactivations in a single cell: a very rare event, accounting for the fact that this form of the disease affects only one eye.

Colorectal Cancer: A Multistage Progression

There is a well-developed genetic model for the development of colorectal (CR) cancer (Figure 5). Mutations in several genes are required for the fully malignant tumor, or carcinoma; benign tumors, or adenomas, from which carcinomas arise, carry fewer genetic changes. The mutations include activation of oncogenes and loss of tumor suppressor gene activity. Fifty percent of CR carcinomas and advanced adenomas have a mutated *ras* oncogene, but in small adenomas this change is rare. Seventy-five percent of carcinomas (but only a few percent of adenomas) demonstrate loss of part of the short arm of chromosome 17; all these deletions, which are of varying extent, include the site of the p53 gene. When the remaining p53 allele is examined, a point mutation is commonly seen; this is consistent with the need for inactivation of both alleles for expression of the recessive phenotype. (There is, however, some evidence that, even in the presence of a wild-type gene product, certain mutant p53 proteins can have some effect on tumor progression, perhaps by reducing the effectiveness of wild-type protein in an oligomeric association.) Chromosome 18 (long arm) also has a deletion in most CR tumors, and a second tumor suppressor gene, DCC has been mapped to this region.

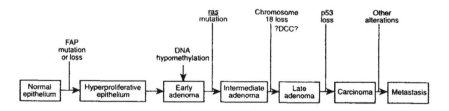

Figure 5. A genetic model for colorectal tumorigenesis. (Redrawn from Fearon and Vogelstein, 1990.)

Familial adenomatous polyposis (FAP) is a rare inherited autosomal dominant syndrome associated with a change in a gene on chromosome 5 (long arm); this apparently allows hyperproliferation of the colon epithelium. Deletions of this region are common in non-FAP colorectal carcinomas, too, and they tend to occur at an earlier stage than the other common deletions (though the order of all these changes is not invariant; it is the total cumulative effect that matters). From an analysis of genetic alterations in many CR tumors, it appears that at least five changes are necessary for cancer formation. The carcinoma can become more invasive as further tumor suppressor genes on other chromosomes are inactivated. A similar model is likely to hold for other common cancers, among them breast cancer, for which a gene *BRCA1* on the long arm of chromosome 17 has been linked with familial predisposition.

Genetic Instability as a Predisposing Factor

An inherited predisposition to cancer may involve a mutation that somehow induces a high rate of mutagenesis in other genes. Xeroderma pigmentosum (XP), and other repair-defective conditions, are associated with high mutation rates in cell culture and with high cancer incidence. Bloom's syndrome is an inherited disorder with a high incidence of cancer and, in cultured cells, very high levels of spontaneous and induced chromosome breakage and sister chromatid exchange; the gene defective in Bloom's syndrome has recently been identified as coding for a helicase—an enzyme likely to be involved in cellular manipulations of DNA during S-phase.

There is a second inherited susceptibility to CR cancer, more common than FAP, known as hereditary nonpolyposis colorectal cancer (HNPCC). It is typically of early onset but otherwise difficult to distinguish from sporadic CR cancer. Our understanding of HNPCC has recently been enhanced with the exciting discovery that cells from HNPCC tumors are deficient in mismatch repair, and have high rates of spontaneous mutation, particularly deletions/insertions in $(CA)_n$ and other simple repeated sequences. Non-tumor cells from affected individuals have one defective copy of a mismatch repair gene, but because the other copy is normal, they show normal mismatch repair. As with retinoblastoma, chance mutation of the normal allele is necessary for the disease to appear. A cell carrying mutations in both alleles has a "mutator phenotype" and subsequent mutations in the crucial oncogenes and tumor suppressor genes occur much more frequently than usual, resulting almost inevitably in cancer.

Alterations in trinucleotide repeat numbers ("microsatellite instability"), apparently occurring through replication slippage with failed mismatch repair, are characteristic of certain human genetic diseases; Huntington's disease, for example, has recently been shown to have just such a cause.

Translocations and Cancer

Very specific chromosome translocations are linked to certain types of cancer. In almost all cases of chronic myelogenous leukemia, reciprocal exchange has occurred between chromosomes 9 and 22. The distinctive shortened chromosome 22, known as the Philadelphia chromosome, is diagnostic for the disease. The significance of the change is that the *abl* proto-oncogene is found on the fragment of chromosome 9 that is translocated to chromosome 22. The protein product of the *abl* gene in its new position is linked with the product of an adjacent chromosome 22 gene, and the fusion protein, through loss of sensitivity to the normal regulatory mechanism, becomes capable of transforming cells to the cancerous phenotype.

Translocation of the *myc* proto-oncogene (normally on chromosome 8) to a site on chromosome 14 puts *myc* under the same control as immunoglobulin genes, resulting in overproduction of *myc* product and transformation of B-lymphocytes to cause Burkitt's lymphoma.

PATTERNS OF MUTATION AT THE SEQUENCE LEVEL

Fine-scale analysis of the molecular changes that underlie genetic disease and cancer can shed light on the mechanism of mutagenesis and on the agents responsible. There are often hotspots for mutagenesis in a particular gene, reflecting both the susceptibility of bases at that position to mutagenic damage and the importance of the corresponding amino acid sequence for the function of the gene product. For example, mammary tumors induced in rats by the methylating agent methylnitrosourea characteristically display a change in the second G of codon 12 of the *H-ras* oncogene.

Eight out of 16 cases of liver cancer from a region of China with known exposure to aflatoxin were found to have a mutation in codon 249 of the p53 gene, and seven of the eight were G→T transversions, as expected from operation of the A-rule on an adducted guanine. This codon is not commonly affected in liver cancer from regions where aflatoxin exposure is not a problem. Still looking at the p53 gene, G:C→A:T transitions are seen in 50% of colon carcinomas. These transitions are likely to be the result of spontaneous deamination of 5-methylcytosine. In inherited human disease, too, G:C→A:T transitions are common.

Invasive squamous carcinomas of the skin tend to carry mutations in the p53 gene, and the presence in several tumors of a double base change supports a causative role for UV light. A T̂T dimer, as a noncoding lesion, is expected to be nonmutagenic, as the A-rule predicts insertion of two As (correctly) opposite the dimer. But a ĈC dimer, by the same rule, will result in a transition, CC→TT. Such changes are very rare in internal cancers. They are especially common in skin tumors from XP patients whose cells (as discussed in chapter 11) do not remove the lesions.

SUMMARY AND PERSPECTIVE

Mutation is an intermediate biological endpoint. It represents the outcome of a complex interplay between the intrinsic instability of DNA, attack by DNA damaging agents, replication fidelity, and the various reparative processes in the cell. As long as the cell and its progeny survive, the mutation is a reminder of past assaults; but many cells that receive mutations die, so the memory is biased.

The mutation endpoint is a convenient, but not infallible, indicator of potentially carcinogenic effects. Most known animal carcinogens are positive when examined (with or without P450 activation; see Chapter 11) in the Ames test—an assay which measures the ability of chemicals to induce mutations in a special strain of *Salmonella* carrying a mutated marker gene for histidine synthesis. Back-mutation in this gene creates cells that can survive and grow on histidine-free medium; the revertant colonies on the bacterial culture plate are counted.

Assessing the exposure of human populations to environmental mutagens requires a different approach, but one also based on a marker gene; in this case *hprt*. Loss of HPRT activity requires only one mutagenic hit as the gene is present in only one active copy, and consequently significant numbers of *hprt⁻* cells are present in the lymphocyte population, reflecting long-term exposure of stem cells to genotoxic agents. *Hprt⁻* frequency rises with age, and is elevated in smokers compared with nonsmokers (Figure 6). It is relatively easy to examine the mutations

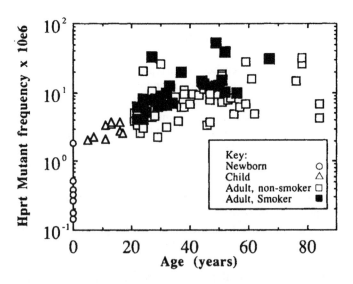

Figure 6. *Hprt* mutant frequency in T-lymphocytes from normal donors. (Kindly provided by Dr. Jane Cole, MRC Cell Mutation Unit.)

in selected *hprt⁻* clones at the DNA sequence level for information about the likely origin of the mutation.

Lymphocytes are convenient to isolate and easy to culture, and they give a fair indication of general mutagen exposure, but they are no substitute for the target cells specific for a particular type of cancer. A promising new approach is the use of oligonucleotide probes for oncogene mutations to pick up tumorigenic changes in DNA isolated from colon epithelial cells recovered from feces. Altered *ras* genes have been detected in this way in patients known to have CR cancer, and it should be possible to monitor those at risk of the disease.

ACKNOWLEDGMENTS

I wish to acknowledge the financial support of the Scottish Office Agriculture Environment and Fisheries Department, and to thank colleagues Dr. Neva Haites and Dr. Susan Duthie for their helpful comments and advice during preparation of this chapter.

REFERENCES

Ames, B.N., Shigenaga, M.K., & Park, E.-M. (1990). DNA damage by endogenous oxidants as a cause of aging and cancer. In: Oxidative Damage and Repair: Chemical, Biological and Medical Aspects. Proceedings of the International Society for Free Radical Research, November 14–20, 1990, Pergamon Press Inc., Oxford.

Fearon, E.R., & Vogelstein, B. (1990). A genetic model for colorectal tumorigenesis. Cell 61, 759–767.

Siminovitch, L. (1976). On the nature of hereditable variation in cultured somatic cells. Cell 7, 1–11.

Stark, G.R., & Wahl, G.M. (1984). Gene amplification. Ann. Rev. Biochem. 53, 447–491.

RECOMMENDED READINGS

Cairns, J., Overbaugh, J., & Miller, S. (1988). The origin of mutants. Nature (London) 335, 142–145.

Cole, J., Waugh, A.P.W., Beare, D.M., Sala-Trepat, M., Stephens, G., & Green, M.H.L. (1991). HPRT mutant frequencies in circulating lymphocytes: Population studies using normal donors, exposed groups and cancer-prone syndromes. In: Trends in Biological Dosimetry (Gledhill, B.L., & Mauro, F., eds.), pp. 319–328, Wiley, New York.

Friedberg, E.C., Walker, G.C., & Siede, W. (1995). DNA Repair and Mutagenesis. ASM Press, Washington.

Jones, P.A., Buckley, J.D., Henderson, B.E., Ross, R.K., & Pike, M.C. (1991). From gene to carcinogen: A rapidly evolving field in molecular epidemiology. Cancer Res. 51, 3617–3620.

Leach, F.S. et al. (1993). Mutations of *mut S* homolog in hereditary nonpolyposis colorectal cancer. Cell 75, 1215–1225.

Sidransky, D., Tokino, T., Hamilton, S.R., Kinzler, K.W., Levin, B., Frost, P., & Vogelstein, B. (1992). Identification of *ras* oncogene mutations in the stool of patients with curable colorectal tumors. Science 256, 102–105.

Strauss, B.S. (1991). The 'A rule' of mutagen specificity: A consequence of DNA polymerase bypass or non-instructional lesions? BioEssays 13, 79–84.

Zarbl, M., Sukumar, S., Arthur, A.V., Martin-Zanca, D., & Barbacid, M. (1985). Direct mutagenesis of Ha *ras*-1 oncogene by *N*-nitroso-*N*-methylurea during initiation of mammary carcinogenesis in rats. Nature (London) 315, 382–385.

Chapter 13

The Polymerase Chain Reaction

KARIN A. EIDNE and ELENA FACCENDA

BASIC METHODOLOGY

The polymerase chain reaction (PCR)[*] (Mullis and Faloona, 1987) has revolution-ized molecular biology. Experiments which could not previously be undertaken

[*]The Polymerase Chain Reaction (PCR) process is covered by US Patents owned by Hoffman-La Roche, Inc.

Principles of Medical Biology, Volume 5
Molecular and Cellular Genetics, pages 271–288.
Copyright © 1996 by JAI Press Inc.
All rights of reproduction in any form reserved.
ISBN: 1-55938-809-9

because of limited tissue or material can now be successfully carried out. Experimental procedures which used to take weeks can now be accomplished in a day. PCR represents a major technological advance for molecular biologists and clinicians alike, enabling the efficient *in vitro* synthesis of large quantities of DNA of known size and sequence for analysis.

PCR exploits features of *in vivo* DNA replication, in which single-stranded DNA is used as a template for the synthesis of a new complementary strand by sequentially linking triphosphate esters of the four deoxynucleosides, deoxyadenosine (A), deoxythymidine (T), deoxyguanosine (G), and deoxycytosine (C) in the order dictated by the template molecule. By chemical convention the opposite ends of a single DNA molecule are termed 5 prime (5') and 3 prime (3'). In the DNA double

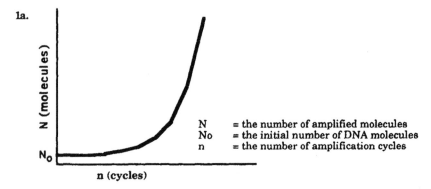

1a.

N (molecules)

N_0

n (cycles)

N   = the number of amplified molecules
N_0 = the initial number of DNA molecules
n   = the number of amplification cycles

1b

Cycle Number	Relative Amount of DNA
1	2
2	4
3	8
4	16
5	32
6	64
7	128
8	256
9	512
10	1 024
20	1 048 576
30	1 073 741 824

Figure 1. PCR amplification of target DNA. (a) Exponential increase in the number of target DNA sequences. The number of amplified molecules can be calculated from the equation $N = N_0 2^n$. (b) Geometric amplification of a DNA strand by PCR. Geometric amplification of the initial DNA results in a doubling of product with each round of the PCR, so that at the end of cycle 1 the amount of target has doubled, by the end of cycle 2 there has been a fourfold increase in DNA, and so on. Theoretically, after 30 cycles of amplification are complete there will be more than 100 million times more target DNA than was added at the outset of the experiment.

helix the two strands run antiparallel, such that the 5' end of one strand pairs with complementary nucleotides at the 3' end of the other strand, and vice versa. The segment of DNA chosen to be amplified is termed the target sequence. The target DNA sequence is selectively amplified using oligonucleotide primers which anneal to the template at the 3' ends of both strands flanking the chosen sequence. *In vitro*, the DNA templates are produced by heating the sample duplex DNA to near boiling which "melts" into single strands. Once denatured, oligonucleotide primers (short stretches of man-made DNA of specifically ordered nucleotide bases) are encouraged to anneal to their complementary regions on the template. This creates the small section of double-stranded DNA necessary for DNA polymerase binding. DNA polymerase enzymatically builds the second strand by incorporating nucleic acid bases at the free hydroxyl group at the 3' end of the primer. New primer binding domains are generated on each of the newly synthesized strands which, when heated to melting temperature, become available for further cycles of primer hybridization (annealing), nucleotide polymerization (extension), and strand separation (denaturation).

As its name suggests, PCR is a chain reaction in which the products of the first amplification cycle become the substrates for the next and so on. Theoretically the number of target DNA molecules increases geometrically, resulting in a doubling of product DNA with every round of amplification (Figure 1). The amount of product increases until either one of the products interferes with progress of the reaction, or one of the substrates becomes limiting.

The Genesis of the PCR

All of the elements required to perform the PCR, including the discovery of DNA polymerase by Arthur Kornberg in the mid-1950s, were in place, but were not associated with each other throughout the 1970s. It was not until early 1983 that PCR was eventually formulated in the mind of Kary Mullis, as he took time off from his work at Cetus Corporation for a regular weekend drive to his woodland cabin in north California. The musings which led to Mullis' realization that reiterative exponential amplification of DNA was possible are documented in an article written by Mullis entitled "The Unusual Origin of the Polymerase Chain Reaction" (Scientific American, April 1990). Although Mullis was extremely excited by his method for high level *in vitro* DNA replication, he was unable to raise the same enthusiasm in his workmates and colleagues within the Cetus Corporation. The first person to appreciate the real significance of Mullis' discovery was the patent attorney at Cetus, who promptly suggested Mullis write a patent disclosure describing his process. Although Mullis' technique was patented, his own findings were not published until 1987. While this milestone paper was in press, the first publication reporting the use of PCR was made by fellow Cetus scientists Henry Erlich and colleagues (including Kary Mullis), and described the prenatal diagnosis of sickle cell anemia (Saiki et al., 1985). The volume of literature using the method of PCR exploded from a total of 3 papers in 1985 to over 3,000 in 1993. Now PCR, in a number of different guises, has become an indispensable tool in the molecular biology laboratory.

The Basic PCR Reaction

Reaction Steps

In the original experiments carried out by Mullis and colleagues it was necessary to add fresh DNA polymerase for every round of amplification because the contemporary enzyme was rapidly and irreversibly denatured by heat. The more recent production of recombinant *Taq* DNA polymerase (cloned from the hot-

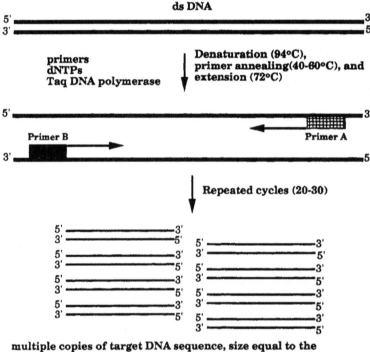

multiple copies of target DNA sequence, size equal to the distance between primers A and B

Figure 2. Basic sequence of steps required for PCR. Initially the DNA is heated almost to boiling to denature double-stranded DNA, forming single DNA strands to which PCR primers will anneal as the temperature is reduced. Once annealed, the temperature is increased to the optimum temperature for the extension of the primers by a DNA polymerase. At the end of many cycles of denaturation, annealing and extension, many more copies of the target DNA sequence will have been synthesized.

spring dwelling bacterium *Thermus aquaticus*), which can withstand repeated exposure to high temperatures (94–95°C), has been of major importance to the development of modern PCR because of its resistance to the degradation caused during the repeated PCR cycles of denaturation, primer annealing, and extension. Latter-day DNA polymerase need only be added at the outset of an experiment, retaining its activity throughout the numerous PCR cycles. Twenty to thirty cycles will normally generate sufficient DNA to allow visualization following electro-phoretic separation on agarose or polyacrylamide gels (Figure 2).

Reaction Components

PCR can be successfully applied to many and varied applications and in this context is extremely user-friendly. A single cycle of DNA polymerization, how-ever, requires the precise interaction of a complex of components and by this criterion is not a simple process.

Taq DNA polymerase. The number of primed template molecules that can be extended in a PCR is equal to the number of *Taq* DNA polymerase molecules present. Therefore, enough enzyme must be added at the outset of the reaction so that the expected number of amplified templates can be extended. *Taq* DNA polymerase extends the growing DNA strand at a rate of approximately 150 nucleotides/second/enzyme molecule. For example, if the target sequence is 1.5 kb, it will take at least 10 seconds for its polymerization. Extension cycles are generally increased beyond this minimum so as to prevent premature termination of strand elongation. *Taq* DNA polymerase is invariably supplied with a buffer containing the correct concentrations of constituents required for polymerization to proceed. The reaction also requires the addition of deoxynucleotide triphosphates (dNTPs), the building blocks of DNA.

As with all other biochemical processes, DNA replication is not perfect. DNA polymerases occasionally, and randomly, incorporate incorrect nucleotides into the extending DNA chain. This is not a major obstacle *in vivo* due to the existence of naturally occurring proofreading and correcting enzymatic machinery. *Taq* DNA polymerase, however, does not have this capability. Under the conditions used in PCR this enzyme has an error rate of approximately one in 1,000 incorporated nucleotides. Any error occurring in the first few rounds of amplification will be maintained throughout the reaction, so for this reason it is important to recognize that PCR bands show significantly higher error rates than non-PCR material.

Template DNA. Template DNA does not have to be thoroughly purified for successful amplification—impurities which inhibit polymerization (e.g., chelating agents) should be removed or diluted out—nor does the amount of DNA template have to be of high order. It has been shown experimentally that the DNA from a single cell is adequate for PCR amplification.

Primers. Primers should ideally be 15 to 30 nucleotides long with annealing temperatures in the range 40–60° C. The primers should be designed to flank the region of DNA to be amplified, and exhibit almost complete homology with the template for high stringency hybridization. High stringency means that the PCR will amplify only the target sequence and unnecessary ghost bands will not be formed. Inter- and intra-primer homology should be avoided to prevent primers annealing to each other, or folding into themselves, thus being effectively unavailable to prime the target sequence.

Control Reactions and Contamination

As contamination with extraneous DNA is a hazard when performing PCR, a negative control (containing no template DNA) should always be included in the set of reactions to evaluate this possibility. A positive control (one which you know works) should also be included as a test for the efficiency of the cycling conditions used in the experiment. Among the best ways to control contamination are to use separate pipettes for pre- and post-PCR liquid handling, and to use commercially available pipette tips fitted with special filters to prevent liquid aerosols containing DNA from contaminating the pipette itself.

Cycling Conditions

Temperature control during cycling is of critical importance, and therefore it is well worth investing in a reliable and efficient automated thermal cycler. It is advantageous to be able to accurately control the rate of temperature change, or ramping, to maintain the fidelity of the PCR. If the reactions are allowed to cool too slowly or too rapidly from denaturing to annealing temperature, it is possible that the primers will anneal randomly, that is, at incorrect positions on the template. This will produce amplified molecules of unpredicted sequence and size, and reduced levels of expected product.

An initial 3–5 minute cycle at 94° C is used to thoroughly denature the duplex DNA generating the single-stranded (ss) template to which the primers and polymerase enzyme can bind. This initial denaturation is rapidly followed by repeated cycles (approximately 30) of denaturation (60 seconds), annealing (60 seconds), and extension (1–2 minutes), at 94° C, 40–60° C, and 72° C, respectively. The annealing temperature (T_a) can be determined by the equation $T_a = T_m - 5° C$, where T_m is the melting temperature of the primer if it were double-stranded DNA. T_m is often calculated automatically by DNA handling computer software, but can also be ascertained manually by the equation

$$T_m = [62.3 + 0.41(G+C\%) - 500]/\text{length (in base pairs)}$$

where G+C% is the percentage composition of deoxyguanylate and deoxycytidylate bases in the primer sequence. A final incubation at 72° C for 3–5 minutes completes extension.

APPLICATIONS

PCR technology is being applied in many areas of scientific research, including disease diagnosis, detection of pathogens, detection of DNA in small samples, comparisons of DNA across phyla and through time, and gene sequencing. Its impact has been extensively felt in most biological and medical disciplines including molecular biological research, forensic science, paleogenetics, mutational analysis of disease etiology, and the Human Genome Project. The source of the DNA is irrelevant, be it a human hair, mummified flesh, bones preserved under boglike conditions, fresh blood or tissue specimens, or cells recovered from amber entombed parasites from the Jurassic period (Cano et al., 1992a,b, 1993).

PCR and its Uses as a Diagnostic Tool

PCR is being used for the rapid detection of pathogens, especially those whose *in vitro* culture is either impossible and/or time consuming, or for whom antibody-based tests are insensitive. PCR also increases the number of specimens that can be analyzed for any organism at any one time. For example, *Treponema pallidum*, the causative agent in syphilis, which will not grow on artificial media, has been detected in patients by PCR. PCR-based detection of *Mycoplasma genitalium* is being used to investigate the pathogenicity of the organism and its main tissue tropism. *Mycobacterium tuberculosis* takes up to six weeks to detect using conventional methods, whereas PCR can detect as few as 10 organisms in only a few hours. The presence of many more organisms, including *Mycobacterium leprae*, *Rickettsia typhi*, and several *Chlamydia* variants can be detected using PCR-based approaches.

Viral pathogens can also be detected by PCR-based methods. HIV was identified by PCR. In this way DNA from peripheral blood samples can be amplified using primers specific for genes belonging to the virus, with an HIV-positive sample producing a DNA fragment of the correct size. The detection of the hepatitis B (HBV) retrovirus by PCR is 10-fold more sensitive than traditional HBV tests. PCR amplification of gastroenteritis causing adenoviruses directly from diluted stool samples proves to be a rapid and reliable detection system. PCR is no longer used only for the detection of pathogens, but is also important as a sensitive diagnostic tool.

Prenatal Sex Determination

Early prenatal diagnosis of the sex of an embryo is very important since certain sex-linked conditions will only be inherited by infants of one sex. Male sex-determination, for detecting inherited disorders, is possible because of the unique DNA sequences carried on the Y-chromosome. A PCR-based test for these sequences can be carried out after *in vitro* fertilization, by the removal of a single cell from the pre-implantation embryo. The technique is shown in Figure 3. The production

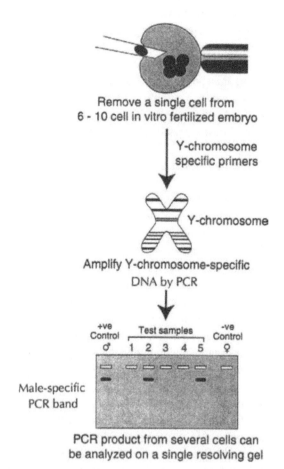

Figure 3. Sex determination of a preimplantation embryo. PCR experiments are carried out using DNA obtained from a single cell harvested from a 6–10 cell *in vitro* fertilized embryo. To confirm a male embryo, PCR primers which anneal to specific sites on the Y-chromosome are used. The appearance of a PCR band on the resolving gel indicates that the embryo is male, no PCR band indicates that the embryo is female.

of a PCR fragment of the expected size indicates that the embryo is male, and by elimination, the remaining female embryos will go on to be used for implantation.

It may be possible to detect the presence of disorders such as Down's syndrome and cystic fibrosis by noninvasive PCR analysis of fetal cells collected from a maternal peripheral blood-sample. Fetal cells which detach from the placenta gain access to the mother's circulation, but are only present in very low numbers (approximately one fetal cell per 1 million maternal cells). There are no current methods able to detect the origin of these different cells, but PCR could possibly amplify just the fetal DNA segment which carries the genetic defect being screened. A prenatal test such as this would negate the risks to the fetus which are repercussions of current amniocentesis and chorionic villus sampling currently in use.

PCR Analysis of Mutations in DNA

PCR has achieved widespread use in the analysis of genetic disease and cancer. Specific mutations in the genes of interest have been shown to relate to a variety of human disorders and can result in genetic dysfunction. It is important to determine the nature of the mutation as this has ramifications pertinent to a patient's diagnosis and the subsequent treatment regime adopted. The target region is usually amplified from genomic DNA or cDNA and examined for mutations or polymorphisms by DNA sequencing, hybridization with allele-specific oligonucleotides, restriction analysis, or enzymatic/chemical cleavage.

Several variations of the basic PCR are now used to identify mutations in genes and consequently many human genes have now been cloned and sequenced. The detection of altered DNA sequences involved in genetic diseases used to be too big a task for routine diagnosis. PCR and automated DNA sequencing have redressed this situation making it possible to rapidly define disease-causing nucleotide alterations in a significant percentage of cases. While Southern blot analysis (the traditional method of detecting genetic mutations) can take up to two weeks, PCR results can be produced in one day. This rapid time-scale is of obvious advantage in prenatal diagnosis of genetic defects. Prospective parents now have a longer period of time in which to make their decision, while still permitting a clinically safe termination of pregnancy. Duchenne muscular dystrophy, Lesch-Nyhan syndrome, β-thalassemia, and hemophilia A are just a few of the genetic diseases which can now be rapidly diagnosed by PCR.

PCR amplification is used in the monitoring of cancer therapy. The pattern of cancer-causing DNA lesions is monitored as treatment with cytotoxic drugs proceeds. When the presence of malignant cells can no longer be detected, therapy can be curtailed or resumed if cancer cells reappear. PCR techniques are capable of detecting a single cancer cell present in a sample of 1 million normal cells. This is 10,000 times more sensitive than the traditional method of detection using Southern blotting.

Mutations in Receptor Proteins

A multitude of receptors have been successfully cloned and sequenced, making possible the identification of diseases linked to abnormal receptor function, be it in the form of modified ligand interactions, or altered intracellular signaling. One of the first cloned human receptors was the low-density lipoprotein receptor (LDL-R). This receptor is membrane bound and is responsible for the uptake of cholesterol into cells which is subsequently used for membrane synthesis. It was hypothesized that a defect in this receptor may be a possible cause of cases of familial hypercholesterolemia (FH). In fact, in Caucasian patients, at least three point mutations of the LDL-R have been associated with FH. In addition, a mutation which introduces a premature stop-codon has been identified (Leren et al., 1993). This mutation truncates the receptor protein such that the carboxy terminal membrane spanning domain is not transcribed, resulting in loss of membrane anchorage.

G-protein coupled receptors exhibiting altered signal transduction have recently been reported as expressing genetic mutations. For example, mutations in the thyroid stimulating hormone (TSH) receptor have been associated with hyperthyroidism (Parma et al., 1993). PCR was used to amplify DNA extracted from tissue obtained following surgical removal of hyperfunctioning pituitary adenomas from patients. Sequence analysis of these products showed the existence of mutations of key amino acids in the third cytoplasmic loop of the receptor. Expression in an *in vitro* eukaryotic system showed that the receptor was constitutively activated in the absence of TSH, and unregulated by normal feed-back mechanisms. This mutation may underlie the clinical manifestation of symptoms of hyperthyroidism. A mutated version of the luteinizing hormone receptor (LH-R) has been associated with familial male precocious puberty (Shenker et al., 1993). In this case, the LH-R gene carries a single-point mutation in the sixth transmembrane helix of the receptor, creating a permanently switched-on receptor, akin to the mutated TSH-R. The permanently activated receptor results in excessive levels of steroidogenesis prior to the male pubertal LH surge. In affected families this untimely secretion of steroid hormones stimulates sexual maturation at ages as young as four years.

PCR together with DNA sequence data is responsible for the accelerating accumulation of documented diseases caused by receptor mutations. Just a few of the mutated receptors implicated in disease etiology identified in 1993 alone include those for insulin, neuropeptide-Y (Y1), adrenocorticotrophin, thyroid hormone, growth hormone, interleukin-6, interleukin-2, and the nuclear androgen- and T-cell receptors. More G-protein coupled receptors involved in human diseases are reviewed by Clapham (1993).

Single-Stranded Conformational Polymorphisms (SSCPs) Detected by PCR

Since PCR alone does not identify the nature of mutations in genetic material, complementary post-PCR protocols must be adapted to analyze the amplification products. There are several methods currently in use to detect single-point mutations in DNA; these include sequencing, temperature gradient gel electrophoresis (TGGE), denaturing gradient gel electrophoresis (DGGE), polymerase chain reaction-restriction length polymorphism (PCR-RFLP), and a novel technique known as polymerase chain reaction-single strand conformational polymorphism (PCR-SSCP; Orita et al., 1989).

Until recently the methods employed to detect mutations were relatively time-consuming and not at all easy to perform. PCR-SSCP is now proving to be better suited to this purpose as this is a simple and rapid procedure, therefore meeting the requirements of any method to be employed for large-scale analyses such as screening and diagnosis. SSCP resolves the products of a PCR in such a way as to identify changes in ssDNA sequence compared to wild-type DNA. TGGE and DGGE are both used to obtain the same information; however, PCR-SSCP is proving to be better suited to the detection of mutations in genomic DNA.

The principle of TGGE, DGGE, and PCR-SSCP is the same, i.e., that changes in nucleotide sequence confer changes in the folding, or secondary structure, of ssDNA. This in turn alters the mobility of the strands within a resolving polyacrylamide gel matrix. TGGE requires a special electrophoresis set-up to generate an increasing temperature gradient which separates or "melts" the double stranded PCR product as it migrates through the gel. DGGE uses specially created polyacrylamide gels containing an increasing gradient of denaturant (usually formamide and urea) towards the distal portion of the gel to dissociate the duplex fragments during migration. The techniques and equipment involved in these two techniques are not standard, and it is often difficult to achieve the conditions required for success. SSCP however, denatures the strands before they are loaded onto the gel and therefore a normal equipment set-up can be used. Although the conformational changes involved are probably very minor, the resolving power of polyacrylamide gels is such that a single base substitution in a sequence of several hundred is detectable. Unfortunately, the sensitivity of PCR-SSCP means that this type of analysis may be more sensitive to the replication errors that occur during PCR.

The sequence of operations for PCR-SSCP involves an initial PCR amplification of target DNA from either genomic or cDNA which is suspected to contain one or more mutations. γ-^{32}P-labeled primers are used to increase the sensitivity of detection. Ideally PCR fragments of 200–300 base pairs should be generated, but longer fragments can be shortened by cutting with restriction enzymes. Next, the PCR products are denatured, then loaded and run on a neutral polyacrylamide gel with the aim of resolving the mutated DNA from the unmutated DNA. Finally, the gel is exposed on X-ray film, producing an autoradiogram on which the DNA bands are visible. The amount of detectable product in isotopic-PCR-SSCP is much smaller than that detected by some other PCR-based techniques which use indirect staining methods. As little as 5 μl reactions (using labeled primers) are satisfactory for most sequences, limiting exposure of laboratory workers to dangerous radioactive isotopes.

Recently, however, Oto et al. (1993) have successfully substituted a smaller and less expensive minislab gel apparatus (9 × 7 × 0.1 cm) for the more commonly used 20 × 40 × 0.02 cm gels, as well as using a sensitive silver stain to reveal the positions of DNA bands in the gel, instead of using radioisotopes to label the PCR primers. These adaptations help to make PCR-SSCP less costly on reagents and reduce the use of radioactivity in the laboratory, while maintaining satisfactory results.

The results of PCR-SSCP analysis merely indicate that a mutation exists. It is not possible to predict from the direction of the shift in mobility what type of base change is involved. The PCR fragment must then be eluted from the gel and subjected to DNA sequencing to determine the exact nature of the base alteration, and its position within the sequence.

Mutations in the p53 gene which have been linked to breast carcinoma have recently been identified using PCR-SSCP, as have conformational polymorphisms in the KIT proto-oncogene. The latter, which encodes a tyrosine kinase receptor crucial in hematopoiesis, can cause myelodysplasia (MDS).

CLONING STRATEGIES

PCR-amplified DNA can be directly cloned into plasmid vectors. Commercially available kits use a method of insertion which utilizes the fact that some DNA poly-merases add dNTPs, preferentially deoxyadenosine (A), to the 3'-end of the polymerized chain independent of the original template sequence. The plasmid is linearized by restriction enzyme digestion and then treated in such a way as to generate single T overhangs at the cut 3' ends of the plasmid vector. The PCR fragment anneals to the plasmid via the complementarity of the single terminal A and T residues, and is covalently linked through the action of the enzyme DNA ligase. Once ligated the plasmid-insert complex is transformed (by electroporation or chemical/heat-shock methods) into appropriate bacterial cells whereupon the DNA is replicated, and can be purified for sequencing and other uses.

To overcome the difficulties often experienced when cloning, the primers used for PCR can be designed to incorporate a restriction enzyme site. The PCR product can then be digested and cloned as before, with the vector of choice also digested following incubation with the same restriction enzyme. Since the 3' end of the primer is important for extension by *Taq* DNA polymerase a restriction site is created at the 5' end of the primer by incorporating the appropriate bases.

Degenerate Primers for DNA Amplification and cDNA Cloning

Because degeneracy exists in the genetic code (i.e., several codons can code for the same amino acid, except methionine) it is impossible to determine the exact nucleic acid sequence of a protein of known amino acids. However, for PCR, amplification primers can be designed to cover all of the possible nucleic acid sequences that could encode a given polypeptide. Such primers are termed degenerate primers, and can be used for PCR when only a small portion of the protein sequence is known, when looking for protein isoforms, or receptor sub-types (Eidne, 1991). Degenerate PCR products are then cloned and sequenced in the same manner as other PCR fragments.

Automated Fluorescent DNA Sequencing

Following amplification and isolation, the PCR product can either be directly sequenced or sequenced after subcloning into a plasmid vector. Traditional dideoxy terminator DNA sequencing has been adapted so that during the PCR stage only one strand of the duplex template is amplified. This is known as asymmetric PCR (Gyllensten and Erlich, 1988; Mihovilovic and Lee, 1989). A single primer, annealing to only one strand of the target DNA, is used in the PCR and radioactive dideoxynucleotide (ddNTP) terminators are used (Sanger et al., 1977). When one of these ddNTP terminators is incorporated into the extending DNA strand, the polymerase ceases to function, and extension stops. During the cycles of PCR, fragments are produced that are lengthened to varying extents and terminated by a radiolabeled ddNTP. Four tubes are used to amplify the same target but each contains only one radiolabel, either A, G, C, or T. The four reactions are run separately on a polyacrylamide gel from which an autoradiogram is produced. The sequence of the DNA is read from the pattern of bands revealed on the X-ray film representing the gel.

The method that is used for automated fluorescent DNA sequencing is a further adaptation in which fluorescent dyes are used to label the ddNTPs rather than radioisotopes (A, T, C, and G are labeled with differently fluorescing dyes). Along with deoxynucleotides, the fluorescently labeled ddNTP terminators are added to the sequencing reaction. The PCR products arising are all terminated by a fluorescent dye corresponding to the complementary nucleotide at that position in the template. Some methods allow the whole sequencing reaction to be carried out in one tube, and run in only one lane of a large polyacrylamide gel, which is a great saving over other traditional and automated sequencing methods which use four tubes and four lanes.

Once PCR is complete, the DNA products are extracted to remove unincorporated dyes and other reaction components, and then run on a denaturing polyacrylamide gel. It is therefore ssDNA which migrates through the gel. By arranging these fragments according to size, the smallest fragments migrate most quickly, while the larger fragments are size retarded. The automatic sequencer uses a laser to scan across the gel, continuously recording the fluorescent pattern, as the fragments migrate down the gel. The sequencer then constructs the sequence of the DNA from the order of the fluorescent bands it detected. The 5'-to-3' sequence of nucleotides in the template strand is computed by inverting the actual sequence obtained since the polymerase generated fragments are complementary to the template and generated in the 3'-to-5' direction. The data are collected into files, and can be analyzed using computer programs which handle this information and which give the actual sequence of the DNA under investigation.

Amplification of DNA to Determine Genomic Structure

Genes consist of two types of regions, exons and introns. Exons represent the DNA which will eventually code for protein, but these are interspersed with intronic regions which, although transcribed into RNA, do not code for amino acids, and are spliced out during the generation of mature, translatable mRNA molecules. To determine the exon-intron configuration of a gene from its cDNA, the protocol outlined in Figure 4 is followed. Two sets of PCR reactions are run, one with primers 1 and 2, and the other with primers 1 and 3. The PCR products are then run on an ethidium staining agarose gel for comparison of the product sizes amplified from the two different templates. If intron sequence is present the PCR bands from genomic DNA will be longer than those from the cDNA with the difference in length being the length of the intronic structure.

Amplification of RNA by Reverse Transcriptase-PCR

Established methods of RNA analysis include *in situ* hybridization, Northern blots, S-1 nuclease assays, and RNase protection studies. The detection limits of these methods require the presence of 10^5–10^6 target molecules, except *in situ*

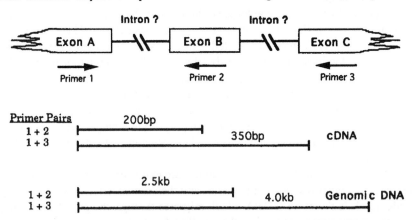

Figure 4. Determining exon/intron structure of a gene. The strategy used to determine the exon-intron components of a gene from its cDNA requires knowledge of the size and sequence of the cDNA clone representing the gene. Two sets of PCR reactions are run, one with primers 1 and 2, and the other with primers 1 and 3. If intron sequence is present the PCR bands from genomic DNA will be longer than those from the cDNA. The difference in length will be the length of the intron(s).

hybridization, which can detect as few as 10–100 RNA molecules in a single cell. Amplification of RNA is achieved by a method which combines reverse transcription and PCR, called RT-PCR. This method is thousands of times more sensitive than the traditional RNA blot techniques, sensitive enough to detect very rare mRNAs.

For PCR amplification to take place, a complementary DNA (cDNA) copy of the RNA must first be synthesized. This is achieved by a single incubation of about an hour with viral reverse transcriptases, Avian myeloblastosis virus RT (AMV RT), or Moloney leukemia virus RT (M-MLV RT). The RT incubation takes place in the presence of RNasin, a ribonuclease inhibitor, which protects the mRNA from degradation by RNases during cDNA synthesis. RNases are ubiquitous and are able to rapidly degrade the RNA template if their action is not inhibited. The cDNA is then amplified in a normal PCR cycle sequence by the addition of appropriate primers and *Taq* DNA polymerase. The DNA products of the PCR are then analyzed by customary methods.

Given that *Taq* DNA polymerase also exhibits significant RT activity at 68° C (Schaffer et al., 1990; Singer-Sam et al., 1990), the quantitation and amplification of mRNA can be further simplified to a one enzyme reaction. The *Taq* DNA polymerase RT generates the first strand cDNA in minutes. The reaction tube is then directly subjected to PCR conditions to allow amplification to occur.

Quantitation of mRNA

The goal of quantitative PCR is to determine from the final amount of product, either the absolute number of starting target molecules, or the relative starting levels among a range of samples. Several methods can be used to measure the total PCR product. These include the use of radioactively labeled primers; hybridization with a radioactively labeled probe (for example, a Southern blot, which is quantitated by counting the level of radioactivity in the hybridization band by measuring the intensity of the image produced on X-ray film by densitometric scanning); measurement of ethidium bromide luminescence on gel electrophoretic resolution; HPLC; or *in vitro* transcription incorporating radioactively labeled ribonucleotide substrates. To obtain truly quantitative data, a DNA target other than that to be quantified must be included in the reaction as a standard control of amplification.

Quantitation of mRNA from RT-PCR is most usefully achieved by a technique called competitive PCR. This method gets its name because the standard and target sequences actually compete for the same primers and therefore for amplification. One method involves the construction of exogenous standards which possess the same priming sites as the target DNA, but contain intervening sequence that is different, usually in length, so that the two products can be easily distinguished by gel electrophoresis. This method produces a standard sequence which mimics the primer binding properties of the target sequence. Quantitation is performed by comparing the amounts of the PCR products amplified from the two templates. Using this technique RT-PCR can be utilized to measure mRNA production in response to stimulation, or to follow inducible transcription of specific genes.

PCR-Based Evolutionary and Phylogenetic Studies (Paleogenetics)

Information about the structure and sequence of genetic material is being used to help identify evolutionary links between different species. Family trees similar to those produced by classical comparative methods are now being produced from molecular biological studies of genetic variation. The genes that are sequenced from ancient DNA must have been passed through lineages that have survived to the present, meaning that relatedness can be estimated by the percentage homology between genes for the same/functionally-similar proteins from different species, or from the same species at different time-points.

Mitochondrial genes have been of special importance as this part of the genetic code, inherited maternally, is not rearranged during meiosis, does not follow Mendelian patterns, and changes over relatively short periods of time owing to a high point-mutation rate. DNA sequence analysis of PCR-amplified mitochondrial DNA from various human sources, including a 7,000 year-old brain, has confirmed the African origin of mitochondrial DNA, suggesting that the common ancestor of modern man was an African woman (called "Eve") who lived 200,000 years ago (reviewed by Wilson and Cann, 1992). Despite criticisms of statistics used in the analysis of mitochondrial DNA mutation rates and their significance (Templeton, 1992), this remains as the most informative marker for uncovering human migration patterns of relatively recent times.

Until the advent of PCR, molecular studies of species relationships had been limited by the necessity of being able to obtain genetic sequence data only from DNA obtained from living organisms. Now, the relationships between extinct and living creatures can be investigated as DNA can be amplified from almost any source, living, recently dead, or ancient. Even if the ancient DNA is partially degraded the PCR is able to synthesize fragments which overlap and hybridize to each other, creating longer templates for subsequent cycles of amplification.

DNA has been amplified from such unlikely sources as museum skins, mummified human bodies, and fossilized plant material. Until recently the oldest material, dating back some 25–40 million years, was obtained from bees and termites trapped and preserved inside pieces of amber (Cano et al., 1992a,b). This has been far surpassed by the extraction, amplification, and sequencing of DNA from a weevil, also caught in amber, 80 million years prior to the demise of these bees and termites (Cano et al., 1993). These findings make the oldest surviving DNA approximately 120–135 million-years-old.

Forensic Science and DNA Fingerprinting

DNA sequencing is now the most powerful tool of the forensic scientist. The analysis of DNA samples used to depend on their quality and quantity. With the advent of PCR, results can be obtained from very small starting quantities, and even from partially degraded DNA, in a very short time, and without technical complications. DNA typing of HLA-DQα (which can distinguish between two individuals

91% of the time) from human hairs has been successfully achieved and is of great help to detectives when the only forensic clue at the scene of a crime is, for example, a single hair.

Maternal relationship can also be determined by analysis of the D-loop mitochondrial DNA which is maternally inherited. DNA fingerprinting is commonly used to solve disputes over paternity, and for correcting sample mix-up. These types of analysis require large amounts of DNA and hybridization probes, both of which are provided, of course, by PCR. Using PCR to amplify just the human DNA in mixed samples eliminates the problems of contamination by fungal or bacterial DNA that are often experienced in DNA fingerprinting.

SUMMARY

The polymerase chain reaction has become an essential molecular biological tool since its inception in the early 1980s. High level *in vitro* synthesis of selected DNA sequences is routinely used in basic research and in clinical diagnosis, giving results which are both more accurate and more quickly achieved than was previously possible. As the application of PCR widens, the method itself is being adapted to suit different requirements. Rapid detection of infectious diseases, such as tuberculosis (whose incidence is rising in developed countries) and HIV, is essential and is made possible by PCR. Prenatal diagnosis should become less of a risk for the fetus if a maternal blood sample will yield cells originating from the developing baby. It is possible that a single isolated cell may be used for PCR to detect fetal sex, if there is a risk of heritable sex-linked disease, or the presence of other genetic abnormalities such as β-thalassemias, cystic fibrosis, or Down's syndrome.

The remarkable power of PCR is also being used by molecular geneticists to unlock the wealth of information held in ancient DNA, retrievable from amber-entombed insects, mummified animal remains, and fossilized plants. This new branch of evolutionary biology is attempting to answer some of the historical questions brought up by archaeologists, such as where *Homo sapiens* arose, the patterns of migrations leading to colonization of new lands, and the order in which subsequent populations arrived and superseded the native inhabitants.

REFERENCES

Cano, R.J., Poinar, H.N., & Poinar, G.O. (1992a). Isolation and partial characterization of DNA from the bee *Proplebeia dominicana* (Apidae: Hymenoptera) in 25–40 million year old amber. Med. Sci. Res. 20, 249–251.

Cano, R.J., Poinar, H.N., & Poinar, G.O. (1992b). Enzymatic amplification and nucleotide sequencing of portions of the 18S rRNA gene from the bee *Proplebeia dominicana* (Apidae: Hymenoptera) isolated from 25–40 million year old Dominican amber. Med. Sci. Res. 20, 619–623.

Cano, R.J., Poinar, H.N., Pieniazek, N.J., Acra, A., & Poinar, G.O. Jr. (1993). Amplification and sequencing of DNA from a 120–135-million-year-old weevil. Nature 363, 536–538.

Clapham, D.E. (1993). Mutations in G protein-linked receptors: Novel insights on disease. Cell 75, 1237–1239.

Eidne, K.A. (1991). The polymerase chain reaction and its uses in endocrinology. Trends in Endocrinology and Metabolism. 2, 169–175.

Gyllensten, U.B., & Erlich, H.A. (1988). Generation of single-stranded DNA by the polymerase chain reaction and its application to direct sequencing of the HLA-DQA locus. Proc. Natl. Acad. Sci. USA 85, 7652–7655.

Mihovilovic, M., & Lee, J.E. (1989). An efficient method for sequencing PCR amplified DNA. BioTechniques 7(1), 14–18.

Leren, T.P., Solberg, K., Rodningen, O.K., Rosby, O., Tonstad, S., Ose, L., & Berg, K. (1993). Screening for point mutations in exon 10 of the low density lipoprotein receptor gene by analysis of single-strand conformational polymorphisms: Detection of a nonsense mutation - FH (469-> stop). Hum. Genet. 92, 6–10.

Mullis, K.B., & Faloona, F.A. (1987). Specific synthesis of DNA in vitro via a polymerase-catalyzed chain reaction. Methods Enzymol. 155, 335–350.

Mullis, K.B. (1990). The unusual origin of the polymerase chain reaction. Scientific American April, 36–43.

Orita, M., Iwanaha, H., Kanazawa, H., Hayashi, K., & Sekiya, T. (1989). Detection of polymorphisms of human DNA by gel electrophoresis as single-strand conformation polymorphisms. Proc. Natl. Acad. Sci. USA 86, 2766–2770.

Parma, J., Duprez, L., Van Sande, J., Cochaux, P., Gervy, C., Mockel, J., Dumont, J., & Vassart, G. (1993). Somatic mutations in the thyrotropin receptor gene cause hyperfunctioning thyroid adenomas. Nature 365, 649–651.

Saiki, R.K., Scharf, S, Faloona, F., Mullis, K.B., Horn, G.T., Erlich, H.A., & Arnheim, N. (1985). Enzymatic amplification of β-globin genomic sequences and restriction site analysis for diagnosis of sickle cell anemia. Science 230, 1350–1354.

Sanger, F., Nicklen, S., & Coulson, A.R. (1977). DNA sequencing with chain-terminating inhibitors. Proc. Natl. Acad. Sci. USA 74, 5463–5468.

Shaffer, A.L., Wojnar, W., & Nelson, W. (1990). Amplification, detection and automated sequencing of gibbon interleukin-2 mRNA by Thermus aquaticus DNA polymerase reverse transcriptase and polymerase chain reaction. Anal. Biochem. 190(2), 292–296.

Shenker, A., Laue, L., Kosugi, S., Merendino, J.J. Jr., Minegishi, T., & Cutler, G.B., Jr. (1993). A constitutively activating mutation of the lutenizing hormone receptor in familial male precocious puberty. Nature 365, 652–654.

Singer-Sam, J., Robinson, M.O., Bellve, A.R., Simon, M.I., & Riggs, A.D. (1990). Measurement by quantitative PCR of changes in HPRT, PGK-1, PGK-2, APRT, MTase, and Zfy gene transcripts during mouse embryogenesis. Nucleic Acids Res. 18(5), 1255–1259.

Templeton, A.R., Hedges, S.B., Kumar, S., Tamura, K., & Stoneking, M. (1992). Human origins and analysis of mitochondrial DNA sequences. Science 255, 737–739.

Wilson, A.C., & Cann, R.L. (1992). The recent African genesis of humans. Scientific American April, 22–27.

RECOMMENDED READINGS

Gelfand, I.M., Sninsky, J.J., & White, T.J. (eds.) (1990). In: PCR Protocols, a Guide to Methods and Applications, Academic, New York. In: PCR: Methods and Applications. Cold Spring Harbor Laboratory Press, Cold Spring Harbor.

Chapter 14

Molecular Cloning

KARIN A. EIDNE

Principles of Medical Biology, Volume 5
Molecular and Cellular Genetics, pages 289–305.
Copyright © 1996 by JAI Press Inc.
All rights of reproduction in any form reserved.
ISBN: 1-55938-809-9

INTRODUCTION

General Aspects of Molecular Cloning

It was during the 1950s and the early 1960s when the genetic code was described and the structure of DNA elucidated. However, it was not until the 1970s that experimental techniques became sophisticated enough for the gene to be studied in any great detail and the past two decades have witnessed remarkable accomplishments in the manipulation and cloning of DNA, generally known as recombinant DNA technology. Recombinant DNA technology has fulfilled its promise many times over as a wealth of information in many areas of biology has been generated. This technology has also spawned the new biotechnology industry which has been applied to produce drugs and vaccines.

New developments in cloning, sequencing, gene engineering, and mutagenesis have increased our understanding of how the structure of a gene relates to its function. The ability to introduce genes and targeted mutations into organisms ranging from plants to higher mammals has altered the way in which experiments are being carried out in every field of biology. Examining the sequence data obtained from a cloned gene can reveal invaluable information, generating insight into how organisms work. This information can ultimately be applied to advance our knowledge of how the cell operates as well as the genetic basis for human disease. Initiatives like the Human Genome Mapping project will generate a wealth of data of this nature and should improve our understanding of the genetic basis for human disease.

This chapter will provide an overview of the basic molecular cloning techniques. This is a very intensive area of study and there has been a vast number of publications describing the different permutations of these techniques in the past decade or so.

THE ISOLATION OF GENES BY MOLECULAR CLONING

A clone is defined as a population of identical cells, generally those which contain identical recombinant DNA molecules. The basic principle behind molecular cloning is to introduce into a host cell a single DNA molecule which is able to replicate in parallel with the host cell genome, but remain nonintegrated (i.e., existing as an independent element or episome). Three basic ingredients are required for cloning: host cells, usually *Escherichia coli*; DNA and possibly RNA; and enzymes.

Host Cells

Cloning is generally carried out in *E. coli* because of the short generation time and also because a wealth of knowledge exists regarding this bacterium. Early strains did not take up DNA very efficiently and also were unable to grow very well. Subsequent strains have been developed which grow extremely well, many with specialized properties useful for certain cloning experiments. Some of these properties can facilitate the identification, functional analysis, and sequencing of cloned DNAs. DNA can be introduced into the bacterial cells by several means. This process, called transformation, can be carried out by briefly heating the cells (heat shock) or by applying an electric current to the cells (electroporation). Cells can also be infected by phage carrying DNA particles (infection). Yeast, fungi, insect, plant, and mammalian cells are also used as host organisms in cloning experiments.

Genomic DNA and cDNA

The genome is defined as the complete set of genes of a particular organism. Complementary DNA (cDNA) is a stretch of DNA which has faithfully copied a stretch of RNA. DNA (genomic and cDNA) is double-stranded, while RNA is single-stranded. DNA is composed of four bases: adenine (A), cytosine (C), guanine (G), and thymine (T). These bases are attached to the 1'- carbon residue of a deoxyribose ring while a three phosphate moiety is attached to the 5' position of the ring. RNA also incorporates four bases except that T is replaced by uracil (U). Messenger RNA (mRNA) is susceptible to degradation by RNAses and as such can be difficult to manipulate. The cDNA prepared by an enzymatic method will be representative of the mRNA present in the initial preparation. Genomic DNA is made up of exons (coding DNA) and introns (noncoding DNA), while the processed mRNA and consequently cDNA has had the intervening intronic sequences spliced out by the cell's transcriptional machinery. The cell's machinery can splice some gene transcripts differently in that the same gene will produce distinct RNAs encoding several forms of the protein. These alternatively spliced forms of the protein are perhaps important in different cell types or at different stages of development. Genes from all eukaryotic animals appear to have introns which vary in number and size from gene to gene. Prokaryotic genomes like those of *E. coli* and yeast are 4.7 and 14 megabases, respectively. These genomes are much smaller than that of humans (3,000 megabases). Prokaryotic genomes can be used as model genomes for humans as many of the genes will be common to all eukaryotic cells. A recent finding has suggested that the relatively small genome of the pufferfish (400 megabases) should provide a more suitable model for the human genome (Brenner et al., 1993).

Enzymes Which Cut, Join, and Copy DNA

The discovery of enzymes, restriction nucleases, which would cut DNA at specific points to produce discrete fragments, as well as other DNA modifying enzymes, provided the primary tools enabling scientists to manipulate and to clone DNA. The initial discovery that the bacterium *Hemophilus influenzae* was able to degrade DNA from foreign cells but not DNA extracted from the cells of *H. influenzae* itself (Smith and Wilcox, 1970) led to the production of the HindIII enzyme. Since then more than 150 of these restriction enzymes have been isolated from a variety of bacterial strains and are commercially available. Many are routinely used in molecular cloning.

Restriction enzymes unambiguously recognize specific sequence motifs of four to eight base pairs (bp) of double-stranded DNA (the recognition sequence) and cut DNA at specific sites within or adjacent to the recognition sequence (Table 1). For a comprehensive list of all known restriction enzymes and their cleavage sequences see Roberts (1989). The number of residues in the recognition sequence determines the frequency with which the enzyme cuts. Statistically, in a stretch of DNA with symmetrically distributed sites, a four bp recognition sequence will occur once in every 256 bases, a six bp sequence once in every 4,096 bases and an eight bp

Table 1. The Recognition Sequences of Some Frequently Used Restriction Enzymes and the Microorganisms From Which They Originate

Enzyme	Microorganism	Sequence	Termini Generated
AluI	*Arthrobacter luteus*	5'..AG|CT..3' 3'..TC|GA..5'	blunt-end
BamHI	*Bacillus amyloliquefaciens*	5'..G|GATCC..3' 3'..CCTAG|G..5'	5'-overhang
Bgl/II	*Bacillus globigii*	5'..A|GATCT..3' 3'..TCTAG|A..5'	5'-overhang
EcoRI	*Escherichia coli*	5'..G|AATTC..3' 3'..CTTAA|G..5'	5'-overhang
EcoRV	*Escherichia coli*	5'..GAT|ATC..3' 3'..CTA|TAG..5'	blunt-end
Hae III	*Hemophilus aegyptius*	5'..GG|CC..3' 3'..CC|GG..5'	blunt-end
HindIII	*Hemophilus influenzae* R_d	5'..A|AGCTT..3' 3'..TTCGA|A..5'	5'-overhang
KpnI	*Klebsiella pneumoniae*	5'..GGTAC|C..3' 3'..C|CATGG..5'	3'-overhang
NotI	*Nocardia otitidis-caviarum*	5'..GC|GGCCGC..3' 3'..CGCCGG|GCG..5'	5'-overhang
PstI	*Providencia stuarti*	5'..CTGCA|G..3' 3'..G|ACGTC..5'	3'-overhang
PvuII	*Proteus vulgaris*	5'..CAG|CTG..3' 3'..GTC|GAC..5'	blunt-end
TaqI	*Thermus aquaticus*	5'..T|CGA..3' 3'..AGC|T..5'	5'-overhang

Note: The sequence shown is that of double-stranded DNA. Almost all cleavage sequences are palindromes, i.e., both strands read the same in either the 5' or the 3' direction.

sequence once in every 65,536 bases Thus, the four bp recognition sequence enzymes can cut DNA frequently generating many small DNA fragments, while an eight bp enzyme will only cut DNA rarely, producing very large DNA fragments. Two different enzymes may recognize the same cleavage sequence; these enzymes are known as isoschizomeric. Restriction enzyme cleavage generates three classes of DNA termini: 5'-overhang sticky-ends, 3'-overhang sticky ends, or blunt-ends (Table 1).

When constructing a recombinant DNA molecule, the step which joins DNA fragments together is catalyzed by the enzyme, DNA ligase. Although the enzyme most used, T4 DNA ligase, is isolated from *E. coli* infected with T4 phage, ligations are performed inside all living cells and are part of the DNA repair process. This process takes place, for example, when the phosphodiester bond between adjacent nucleotides is missing as a result of DNA replication, recombination, or breakages. Ligation reactions can be carried out when the DNA to be joined has either compatible overlapping fragments (sticky ends) or nonoverlapping fragments (blunt ends). The sticky-ended ligation is much more efficient than the blunt-ended one as the compatible ends can form hydrogen bond contacts with each other thus facilitating the enzymatic step.

In addition to enzymes which can cut and join DNA, there are enzymes which synthesize DNA. These include: DNA polymerase I (has 5'- to 3'- polymerase activity); reverse transcriptase (copies single-stranded RNA or DNA templates); and terminal transferase (adds residues to any free 3' terminus). Enzymes which degrade DNA include: S1 nuclease (degrades single-stranded DNA); endonuclease III (progressively degrades one of the two strands of double-stranded DNA); and DNAse I (introduces random nicks into double-stranded DNA). The specific properties of all the above mentioned enzymes can be utilized to manipulate DNA in genetic engineering experiments.

Cloning Vectors

Cloning vectors derived from naturally occurring plasmids were the first primitive tools that molecular biologists used for cloning genes. Plasmid vectors are double-stranded, circular stretches of DNA ranging in size from approximately 3 kb to 10 kb. They have the ability to carry fragments of foreign DNA of up to about 5 kb in length. The most widely applied vectors are compatible for use in *E. coli* host cells. Cloning vectors are also available for use in yeast, insect cells, fungi, plants, and mammalian cells. The simplest vectors are those based on plasmids obtained from bacteria. One of the original cloning vectors, pBR322, was designed to carry several features, e.g., restriction enzyme sites for the insertion of foreign DNA; antibiotic resistance genes which select for the cloned cells carrying the plasmid; and an origin of replication, *ori*, allowing efficient replication of plasmids to hundreds of copies per cell. More sophisticated vectors, many of them based on pBR322, have since been developed. Each of these vectors displays additional features which make them a better choice for certain cloning experiments. Vectors are now available which incorporate additional useful characteristics like reporter genes (allowing the expression of the gene to be detected in some systems); the ability to make single stranded DNA (useful for sequencing and for mutagenesis studies); RNA transcription sites such as T3, T7, and SP6 (for the transcription of RNA); a polylinker or multiple cloning site (a range of restriction enzyme cloning sites which facilitate the insertion or the removal of the cloned gene from the vector); and a *lac Z'* gene (for the synthesis of β-galactosidase which allows clones carrying both the plasmid vector and foreign DNA to be distinguished from clones which contain only plasmid vector DNA).

A range of bacteriophage vectors like lambda phage have been developed which allow the cloning of larger fragments of DNA (about 5 kb to 15 kb). The lambda DNA vectors, e.g., λgt10 and λgt11, are long linear stretches of DNA, much bigger than plasmid vectors, around 40 kb in size. These phage are packaged *in vitro* to form phage particles. The phage particles then infect the host bacterial cells thereby introducing their DNA into the cells. Usually the phage infect a lawn of bacterial cells and the infected cells lyse, forming a visible plaque on the agar plate.

A hybrid series of vectors, phagemids, which combine some of the features of plasmids and phage vectors have now been developed. As the size of DNA that can be carried in plasmid, lambda, and phagemid vectors is limited, cosmid vectors which can accommodate very large fragments of DNA, up to 45 kb in length, have also been developed. Cosmids, also a hybrid between phage and lambda vectors, carry *cos* sites (lambda phage sites which are required by enzymes for packaging DNA into the phage protein coat), an antibiotic resistance gene, Tet^R and the origin of replication, *ori*. These *cos* sites consist of single-stranded complementary stretches of DNA at each end of the lambda DNA. Following replication, many copies of phage DNA are joined together at the *cos* sites to form concatemers. Recombinant phage DNA can be produced by ligating genomic DNA and vector together, which is then packaged by packaging enzymes which recognize any two *cos* sites 35 to 45 kb apart.

Yeast artificial chromosomes (YAC) vectors can be used to clone huge fragments of DNA, hundreds of kilobases in length. Stretches of DNA 400 to 500 kb are ligated into these vectors which are then used to transform spheroplasts (yeast cells denuded of their cell walls). YACs incorporate features like a yeast centromere and two sets of telomeres, which are needed for replication of the vector as a linear artificial chromosome in yeast cells.

The Strategy of Gene Cloning

To carry out detailed studies of the structure of a gene of interest, large amounts of a homogeneous preparation of the gene will need to be obtained. Among the first genes to be cloned were those encoding very abundantly expressed RNA structural species like the 45S ribosomal RNA. Multiple copies of these types of genes are expressed in the genome making it a much simpler task to detect and isolate the relevant clones. In contrast, isolation of genes encoding proteins which are not abundantly expressed can present major difficulties. If the chromosomal or genomic DNA contains only a single copy of the gene, this will represent a very small amount of the total DNA in that cell. However, if the gene is expressed in appreciable amounts in a tissue or a cell type, using mRNA as a template to produce cDNA may circumvent some of the difficulties experienced in cloning genes.

When attempting to isolate a particular cDNA clone by carrying out cDNA library screens, the starting number of clones plated out on agar plates depends on the abundance of the particular mRNA expressed in the source tissue or cells. The sequence of interest will usually be present in the mRNA in an abundance of between 1 in 10^3 and 1 in 10^4, i.e., the protein will be expressed by between 0.1% and 0.01% of total mRNA. Between 10^4 and 10^5 clones will probably need to be screened in order to isolate the clone of interest. Very rare mRNA will be present somewhere between 1 in 10^5 and 1 in 10^6 in the source tissue. In situations like these, the task of screening is increased by an order of magnitude; in order to maximize the chances of finding the clone of interest more than 10^6 clones may need to be screened.

The basic principles of isolating cDNA clones and genomic clones are essentially similar. Several different procedures by which a gene can be cloned are available. When designing a gene cloning experiment, there are a number of basic steps to follow. To illustrate the techniques of molecular cloning, a series of possible experiments describing the cloning of a representative gene, gene A, will be outlined in the next two sections.

Preparation of a cDNA or a Genomic DNA Library

A cDNA library is a collection of cloned DNA fragments which are representative of the mRNA present in a particular cell or tissue preparation, while a genomic library is a collection of cloned DNA fragments sufficient in number to be likely to represent every single gene present in a particular organism. When preparing a cDNA library, the starting point is to obtain sufficient amounts of tissue or cells which are thought to express gene A. First, polyA$^+$-enriched mRNA is prepared from the tissue and used as a template for the synthesis of cDNA using the enzyme reverse transcriptase. A cDNA library is then constructed, where the synthesized cDNA fragments are individually inserted into an appropriate bacteriophage or plasmid vector. The vector acts as a vehicle that transports the gene into a host cell. Each of the million or so bacteriophage or plasmids in the library contain a unique inserted DNA fragment. The host cell and the DNA:vector construct then multiply producing many identical copies of the foreign gene. Following many cell divisions, a colony of identical host cells (a clone) is produced. Each cell is likely to contain at least one copy of a foreign DNA gene.

A genomic library is produced by enzymatically digesting genomic DNA from a particular organism into random fragments. This collection of fragments should be representative of the entire genome. As with the cDNA library, the genomic fragments are individually inserted into an appropriate plasmid or bacteriophage vector and thus make up the recombinant genomic library.

Identification of Recombinant Clones

Types of Library Screening

If the protein encoded by the gene to be cloned has been purified and partial protein sequence information is available, then it is fairly straightforward to design oligonucleotide probes with which to screen the library. These oligonucleotide probes are based on information contained within nucleotides coding for the protein sequence. This method of hybridization probing is the most straightforward method for identification of a recombinant gene from a library.

Another method of cloning is expression cloning, where the method of identification is based on detection of the translation product of the cloned gene. The purified protein can be used to generate antibodies to screen expression libraries (immunological screening). For those proteins whose purification is difficult, alternative cloning strategies are necessary in order to isolate the genes. These may include methods like the polymerase chain reaction (PCR), the *Xenopus* oocyte expression system, and functional assay in eukaryotic cells.

Library Screens

The libraries have no catalogue list of clones and genes, so to establish which of these plasmids or bacteriophage in the library contains gene A, the entire library must be screened. A bacteriophage library screen can be carried out as follows (Figure 1). The library (10^5 to 10^6 phages) is spread out on agar culture plates covered with a lawn of host bacteria. Each separate plaque comprises an individual clone of DNA. A small portion of the DNA from each phage plaque is transferred to a filter membrane. The DNA on the filter membrane is then hybridized to a radiolabeled gene A-specific probe. The probe will specifically bind to the plaque DNA attached to the membrane and, following autoradiography with X-ray film, will allow the identification of any positive plaques. These positive plaques can be isolated from the agar plate and following a few more rounds of screening (secondary and tertiary screens), a plaque-pure, homogeneous DNA clone containing gene A obtained. Because of their size, bacteriophage vectors are difficult to manipulate for further investigation. The DNA insert can be excised from this

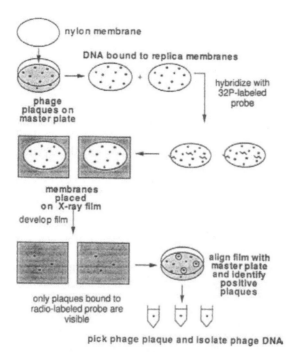

nylon membrane

DNA bound to replica membranes

hybridize with
32P-labeled
probe

phage
plaques on
master plate

membranes
placed
on X-ray film

develop film

align film with
master plate
and identify
positive
plaques

only plaques bound to
radio-labeled probe are
visible

pick phage plaque and isolate phage DNA

Figure 1. Library screening. Several thousand phage plaques (bacteriophage vector DNA library) or colonies (plasmid vector DNA library) are plated out on agar plates. The agar plates are overlaid with a filter (nylon or nitrocellulose) in order to create an exact replica of the plaque (or colony) pattern. The filters, to which DNA from the plaques or colonies is now tightly adhered, are incubated with a radiolabeled probe. The probe specifically binds to complementary sequences from the transferred DNA on the filters. Following washes to remove unbound probe, the filters are exposed to autoradiographic X-ray film. Positive clones, which appear as dark spots on the film, are identified on the agar plates and isolated.

vector using a restriction enzyme and inserted into a smaller, plasmid vector. The cloned gene A can now be propagated to produce large amounts to facilitate future study.

Polymerase Chain Reaction (PCR) Gene Cloning

The tools of cloning have been expanded by the recent development of the versatile PCR (Mullis and Faloona, 1987) (for further information see chapter 13). If some prior knowledge about the sequence of interest is available, PCR can be applied to clone that particular gene. It can be used to generate sufficient amounts of a stretch of DNA (a partial gene) producing a suitable probe for conventional library screening in order to isolate the full-length gene. It can also be used in some situations to directly isolate a full-length stretch of DNA, thereby circumventing the need to construct time-consuming DNA libraries. For example, members of the G-protein coupled receptor family have been isolated by the application of PCR techniques. New members of the G-protein coupled receptor family which shared homology in the conserved regions, but had diverse sequence outside these regions, have been identified and cloned (Libert et al., 1989, Eidne et al., 1991). The amino acid sequences common to this receptor family were used to design degenerate oligonucleotide primers. The PCR was performed using these primers which were based on conserved regions within the molecule.

Expression Cloning

Expression Cloning Using Xenopus Oocytes

Xenopus oocytes can be used as part of a cloning strategy to clone certain genes, e.g., DNA encoding receptors, channels, and gap junction proteins. The *Xenopus* oocyte translation system, where exogenous membrane proteins can undergo transient expression, has proved to be a critical tool in the elucidation of the molecular structures of many G-protein coupled and ion channel receptors. Gurdon et al. (1971) was the first to observe that foreign mRNAs injected into the oocytes of the South African clawed toad, *Xenopus laevis*, could lead to synthesis and translation. The oocyte is a self-contained system, capable not only of translation of exogenous mRNAs but also of posttranslational modifications such as phosphorylation, glycosylation, subunit assembly as well as the ability to insert the molecule into the membrane.

In order to express proteins like ion channel receptors (e.g., the GABA receptor; Blair et al., 1988) and G-protein coupled receptors (e.g., the gonadotropin-releasing hormone receptor; Yoshida et al., 1989) in *Xenopus* oocytes, intact mRNA needs to be injected into the oocyte's cytoplasm. For example, during the cloning of the cDNAs encoding the 5HT1c receptor (Julius et al., 1988) and the substance K receptor (Masu et al., 1987), mRNAs were transcribed *in vitro* from pools of cDNA clones, oocytes were then injected with these mRNAs and evaluated for expression of the receptor protein using electrophysiological techniques. To be able to transcribe RNA from the cloned cDNAs, it is important that the cDNAs primarily be inserted into a vector containing suitable transcription promotors (SP6, T3, or T7). cDNA libraries are therefore constructed in vectors containing promotors which flank the cloning site. The cDNA library is transcribed and the resultant transcription products injected into the oocyte. Following a receptor positive response, the library is split into a number of pools and each pool examined for a response. An active pool is progressively subdivided into smaller pools until a single clone is found which is capable of producing a positive response.

Immunological Screening

Expression cloning vectors like lambda gt11 are available which can promote the bacterial host to express a foreign protein. This protein will not normally be expressed by the host cells. Antibodies are required for immunological detection methods and the antibody should be specific for the protein in question. Recombinant colonies are screened with the labeled antibody (or labeled protein A, a bacterial protein which itself binds to the antibody) instead of with a labeled DNA probe. In this way, the recombinant clone can be identified and isolated.

Functional Assay in Eukaryotic Cells

Proteins which are functional in eukaryotic cells but not in bacterial cells have been cloned using a functional assay as part of the detection system. A variation of this method is called panning (Seed and Aruffo, 1987) and certain receptors (e.g., activin receptor; Mathews and Vale, 1991) have been isolated by this method. Briefly, mammalian cells which do not express the protein endogenously are transfected with a cDNA library prepared from a tissue source expressing the protein of interest. The vector carrying the inserted sequences has a promotor which allows it to direct the expression of the cDNAs.

GENE ANALYSIS

Following the isolation of a specific gene, the next stage is to obtain as much information as possible about this gene. Determination of the gene's DNA sequence will provide much useful information. Techniques are also available which allow one to determine which tissues are producing this gene and also on which chromosome the gene is located. It is also possible to study the expression of the mRNA encoded by the gene at a cellular level.

DNA Sequencing

DNA sequencing, the ability to determine the precise order of the nucleotides in a fragment of DNA, is probably the most valuable technique available in molecular biology. Two very different techniques for carrying out this procedure were developed at around the same time. In the United Kingdom, Sanger and Coulson developed the chain termination method (Sanger et al., 1977a), while in the United States, Maxam and Gilbert developed a method based on the chemical degradation of DNA chains (Maxam and Gilbert, 1977). DNA sequencing has now become more and more automated and machines which run and analyze sequencing gels are now available.

The first DNA molecule to be sequenced was the 5 kb chromosome of the bacteriophage, Φx174 (Sanger et al., 1977b). The number of nucleotides in a fragment of DNA accurately sequenced per sequencing experiment is usually only about 400. This means that overlapping fragments of DNA will need to be sequenced, aligned, and merged before the final sequence can be obtained. Computers are very useful for carrying out this task.

The Use of Computers in DNA Analysis

Computers have simplified the laborious task of analyzing long stretches of DNA sequence. Raw sequence data produced from either manual or automated DNA sequencing projects can be loaded directly onto a computer for further analysis. Computer software designed to manage DNA sequences, assembles the data from each sequencing experiment and compiles the nucleotides into a contiguous stretch. These stretches of nucleotides can be compared and aligned with other DNA sequences. Areas of homology can be identified as can restriction enzyme recognition sites.

The computer can search through the sequence to try and identify open reading frames (ORFs) (Figure 2). These are regions of the sequence which can potentially code for proteins. Triplet codons are translated into their corresponding amino acids and the start codons (ATG) and stop codons (TAA, TGA*, TAG) identified. The nucleotide sequence is converted into amino acids in both orientations which allows for six possible reading frames. When a long stretch of contiguous nucleotides uninterrupted by stop codons can be translated into protein, this is known as an ORF.

A series of As or Ts in the nucleotide sequence could indicate a polyA tail, identifying the 3' end of the molecule. Both the nucleotide and the protein sequences can be compared with other known sequences by carrying out a database search on the computer. These databases, updated on a daily basis, currently catalogue over

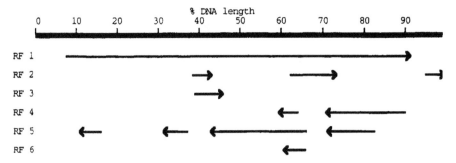

Figure 2. Six-phase open reading frame (ORF). Diagram showing locations of stretches of DNA longer than 75 bp with no stop codon in frame. Arrows indicate the direction of read. There are six reading frames (RF) and RF 1 starts on the first nucleotide.

200,000 DNA sequences; this number doubles every two years. Information obtained in this way will indicate whether or not the sequence has been previously identified, or shares homology with other molecules. If the sequence of interest is identical to a known sequence, then this provides conclusive evidence as to the identity of the cDNA clone.

Analysis of DNA by Agarose Gel Electrophoresis

Following digestion by restriction enzymes, DNA molecules of different sizes can be easily separated by means of electrophoresis. DNA is negatively charged and can be driven through porous agarose gels in an electrical field. The smaller DNA fragments will migrate at a faster rate than the larger DNA fragments. Depending on the length of the agarose gel, the time taken to run, and the intensity of the electrical field, the DNA will separate out into discrete bands. The DNA can then be visualized following staining with a fluorescent dye, ethidium bromide. Restriction maps are obtained by mapping the positions where the DNA digested by a range of enzymes electrophoretically migrates on a calibrated gel. Different patterns will be obtained for DNA from different sources.

Very large DNA molecules (> 5000 kb) can be separated on gels by means of pulsed field electrophoresis (Schwartz and Cantor, 1984), an electric field which alternates rapidly between two pairs of electrodes, each pair set at a 45º angle. The DNA molecules in the gel have to continually change direction following each pulse. The shorter DNA molecules can reorientate themselves faster than the larger ones, the net result being that they move faster. Thus, the alternating fields allow the DNA to separate according to size.

DNA and RNA Hybridizations

Analysis of the expression of a gene, the organization of a gene, or the localization of a particular sequence within a gene can usually be achieved by using techniques based on the ability of labeled nucleic acid probes to hybridize with specific DNA or RNA sequences.

Southern and Northern Hybridizations

To localize a specific DNA sequence from within a particular gene, Southern blotting, first described by Southern in 1975, is carried out. Briefly, DNA is digested with restriction enzymes and then electrophoretically separated according to size on an agarose gel (Figure 3). The separated DNA is then transferred to a filter membrane (nylon or nitrocellulose). The membrane is then incubated with a radiolabeled probe which will specifically hybridize to any stretches of complementary DNA. The membrane is subjected to a series of washes to remove unbound

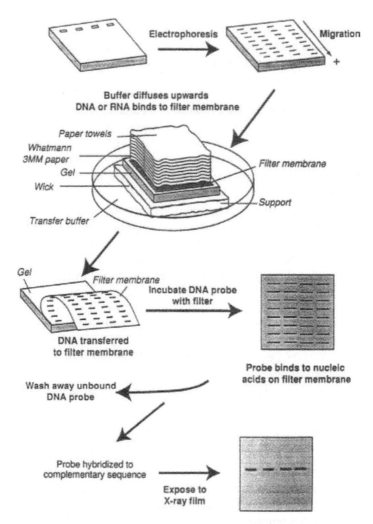

Figure 3. RNA and DNA blot analysis. Agarose gel electrophoresis is carried out to separate by means of size either RNA or DNA. The separated RNA (Northern blotting) or DNA (Southern blotting) is then transferred or blotted onto a filter membrane (nitrocellulose or nylon) by placing the filter directly onto the gel followed by a stack of paper towels. By a process of diffusion, the nucleic acids become firmly attached to the filter, thereby making an exact replica of the gel. The filter is then incubated with a radiolabeled probe which is designed to specifically recognize the desired sequence. Following several washes which remove any unbound probe, the filter is exposed to autoradiographic X-ray film. The bands visualized on the gel are complementary to the probe used.

probe and then exposed to an X-ray film. The autoradiographic results will show a pattern of bands corresponding to the fragment of DNA complementary to the probe. Results obtained from Southern blotting can be used to detect the presence of gene deletions or rearrangements, particularly those found in a variety of human disorders. Zoo blots, Southern blots of genomic DNA extracted from an array of species, can provide useful information on the degree of evolutionary conservation of a particular gene. Noah's Ark blots are similar, except that they contain pairs of genomic DNA from both the male and female of the species and can provide information on genes which are only expressed, or expressed to a varying degree, in one sex, e.g., the male infertility factor (Ma et al., 1993). Southern blotting of chromosomal DNA molecules, separated by pulse-field electrophoretic techniques permits the localization of the chromosome carrying a particular gene.

Northern blotting is a technique for transferring bands of RNA from an agarose gel to a nitrocellulose or nylon membrane. RNA is separated on a gel and transferred in a fashion similar to that described for DNA blotting. After probing with an appropriate probe, the RNA species encoding the gene of interest can be identified. This method can provide information as to the size of the specific RNA, which tissues or cell types are expressing the gene, and the regulatory factors which may affect that expression.

In Situ Hybridization

In situ hybridization is a powerful technique which can be used to localize the expression of mRNAs in different tissues and cell types. It is a useful tool to neuroscientists, among others, as it has enabled further analysis of brain function. For example, *in situ* hybridization has been used to demonstrate the expression of the pituitary receptor for thyrotrophin-releasing hormone in discrete areas of the brain, indicating that this peptide may have a role to play in the central nervous system (Zabavnik et al., 1993). Histochemical tissue sections are incubated in the presence of labeled probes that contain DNA or RNA sequences complementary to the mRNA to be localized. In addition to being able to determine the localization of mRNA in different tissues, morphological data regarding the labeled cells can be obtained. The technique also permits quantitative assessments of changes in levels of expression to be made.

In situ hybridization can also be used to detect DNA on the chromosomes of higher eukaryotes. The Southern blotting technique used to visualize the position of a gene on a chromosome is limited to lower eukaryotes whose chromosomes are relatively small. Under the light microscope, chromosomes can be identified by their shape and banding patterns. Fluorescent *in situ* hybridization (FISH), where the probe is labeled with a fluorescent tag, is a sensitive method which provides a direct visual localization of the exact chromosome bearing the gene of interest as well as its position on the chromosome.

SUMMARY

Once the DNA of interest has been cloned, it now becomes possible to study the function of that gene in depth. DNA sequences which may be important for the different functions can be identified by sequence comparisons with other similar genes. Gene manipulations using techniques like site-directed mutagenesis, allow specific regions of the DNA to be changed and the effect of the change on gene function examined.

The recent exciting advances in molecular cloning technology has led to the development of innovative biological products designed to improve health care. Basic experimental techniques such as cloning, PCR, sequencing, site-directed mutagenesis and transfection have been combined to yield a technology of extraordinary potential. Gene therapy may now become a feasible form of clinical treatment for both acquired and genetic disorders, particularly since accurate animal models of human genetic disorders, e.g., cystic fibrosis, can now be obtained by gene targeting (Dickenson et al., 1993). There have been many attempts to link the genetic and environmental components of complex polygenic disorders. Recently, the tools of recombinant DNA have been applied to cloning some of the genes implicated in human cancers. Studies like these will increase our understanding of the genetic pathways leading to malignancies, coronary artery disease, and mental illness in humans.

REFERENCES

Barrell, B.G., Anderson. S., Bankier, A.T.. deBruijn. M.H.L., Chen, E., Coulsen, A.R., Drouin, J., Eperon. I.C., Nierlich, D.P., Roe, B.A., Sanger, F., Schreier, P.H.. Smith, A.J.H.. Staden, R.. & Young, I.G. (1980). Different pattern of codon recognition by mammalian mitochondrial tRNAs. Proc. Natl. Acad. Sci. USA 77, 3164–3166.

Blair, L.A.C., Levitan, E.S.. Marshall, J., Dionne, V.E., & Barnard, E.A. (1988). Single subunits of the GABA A receptor form ion channels with properties of the native receptor. Science 242, 577–579.

Brenner, S., Elgar, G., Sandford, R., Macrae, A., Venkatesh, B., & Aparicio, S. (1993). Characterization of the pufferfish (Fugu) genome as a compact model vertebrate genome. Nature 366, 265–268.

Davis, L.G., Dibner, M.D., & Battey J.F. (1986). In: Basic Methods in Molecular Biology. Elsevier, New York.

Dickenson, P., Kimber, W.L., Kilanowski, F.M., Stevenson, B.J., Porteous, D.J., & Dorin, J.R. (1993). High frequency gene targeting using insertional vectors. Hum. Mol. Gen. 2, 1299–1302.

Eidne, K.A.. Zabavnik. J., Peters, T.. Yoshida, S.. Anderson, L., & Taylor, P.L. (1991). Cloning, sequencing and tissue distribution of a candidate G-protein coupled receptor from rat pituitary gland. FEBS Lett. 292, 243–248.

Gurdon, J.B., Lane, C.D., Woodland, H.R., & Marbaix, G. (1971). Use of frog eggs for the study of messenger RNA and its translation in living cells. Nature 233, 177–182.

Julius, D., MacDermott, A.B., Axel, R., & Jessell, T.M. (1988). Molecular characterization of a functional cDNA encoding the serotonin 1c receptor. Science 241, 558–564.

Libert, F., Parmentier, M., Lefort, A., Dinsart, C., Vansande, J., Maenhaut, C., Simons, M.J., Dumont, J.E., & Vassart, G. (1989). Selective amplification and cloning of four new members of the G-protein coupled receptor family. Science 244, 569–572.

Ma, K., Inglis, J.D., Sharkey, A., Bickmore, W., Hill, R.E., Prosser, E.J., Speed, R.M., Thomson, E.J., Jobling, M., Taylor, K., Wolfe, J., Cooke, H.J., Hargreave, T.B., & Chandley, A.C. (1993). A Y-chromosome gene family with RNA-binding protein homology—candidates for the azoospermia factor AZF controlling spermatogenesis. Cell 75, 1287–1295.

Masu, Y., Nakayama, K., Tamaki, H., Harada, Y., Kuno, M., & Nakanishi, S. (1987). cDNA cloning of bovine substance K receptor through oocyte expression system. Nature 329, 836–838.

Mathews, L.S., & Vale, W.W. (1991). Expression cloning of an activin receptor, a predicted transmembrane serine kinase. Cell. 55, 973–982.

Maxam, A.M., & Gilbert, W. (1977). A new method of sequencing DNA. Proc. Natl. Acad. Sci. USA 74, 560–564.

Mullis, K., & Faloona, F. (1987). Specific synthesis of DNA in vitro via a polymerase catalysed chain reaction. Meth. Enzymol. 55, 335–350.

Roberts, R.J. (1989). Restriction enzymes and their isoschizomers. Nuc. Acids Res. 17, r347–r387.

Sambrook, J., Fritsch, E.F., & Maniatis, T. (1989). In: Molecular cloning. A laboratory manual Vol. 1, 2, and 3. 2nd edition, Cold Spring Harbor Laboratory Press, Cold Spring Harbor.

Sanger, F., Nicklen, S., & Coulson, A.R. (1977a). DNA sequencing with chain-terminating inhibitors. Proc. Natl. Acad. Sci. USA 74, 5463–5467.

Sanger, F., Air, G.M., Barrell, B.G., Brown, N.L., Coulson, A.R., Fiddes, J.C., Hutchison III, C.A., Slocombe, P.M., & Smith, M. (1977b). Nucleotide sequence of bacteriophage Φx174. Nature 265, 678–695.

Schwartz, D.C., & Cantor, C.R. (1984). Separation of yeast chromosome-sized DNAs by pulsed-field gradient gel electrophoresis. Cell 37, 67–85.

Seed, B., & Aruffo, A. (1987). Molecular cloning of the CD2 antigen, the T-cell erythrocyte receptor, by a rapid immunoselection procedure. Proc. Natl. Acad. Sci. USA 84, 3365–3369.

Smith, H.O., & Wilcox, K.W. (1970). A restriction enzyme from Haemophilus influenzae, I. Purification and general properties. J. Mol. Biol. 51, 379–391.

Southern, E.M. (1975). Detection of specific sequences among DNA fragments separated by gel electrophoresis. J. Mol. Biol. 98, 503–517.

Yoshida, S., Plant, S., Taylor, P., & Eidne, K.A. (1989). Chloride channels mediate the response to gonadotropin-releasing hormone (GnRH) in *Xenopus* oocytes injected with rat anterior pituitary mRNA. Mol. Endocrinol. 3, 1953–1960.

Zabavnik, J., Arbuthnott, G., & Eidne, K.A. (1993). Distribution of thyrotrophin-releasing hormone receptor messenger RNA in rat pituitary and brain. Neuroscience 53, 877–887.

RECOMMENDED READINGS

Hall, S.S. (1989). In: Invisible Frontiers: The Race to Synthesize a Human Gene. Tempus, Washington, D.C.

Sambrook, J., Fritsch, E.F., & Maniatis, T. (1989). In: Molecular Cloning. A laboratory manual. Vol. 1, 2, and 3. 2nd edition, Cold Spring Harbor Laboratory Press, Cold Spring Harbor.

Williams, J.G., & Patient, R.K. (1988). In: Genetic Engineering (Rickwood, D. and Male, D., eds.) IRL Press, Washington D.C.

Chapter 15

DNA Probes in Human Diseases

PAUL M. BRICKELL

Principles of Medical Biology, Volume 5
Molecular and Cellular Genetics, pages 307–329.
Copyright © 1996 by JAI Press Inc.
All rights of reproduction in any form reserved.
ISBN: 1-55938-809-9

INTRODUCTION

Genetic defects make a significant contribution to human disease, in terms both of the misery that they cause and the burden that they impose on health services. A recent survey of the literature estimated that, in Western societies, genetic disease affects between 3.7% and 5.2% of all live births, accounts for up to one third of pediatric admissions to hospitals, and is a major cause of death in childhood (Weatherall, 1991). These figures include not only inherited single-gene disorders (Table 1) and chromosomal abnormalities, but also congenital malformations and common diseases that are thought to have a significant genetic component (Table 2). If the definition of genetic disease is broadened to include genetic defects arising in somatic cells, as well as inherited defects, the scale of the problem becomes even greater. In particular, it has become clear that the development and progression of human tumors involves the accumulation of mutations in the DNA of somatic cells.

Table 1. Some Common Inherited Single-Gene Disorders

Disorder	Frequency/10,000 Births	Protein Encoded by Affected Gene
Autosomal dominant disorders	18–95	
Acute intermittent porphyria	0.1	porphobilinogen deaminase
Hypercholesterolemia	2	low density lipoprotein receptor
Huntington's disease	2	HD gene product
Marfan syndrome	1	fibrillin
Myotonic dystrophy	0.5	myotonin protein kinase
Neurofibromatosis type I	2.5	neurofibromin
Osteogenesis imperfecta	1	α1(I) collagen
Otosclerosis	30	not known
Polycystic kidney disease	10	not known
Tuberous sclerosis	0.8	TSC2 gene product
Autosomal recessive disorders	22–25	
α$_1$-antitrypsin deficiency	1–5	α$_1$-antitrypsin
β-thalassemia	0.5	β-globin
Cystic fibrosis	5–6	CF transmembrane regulator
Phenylketonuria	2–5	phenylalanine hydroxylase
Sickle cell anemia	1	β-globin
Tay-Sachs disease	0.04	hexosaminidase A
X-linked disorders	5–20	
Duchenne muscular dystrophy	3	dystrophin
Hemophilia A	0.5	factor VIII
Hemophilia B	0.1	factor IX
Fragile X mental retardation	9	FMR-1 gene product
Ichthyosis	1	steroid sulphatase

Note: Frequencies are for north European populations and are very approximate, as discussed by Weatherall (1991), from whom the figures are taken.

Table 2. Some Congenital Malformations and Common Diseases
That Are Thought to Have a Significant Genetic Component

Congenital malformations
  Cleft lip
  Cleft palate
  Dislocated hip
  Pyloric stenosis
  Spina bifida

Common diseases
  Cancer
  Diabetes mellitus (particularly the insulin-independent form)
  Epilepsy
  Hypertension
  Ischemic heart disease
  Manic depression
  Schizophrenia

The human genome comprises approximately 3.5×10^9 base pairs of DNA, distributed between 23 chromosomes. There are subtle differences in the sequence of base pairs in different individuals, giving rise to normal biological variation between individuals. The question is, which of these differences are responsible for genetic diseases? This is an important question, because its answer can lead to a better understanding of genetic diseases and to improved strategies for their diagnosis and treatment. The aim of this article is to describe how DNA probes can be used to investigate genetic disease in humans.

WHAT IS A DNA PROBE?

A DNA molecule consists of two strands wound around each other in a double-helix. Each strand has a backbone of repeating sugar-phosphate units, and attached to each sugar residue is one of four bases: adenine (A), cytosine (C), guanine (G), or thymine (T). The two strands are held together by hydrogen-bonding between bases, such that A binds only to T, while C binds only to G. The sequences of the bases on the two strands are therefore said to be complementary. When double-stranded DNA in solution is heated or treated with alkali, the hydrogen bonds that hold the base pairs together are broken and the two strands separate. This is termed denaturation. If the solution is then allowed to cool, or if its pH is lowered, complementary bases pair with each other once again and double-stranded DNA molecules indistinguishable from the starting material reform. This is termed renaturation, or annealing. Single DNA strands from completely different sources will also renature in this way, as long as they have complementary base sequences. Such double-stranded molecules are termed DNA-DNA hybrids. In the same way, DNA-RNA and RNA-RNA hybrids can form between complementary DNA and

RNA strands. The ability of nucleic acids to form hybrids makes it possible to take a purified DNA fragment, to label it and to use it as a probe with which to search for complementary DNA or RNA sequences in complex mixtures of nucleic acids.

HOW ARE DNA PROBES MADE?

Isolating DNA Probes

There are three main classes of DNA probe. These are cloned DNA fragments, synthetic oligonucleotides, and DNA fragments amplified by the polymerase chain reaction (PCR) (see chapter 13). Cloned DNA fragments may be isolated from a genomic library or a cDNA library. The former is a collection of cloned fragments of genomic DNA. The latter is a collection of cloned cDNA molecules, which are double-stranded DNA copies of single-stranded mRNAs. Oligonucleotides are short, single-stranded DNA molecules, typically 15–50 nucleotides long. They are synthesized chemically to have the desired base sequence. PCR can be used to generate double-stranded DNA fragments for use as probes.

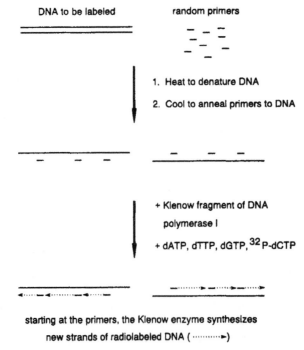

Figure 1. Labeling DNA by the random priming method. Random primers are a mixture of synthetic, single-stranded DNA molecules each consisting of a random sequence of six nucleotides. The labeled probe must be denatured by boiling before use.

Labeling DNA Probes

DNA probes must be labeled so that they can be detected once they have hybridized to complementary nucleic acid sequences. Traditionally, DNA probes have been labeled with a radioisotope that can be detected by autoradiography. The most commonly used isotope is ^{32}P. More recently, a range of nonradioactive labeling and detection systems have been developed. These are safer to use, but less sensitive, than ^{32}P.

Cloned DNA fragments and PCR fragments for use as hybridization probes are usually labeled by the random priming method (Figure 1). In this method, the DNA fragment is used as a template for the synthesis of new DNA strands. If one of the deoxyribonucleoside 5'-triphosphates (dNTPs) in the reaction mixture is labeled, the label becomes incorporated along the whole length of the newly-synthesized DNA strands. The resulting labeled probe will be double-stranded. Cloned DNA fragments can also be used as templates for the synthesis of labeled single-stranded RNA probes. In contrast, oligonucleotides are usually labeled by attaching a ^{32}P-labeled group to their 5' or 3' end. Most frequently, the enzyme T4 polynucleotide kinase is used to transfer the ^{32}P-labeled γ-phosphate group from [γ-^{32}P]ATP to the 5' end of the oligonucleotide.

HOW ARE DNA PROBES USED?

Southern Blotting and Restriction Fragment Length Polymorphisms (RFLPs)

The structure of a gene in a healthy or diseased individual can be analyzed using the Southern blotting technique, which is named after Ed Southern, its inventor. The steps involved in Southern blotting are shown in Figure 2. Genomic DNA can be isolated from any tissue, but peripheral blood leukocytes are the most convenient source in humans. The genomic DNA is cut into a large number of different-sized fragments with a restriction endonuclease. Such enzymes bind to specific short sequences of base pairs in DNA and cut through the two strands at that position (Table 3). The fragments are then separated according to their size by electrophoresis on an agarose gel; the smaller the fragment, the further it will migrate. To permit subsequent hybridization to a DNA probe, the double-stranded genomic DNA fragments in the gel are denatured by soaking the gel in an alkaline solution. The fragments are then transferred, or blotted, on to a nylon membrane, to which they bind covalently. It is necessary to transfer the DNA from the gel to a solid support because attempts to hybridize a probe to DNA in the fragile gel would cause it to disintegrate. Short oligonucleotide probes can be hybridized to DNA in dried gels, but only with some difficulty. There are a number of methods for transferring nucleic acids from agarose gels to nylon membranes, but the most popular is the simple capillary blotting method illustrated in Figure 3.

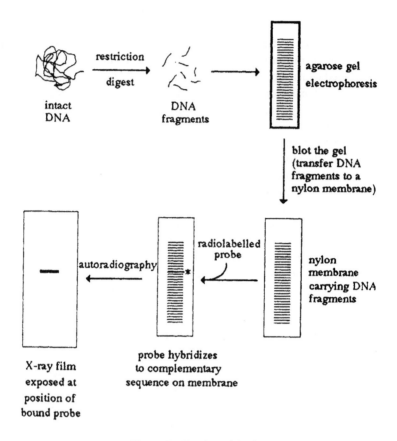

Figure 2. Southern blotting.

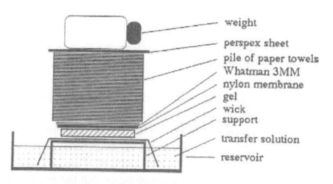

Figure 3. Capillary blotting apparatus.

Table 3. Some Restriction Endonucleases and Their Recognition Sequences

Enzyme	Source	Recognition Sequence
MboI	*Moraxella bovis*	$^{\triangledown}$GATC CTAG$_{\triangle}$
TaqI	*Thermus aquaticus*	T$^{\triangledown}$CGA AGCT$_{\triangle}$
DdeI	*Desulfovibrio desulfuricans*	C$^{\triangledown}$TNAG GANTC$_{\triangle}$
EcoRI	*Escherichia coli* R factor	G$^{\triangledown}$AATTC CTTAAG$_{\triangle}$
EcoRV	*Escherichia coli* R factor	GAT$^{\triangledown}_{\triangle}$ATC CTATAG$_{\triangle}$
PstI	*Providencia stuartii*	CTGCA$^{\triangledown}$G GACGTC$_{\triangle}$
NotI	*Nocardia otitidis-cavarium*	GC$^{\triangledown}_{\triangle}$GGCCGC CGCCGGCG$_{\triangle}$

Note: The arrowheads indicate the positions at which the enzymes cleave the DNA. N, any base.

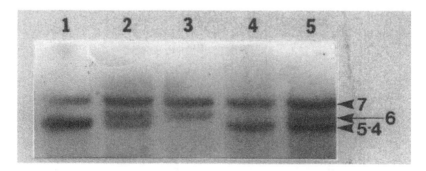

Figure 4. Detection of an RFLP by Southern hybridization. Genomic DNA from five healthy people was digested with *Taq*I and hybridized with a probe for the gene encoding complement component C4. The probe hybridized to a 7 kb fragment found in all people and to a polymorphic fragment that was either 6 kb or 5.4 kb long. Persons 1 and 4 were homozygous for the 5.4 kb allele. Person 3 was homozygous for the 6 kb allele. Persons 2 and 5 were heterozygous.

After blotting, the membrane is incubated with the radiolabeled probe. If the probe is double-stranded, it is first denatured by boiling. After hybridization, excess probe is washed off and the bound probe is detected by autoradiography. Genomic DNA fragments that are complementary to the probe appear as black bands on the film (Figure 4). While this procedure is quite straightforward, care must be taken to perform the hybridization and washing steps under conditions that permit the formation of stable hybrids between the probe and membrane-bound DNA, while minimizing hybridization between mismatched sequences and nonspecific binding of the probe to the membrane.

In most cases, gene probes hybridize to fragments of equal size in all individuals tested. Occasionally, however, a probe hybridizes to fragments of different sizes in different individuals (Figure 4). Such a probe is detecting a genetic difference between healthy individuals. Such a difference is termed a polymorphism, and since the polymorphism is detected as a difference in restriction fragment lengths, it is termed a restriction fragment length polymorphism, or RFLP. RFLPs can arise in

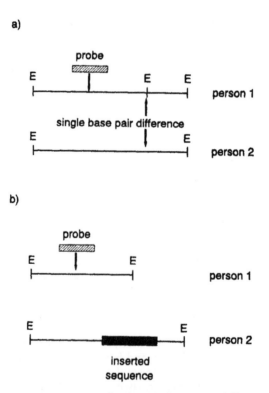

Figure 5. RFLPs can arise as a result of single base pair differences that create or destroy a restriction site (**a**) or as a result of deletions or insertions of DNA (**b**).

two ways. First, a change in a single base pair may create or destroy a restriction enzyme site, which is consequently present in some individuals but not in others (Figure 5a). This will lead to the generation of restriction fragments of unequal lengths. Second, the distance between two restriction sites may vary between individuals as a result of the insertion, or deletion, of a DNA sequence (Figure 5b). A common cause of this is the presence of repeated DNA sequences that are scattered throughout the genome in blocks containing multiple copies of the same sequence. The number of copies of the repeated sequence in a given block can vary between individuals, so that a restriction fragment containing a block of repeats will vary in length between individuals. An RFLP generated thus is called a variable number of tandem repeats (VNTR) RFLP.

The polymorphisms discussed above can also be detected using PCR (see chapter 13). In fact, given that the nucleotide sequence of the region containing the polymorphism is available, it is usually more convenient to use PCR than it is to use Southern blotting. The existence of RFLPs permits the mapping of DNA

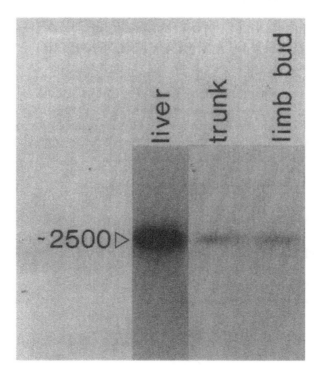

Figure 6. A Northern blot. Equal quantities of RNA from the liver, trunk, or limb buds of a chick embryo were electrophoresed, blotted, and hybridized with a probe for retinoid-X-receptor-γ mRNA, which is approximately 2,500 nucleotides long. The abundance of the mRNA differs between the tissues.

sequences, the tracking of mutant genes through families, and the prenatal detection of inherited diseases.

Northern Blotting

Northern blotting is a modification of Southern blotting that enables the detection of RNA, rather than genomic DNA. It was called Northern blotting as a joke. Briefly, RNA is isolated from tissue and size-separated by agarose gel electrophoresis. Since RNA molecules are single-stranded, they form weak intramolecular and intermolecular hydrogen bonds. The resulting folding and aggregation of the RNA molecules means that they do not migrate true to size. RNA is therefore commonly electrophoresed in gels containing formaldehyde, as a denaturing agent. The electrophoresed RNA is transferred to a nylon membrane which is incubated with labeled probe, washed, and autoradiographed essentially as for a Southern blot. Northern blotting allows you to estimate the size and abundance of an mRNA species in a cell line or tissue (Figure 6).

DNA PROBES AND INHERITED DISEASES INVOLVING A DEFECT IN A SINGLE, KNOWN GENE

If you have a probe for a gene that is known to be mutated in a particular inherited disease, it can be used to determine whether or not an individual carries the mutation. It is also possible to diagnose an inherited disease in a fetus (prenatal diagnosis), since DNA can be isolated from fetal tissue obtained by chorionic villus sampling (Doran, 1990). Each of the techniques described below can be adapted for use in carrier detection, prenatal diagnosis or postnatal diagnosis.

Detecting Gross Gene Deletions

The simplest genetic lesions to detect are large deletions. These can usually be detected by Southern hybridization. For example, it has been found that many individuals with human pituitary growth hormone (hGH) deficiency are homozygous for a deletion that removes the *GH1* gene, which encodes hGH (Vnencak-Jones et al., 1988). This can readily be detected using an hGH cDNA probe or a *GH1* gene fragment probe (Figure 7).

Detecting Point Mutations

RFLPs

More subtle mutations, such as point mutations, are more difficult to detect. Most point mutations that cause disease do not alter restriction sites and so do not create RFLPs. Conversely, the vast majority of RFLPs result from genetic differ-

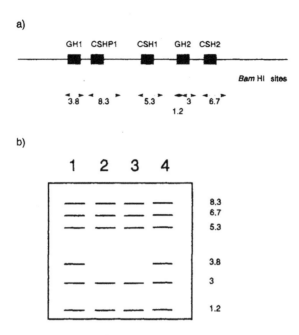

Figure 7. Detection of a human growth hormone gene deletion by Southern hybridization. **a)** Map of the human growth hormone gene cluster, which occupies approximately 50 kb of DNA. Pituitary growth hormone is encoded by the *GH1* gene. The sequences of the five genes are so similar that under the right conditions a *GH1* gene probe will hybridize to them all. **b)** Sketch of a Southern blot of *Bam*HI-digested DNA hybridized with a *GH1* gene probe. Persons 2 and 3 are homozygous for a deletion that removes the *GH1* gene, while leaving the other genes of the cluster intact. The sizes of the *Bam*HI fragments are shown in kb.

ences that have no consequence as far as disease is concerned. Very occasionally, however, a mutation that causes a disease also creates an RFLP. For example, sickle cell anaemia affects people who are homozygous for a point mutation in the β-globin gene that changes the sixth codon from one encoding glutamic acid to one encoding valine. The sequence of the normal β^A-globin gene in this region includes an *Mst*II site (CCTGAGG). The point mutation in the β^S-globin gene destroys this restriction enzyme site, so that *Mst*II can no longer cut there. This makes it very easy to distinguish the normal and mutated forms of the β-globin gene using Southern hybridization, as illustrated in Figure 8 (Chang and Kan, 1982).

As noted above, most point mutations that cause diseases do not create RFLPs. There are two ways in which gene probes have been used to detect such mutations directly. These are the use of allele-specific oligonucleotides (ASOs) and ribonuclease A (RNase A) protection.

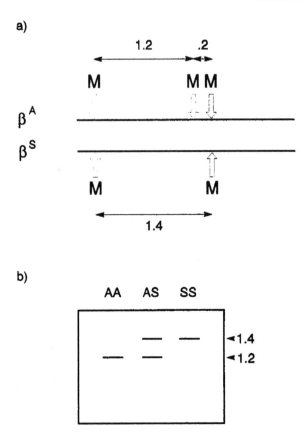

Figure 8. Detection of the β^S^-globin gene by Southern hybridization. **a)** The mutation responsible for sickle cell anemia converts an *Mst*II site in the normal β^A^-globin gene (CCTGAGG) into a sequence that is not recognized by *Mst*II (CCTGTGG). **b)** Genomic DNA from individuals homozygous for the β^A^-globin gene (AA), homozygous for the β^S^-globin gene (SS) or heterozygous (AS) was digested with *Mst*II, blotted, and hybridized with a probe derived from the 1.2 kb *Mst*II fragment. Fragment sizes are shown in kb. M, *Mst*II site.

Allele-Specific Oligonucleotides (ASOs)

In this technique (Figure 9), short oligonucleotides corresponding to the normal and mutant sequences are synthesized, labeled, and used to probe blots of genomic DNA from the individuals being tested. Hybridization and washing conditions are adjusted so that the oligonucleotides form stable hybrids with perfectly-matched complementary sequences in the genomic DNA, while hybrids with a single mismatch are unstable. The "normal" oligonucleotide will therefore give a hybridization signal with DNA from heterozygotes and from individuals homozygous for the normal sequence, while the "mutant" oligonucleotide will give a hybridization signal with DNA from heterozygotes and from individuals homozygous for the mutated sequence (Figure 9).

Each oligonucleotide probe must be short, so that a single base mismatch has a significant effect on the stability of the hybrids that it forms. If the probe is too long, it will be impossible to find hybridization and washing conditions that allow it to discriminate between a perfectly-matched sequence and one with a single mismatch. However, the shorter the oligonucleotide, the more likely it is that its sequence will appear elsewhere in the genome, resulting in spurious hybridization to other fragments. In practice, probes are usually 19–21 nucleotides long.

This technique gives best results if the region of genomic DNA containing the point mutation is first amplified using PCR. The PCR product is then bound to nylon membranes and probed with the labeled normal and mutant oligonucleotides. This avoids the problems with spurious hybridization that arise if a probe's

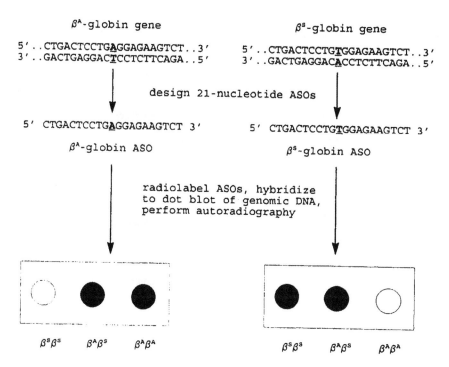

Figure 9. Use of ASOs to detect the β^A-globin and β^S-globin genes. To make the dot blots, genomic DNA or PCR fragments are loaded directly on to a nylon membrane, without first being electrophoresed. Hybridization conditions are controlled so that the ASOs hybridize only to perfectly matched sequences.

sequence occurs elsewhere in the genome, and gives stronger specific hybridization signals, since more target DNA can be applied to the membranes.

ASO is now used for prenatal diagnosis of a range of inherited diseases, including phenylketonuria (phenylalanine hydroxylase deficiency), hemophilia A (factor VIII deficiency), hemophilia B (factor IX deficiency), and α_1-antitrypsin deficiency (Kidd et al., 1984).

Ribonuclease A (RNase A) Protection

In this technique (Figure 10), a labeled, single-stranded RNA probe is synthesized *in vitro* by transcribing cloned DNA that encompasses the region containing the point mutation. This is then hybridized in solution to genomic DNA from the individuals being tested. Where the RNA probe forms a perfectly-matched DNA-RNA hybrid, it is protected from subsequent digestion by RNase A. However, the RNA probe will be cleaved at any sites of mismatch. The products of RNase A

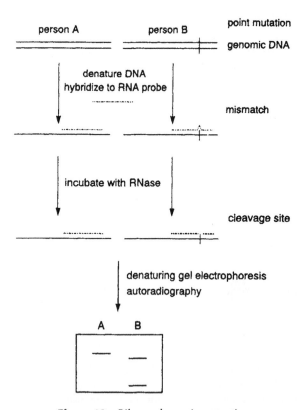

Figure 10. Ribonuclease A protection.

treatment are electrophoresed on a denaturing polyacrylamide gel and the labeled RNA fragments are detected by autoradiography. This technique has been used to identify mutations in a number of genes, including that encoding hypoxanthine phosphoribosyltransferase (HPRT), which is defective in people with Lesch-Nyhan syndrome (Gibbs and Caskey, 1987). Once again, the sensitivity and specificity of the technique are increased if the region of genomic DNA containing the point mutation is first amplified using PCR.

Linked RFLPs

Point mutations responsible for a disease can also be tracked indirectly by looking at linked RFLPs. For example, in the normal human population the β-globin gene is found on a *Hpa*I fragment of either 13 or 7.6 kb, depending on whether a polymorphic *Hpa*I site is present (Figure 11). The polymorphism is said to be linked to the β-globin gene, because it lies on the same chromosome. This polymorphism has no known effect on health, but it turns out that most β^S-globin genes just happen to lie on the 13 kb version. In many families, it is therefore possible to track the β^S-globin gene by probing Southern blots of *Hpa*I-digested genomic DNA with a β-globin gene probe (Kan and Dozy, 1978). This technique can be used for carrier detection and prenatal diagnosis, although there is a very small risk of misdiagnosis resulting from meiotic recombination between the polymorphic *Hpa*I site and the β^S-globin gene in the egg or sperm from which an individual develops. Clearly, prenatal diagnosis of sickle cell anemia is best performed by looking directly at the disease-causing mutation, rather than at this linked RFLP. However, the *Hpa*I RFLP is useful for prenatal diagnosis of other less common β-globin gene mutations that cause β-thalassemia, when the mutation that causes the disease has not been identified. The same is true for a number of

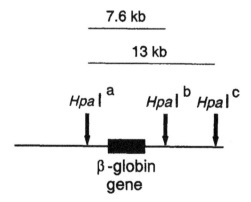

Figure 11. An *Hpa*I RFLP that is closely linked to the β-globin gene. Sites a and c are present in all people. Site b is polymorphic.

of other inherited diseases involving known genes. Diagnostic tests based on ASO or ribonuclease A protection can be set up for commonly-encountered, well-characterized mutations, while rare and/or uncharacterized mutations can be tracked through families using linked RFLPs.

DNA PROBES AND INHERITED DISEASES INVOLVING A DEFECT IN A SINGLE, UNKNOWN GENE

Linkage Analysis

The use of linked RFLPs (or other genetic markers) is the only option for prenatal diagnosis or carrier detection when the genetic disease in question arises from a mutation in a single, unknown gene. A prerequisite is to find RFLPs that are closely linked to the disease locus. This can be done as described below, although the growth of linkage maps of the human genome and the development of PCR-based techniques is making the procedure much more streamlined.

First, probes are taken at random from a genomic library and tested for their ability to detect RFLPs. Next, the RFLP locus is tested for linkage to the disease locus by probing Southern blots of DNA from families with the disease. The RFLP locus and the disease locus are linked (on the same chromosome) if they segregate together in families more often than would be expected by chance. Linkage analysis in humans is performed in the same way as in any other diploid species but is complicated because it is not possible to perform suitable matings to order. Analysis is further complicated because alleles can be mixed up by meiotic recombination. In practice, linkage analysis in humans relies on the availability of DNA from large families in which the disease has been passed through two or more generations. Data from several families can be combined by means of sophisticated statistical treatment performed by computers.

When a probe detecting a linked RFLP has been identified, it can be used to determine the chromosomal location of the disease locus. This can be done by hybridizing the labeled probe to a spread of metaphase chromosomes. The use of DNA probes labeled with a fluorescent marker has greatly enhanced the sensitivity and resolution of this technique, which is called fluorescence *in situ* hybridization, or FISH (Lichter et al., 1990). Alternatively, the chromosomal location of a DNA sequence can be determined by hybridizing the probe to Southern blots of DNA from a panel of somatic cell hybrids. These are hybrids of mouse and human cells that contain only one or a few human chromosomes. More accurate localizations can be performed using radiation hybrids, which retain only small fragments of human chromosomes (Cox et al., 1990). Once the disease locus has been assigned to a chromosome, it becomes easier to find more linked markers, since it is only necessary to test probes that have previously been assigned to the same chromosome or that have been isolated from genomic libraries constructed from that chromosome alone.

Prenatal Diagnosis

Figure 12 illustrates the use of a closely linked RFLP to perform pre-natal diagnosis for Huntington's disease (Meissen et al., 1988). This is an autosomal dominant condition resulting from a mutation in the *HD* gene. It is a progressive and fatal neurological disease that typically manifests itself in early middle age. The RFLP shown in Figure 12 has two alleles of 1.9 and 2.0 kb. Person I-1 is

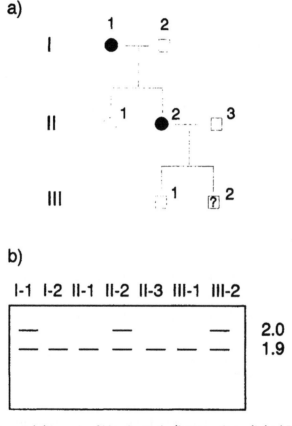

Figure 12. Prenatal diagnosis of Huntington's disease using a linked RFLP. **a)** Three generations of a family with the disease. Affected individuals are shown as black symbols. III-2 is the fetus for whom prenatal diagnosis is being performed. **b)** Sketch of a Southern blot of digested DNA from the family members, hybridized with a probe that detects an RFLP with two alleles, of 1.9 and 2.0 kb.

affected by the disease and is heterozygous for the RFLP. I-2 is unaffected by the disease and is homozygous for the 1.9 kb allele. II-1 and II-3 have inherited the 1.9 kb allele from I-1 and are unaffected, while II-2 has inherited the 2.0 kb allele from I-1 and has the disease. This indicates that in I-1, the 2.0 kb allele of the RFLP lies on the same chromosome as the defective *HD* gene, while the 1.9 kb allele lies on the same chromosome as the normal *HD* gene. It would be predicted from the result of the Southern blot that child III-1 has two normal *HD* genes, while the fetus III-2 has a defective *HD* gene and is therefore at risk of developing Huntington's disease. Accurate prenatal diagnosis for Huntington's disease requires the use of a panel of RFLPs and will be greatly improved by the recent isolation of the *HD* gene (Huntington's Disease Collaborative Research Group, 1993), since it will now be possible to detect *HD* gene mutations directly, using the techniques described in the section titled "DNA Probes and Inherited Diseases."

A similar approach can be used to perform prenatal diagnosis for autosomal recessive conditions, such as cystic fibrosis (CF; Farrall et al., 1986). CF is an autosomal recessive disorder that affects approximately 1 in 2,000 live births in Caucasian populations. Approximately 1 in 20 people carry a defective *CF* gene. The disease is characterized by abnormal chloride ion transport by epithelial cells, resulting in the secretion of viscous mucus. A number of organs are affected, but the most serious effects are found in the lung. Obstruction of airways by mucus can lead to recurring infections and lung damage, leading to death. There are three steps to the diagnostic procedure:

1. Determine whether the parents are heterozygous for the linked RFLP. If they are not, the test will not work and another linked RFLP must be tried. It is possible to use different linked RFLPs for each parent.
2. For each parent, work out which allele of the RFLP is on the chromosome carrying the mutant *CF* locus. This is done by studying other members of the family, including a previously affected child.
3. Analyze fetal DNA, obtained by chorionic villus sampling, to determine whether each parent passed on a normal or mutant *CF* allele.

There are a number of problems with this kind of analysis:

1. The combination of the alleles at the RFLP and disease loci in the two parents may not permit any conclusion about the fetus, in which case the family is said to be uninformative for that RFLP. The solution is to use other RFLPs for which the parents are heterozygous and to see if they are informative for the family in question.
2. Recombination between the RFLP and the disease locus during meiosis may result in false positive or false negative diagnoses. Obviously, the more closely linked the RFLP and disease loci, the smaller the recombination rate and the fewer the errors. To minimize such errors, it is usual to use two RFLPs, one on each side of the disease locus (called bridging markers or

flanking markers). In this way, only extremely rare double recombinants will give false results.

Once again, the isolation of the *CF* gene has now made it possible to perform prenatal diagnosis by looking directly for disease-causing mutations in the *CF* gene. It has also become feasible to devise programs for screening the general population for carriers of the most common *CF* gene mutations (Collins, 1992).

Identification of Disease Genes

As we have seen, probes that detect linked RFLPs are extremely valuable as markers for carrier detection and prenatal diagnosis of inherited disease. However, perhaps their greatest power lies in their use to identify previously unknown disease genes. For a large number of inherited single-gene disorders, nothing was known about the defective protein until the defective gene was identified. In recent years, a number of such genes have been cloned, starting only with a knowledge of their general chromosomal location (Table 4). This approach to identifying disease genes is often called positional cloning, or reverse genetics.

As noted above, CF is one example of an inherited disease that has yielded to this approach. The steps in the positional cloning of the *CF* gene were as follows (Collins, 1992):

1. Linkage analysis was performed on a large number of families with CF to find RFLPs that were linked to the *CF* gene.
2. An RFLP that was loosely linked to the *CF* gene was found. It was shown by *in situ* hybridization that this RFLP mapped to the long arm of chromosome 7.
3. Two RFLPs that were tightly linked to the *CF* gene, named *MET* and *D7S8*, were found. These RFLPs lay on either side of the *CF* gene and mapped to chromosome 7q31-32.

Table 4. Some of the Human Disease Genes That Have Been Identified by Positional Cloning

Duchenne muscular dystrophy
Chronic granulomatous disease
Retinoblastoma
Wilm's tumor
Cystic fibrosis
Neurofibromatosis type I
Choroideremia
Adenomatous polyposis coli
Fragile X-associated mental retardation
Myotonic dystrophy
Huntington's disease
Tuberous sclerosis

4. The region of DNA between the *MET* and *D7S8* markers was systematically cloned. This was a very laborious procedure, involving cloning techniques called "chromosome walking" and "chromosome jumping". In all, some 500 kb of DNA around the *CF* gene was cloned.

5. Genes within the cloned region were identified by the technique of "zoo blotting". This technique is based on the idea that exon sequences are evolutionarily conserved while intron and intergenic sequences are not. Short probes derived from the cloned *CF* gene region were therefore incubated with Southern blots of genomic DNA from a range of animal species. Probes that hybridized to several of the DNA samples were judged to contain exon sequences. Five such probes were found, but four proved not to be derived from the *CF* gene. Two were excluded by linkage studies, one was a pseudogene and one failed to detect mRNA in cells affected in CF patients.

6. The fifth probe was shown to hybridize to mRNA from cultured sweat gland cells, a cell type that is known to be affected in CF patients. This mRNA was cloned.

7. Using the cloned mRNA as a probe, it was shown that there was no gross abnormality in the most common mutant *CF* gene alleles. It was therefore necessary to determine and compare the nucleotide sequences of the *CF* genes from a normal individual and from a CF patient. This analysis showed that the most common mutant *CF* gene allele has a 3 bp deletion, resulting in the deletion of a phenylalanine residue from CF transmembrane regulator (CFTR), the protein encoded by the *CF* gene.

8. The isolation of the *CF* gene has made it possible to study the structure and function of the CFTR protein, to investigate the biochemistry of the defect in CF patients, and to plan new therapeutic approaches (Collins, 1992).

DNA PROBES AND MULTIFACTORAL DISEASES WITH A GENETIC COMPONENT

As noted above (Table 2) a number of common diseases have been shown to have a significant genetic component. The development of these diseases may involve the interaction of a number of genes with a number of environmental factors. Application of the techniques described above has led to the identification of some of the genes that are involved in some of these conditions (Weatherall, 1991).

DNA PROBES AND CANCER

The application of recombinant DNA techniques to tumor biology has led to the identification of a large number of cellular proto-oncogenes. Mutation of these genes can contribute to the development of malignant tumors. In some cases, mutation can convert a proto-oncogene into an active, dominant oncogene that contributes to tumor development even in the presence of a normal, inactive allele. For example, a large proportion of human colon carcinomas have an activating point mutation in one of their two Ki-*ras* proto-oncogenes. This point mutation can be detected by Southern hybridization, using ASOs for the normal and mutated Ki-*ras* genes (Yasui et al., 1992).

Tumor development also involves the deletion of tumor suppressor genes. For example, loss of the *FAP* gene on chromosome 5, the *DCC* gene on chromosome 18, and the *p53* gene on chromosome 17 have all been implicated in the development of colon carcinomas. Southern hybridization can be used to detect the changes in restriction fragment sizes that result from the deletion of such genes (Yasui et al., 1992). It is hoped that such DNA-based tests might be valuable in some situations for diagnosis, for the assessment of prognosis, and for the detection of residual cancer cells after treatment.

DNA PROBES AND MEDICAL MICROBIOLOGY

In all of the above examples, gene probes have been used to detect specific human genes. However, another major use for gene probes is to detect the nucleic acid genomes of microorganisms in tissues or biological fluids. For example, gene probes can been used to detect viral DNA in sections of formalin-fixed, paraffin-embedded biopsy material. This technique, which is called *in situ* hybridization, can be a valuable diagnostic procedure in cases where there are no available serological tests for the virus, where virus cannot be isolated from clinical samples by tissue culture methods, where the virus infects only a small number of cells in a sample, or where the virus establishes a latent infection with only minimal viral gene expression. Examples of the diagnostic use of *in situ* hybridization include detection of human papillomavirus, cytomegalovirus, and HIV (Syrjänen, 1992).

Gene probes have also been used to distinguish pathogenic and nonpathogenic strains of bacteria. For example, enterotoxigenic *Escherichia coli* (ETEC) can be distinguished from normal commensal *E. coli* by growing the two types of bacteria as colonies on nylon membranes laid on top of agar, breaking open the bacteria and denaturing their DNA with sodium hydroxide solution, fixing the DNA to the membrane by UV irradiation, and hybridizing the immobilized DNA with a DNA probe for sequences that are found only in ETEC (McCreedy and Chimera, 1992). This procedure can be quicker and cheaper than diagnostic assays based on the detection of toxins by means of their immunological or cytotoxic properties.

SUMMARY

The availability of DNA probes has revolutionized our understanding of gene structure, expression, and function; has allowed us to explore how structure, expression, and function are deranged in genetic diseases; and has provided new tools for the diagnosis of genetic diseases. In this chapter, we first saw how DNA probes are made and how they can be used in hybridization experiments to detect specific gene sequences. We then saw that DNA probes can be used in a number of ways to analyze and diagnose inherited diseases that result from a mutation in a single, known gene. The analysis of inherited diseases involving unknown genes is possible, but more problematic. However, the power of recombinant DNA technology is such that the list of inherited diseases involving unknown genes is steadily becoming smaller, as more and more disease genes are identified for the first time. Finally, we saw that DNA probes can be used to analyze somatic mutations in tumor cells, and to detect the genomes of infectious agents.

REFERENCES

Chang, J.C., & Kan, Y.W. (1982). A sensitive new prenatal test for sickle cell anemia. New Engl. J. Med. 307, 30–32.

Collins, F.S. (1992). Cystic fibrosis: Molecular biology and therapeutic implications. Science 256, 774–779.

Cox, D.R., Burmeister, M., Price, E.R., Kim, S., & Myers, R.M. (1990). Radiation hybrid mapping: A somatic cell genetic method for constructing high-resolution maps of mammalian chromosomes. Science 250, 245–250.

Doran, T.A. (1990). Chorionic villus sampling as the primary diagnosis tool in prenatal diagnosis. Should it replace genetic amniocentesis? J. Repro. Med. 35, 935–940.

Farrall, M., Law, H-Y., Rodeck, C-H., Warren, R., Stanier, P., Super, M., Lissens, W., Scambler, P., Watson, E., and Wainright, B. (1986). First-trimester prenatal diagnosis of cystic fibrosis with linked DNA probes. Lancet 1, 1402–1405.

Gibbs, R., & Caskey, C.T. (1987). Identification and localization of mutations at the Lesch-Nyhan locus by ribonuclease A cleavage. Science 236, 303–305.

Huntington's Disease Collaborative Research Group (1993). A novel gene containing a trinucleotide repeat that is expanded and unstable on Huntington's disease chromosomes. Cell 72, 971–983.

Kan, Y.W., & Dozy, A.M. (1978). Antenatal diagnosis of sickle-cell anaemia by DNA analysis of amniotic fluid cells. Lancet II, 910–912.

Kidd, V.J., Golbus, M.S., Wallace, R.B., Itakura, K., & Woo, S.L.C. (1984). Prenatal diagnosis of α_1-antitrypsin deficiency by direct analysis of the mutation site in the gene. N. Engl. J. Med. 310, 639–642.

Lichter, P., Tang, C.J.C., Call, K., Hermanson, G., Evans, G.A., Housman, D., &Ward, D.C. (1990). High-resolution mapping of human chromosome 11 by in situ hybridization with cosmid clones. Science 247, 64–69.

McCreedy, B.J., & Chimera, J.A. (1992). Molecular detection and identification of pathogenic organisms. In: Diagnostic Molecular Pathology: A Practical Approach (Herrington, C.S., & McGee, J.O'D., eds.), Vol. II, pp. 173–182, IRL Press, Oxford.

Meissen, G.J., Myers, R.H., Mastromauro, C.A., Koroshetz, W.J., Klinger, K.W., Farrer, L.A., Watkins, P.A., Gusella, G.F., Bird, E.D., & Martin, J.B. (1988). Predictive testing for Huntington's disease with use of a linked DNA marker. N. Engl. J. Med. 318, 535–542.

Syrjänen, S.M. (1992). Viral gene detection by in situ hybridization. In: Diagnostic Molecular Pathology: A Practical Approach (Herrington, C.S., & McGee, J.O'D., eds.), Vol. I, pp. 103–139, IRL Press, Oxford.

Vnencak-Jones, C.L., Phillips, J.A., Chen, E.Y., & Seeburg, P.H. (1988). Molecular basis of human growth hormone gene deletions. Proc. Natl. Acad. Sci. USA 85, 5615–5619.

Weatherall, D.J. (1991). The frequency and clinical spectrum of genetic diseases. In: The New Genetics and Clinical Practice, Third edn. (Weatherall, D.J., ed.), pp. 4–38, Oxford University Press, Oxford.

Yasui, W., Ito, H., & Tahara, E. (1992). DNA analysis of archival material and its application to tumor pathology. In: Diagnostic Molecular Pathology: A Practical Approach (Herrington, C.S., & McGee, J.O'D., eds.), Vol. II, pp. 193–206, IRL Press, Oxford.

RECOMMENDED READINGS

McConkey, E.H. (1993). Human Genetics: The Molecular Revolution. Jones and Bartlett Publishers, Boston.

Strachan, T. (1992). The Human Genome. Bios Scientific Publishers, Oxford.

Watson, J.D., Gilman, M., Witkowski, J., & Zoller, M. (1992). In: Recombinant DNA, Second Ed., Chapters 18, 26, 27. Scientific American Books, New York.

Weatherall, D.J. (1991). The New Genetics and Clinical Practice, Third Edition. Oxford University Press, Oxford.

Chapter 16

Advances in Clinical Genetics

RAM S. VERMA

INTRODUCTION

The laws of inheritance were considered quite superficial until 1903 when the chromosome theory of heredity was established by Sutton and Boveri. Development of the discipline of genetics took place in phases and the positive impact of genetics in medicine remained elusive until 1956 when the correct number of chromosomes in man was discovered by Tjio and Levan (1956). The chromosomal

Principles of Medical Biology, Volume 5
Molecular and Cellular Genetics, pages 331–344.
Copyright © 1996 by JAI Press Inc.
All rights of reproduction in any form reserved.
ISBN: 1-55938-809-9

basis of disease diagnosis became an essential tool after the discovery where an extra chromosome 21 resulted in Down syndrome (Lejeune et al., 1959). Immediately after the invention of banding techniques, human diseases were correlated with chromosomal segments (bands) and hundreds of so-called syndromes were described that were associated with specific genomic aberrations (DeGrouchy and Turleau, 1984).

The field of human genetics has been revolutionized by the chromosomal basis which is presently being coupled with molecular techniques. The most spectacular progress has been made in the area of recombinant DNA technology. A variety of other strategies are now being utilized to decipher the morbid anatomy of the genome (Verma, 1993). The genetic map for individual chromosomes has just begun to emerge. Pinpointing molecular lesions at a single gene with such accuracy was once inconceivable, but has now become routine practice. The message emerging with increasing clarity is single nucleotide mutations can regulate the very fundamental basis of human survival. Diseases at a nucleotide level resulting in enzymatic defects are being unraveled at a rapid pace.

New technology is providing a paradigm for pathogenesis of diseases associated with mitochondria. We are on the verge of a new era where diseases are being redefined in a genetic sense. The knowledge which we are gathering is emerging in futuristic dreams and gene therapy has become a reality.

CHROMOSOMAL BASIS OF HUMAN DISEASES

The technical advances in chromosome preparations and banding techniques allowed dissection of chromosomes facilating the search for defective genes that cause human diseases (Verma and Babu, 1989). Molecular cytogenetics, coupled with routine chromosome analysis, clearly reflects the discovery of an ever-increasing number of morbid sites in the genome. The remarkable discoveries that a minute chromosome segment is associated with specific phenotypes opened new avenues for establishing over 60 recognizable syndromes (DeGrouchy and Turleau, 1984; Jones, 1988; Robinson and Linden, 1993). The severity of birth defects depends upon the size of the genetic material duplicated or deleted. Some chromosomal abnormalities are lethal while a few may survive for a long time. Approximately 1 in 160 liveborn infants have some type of chromosomal abnormality, and in newborn children, half involve the autosomes while the remaining half have sex chromosomal abnormalities. About 50% of first-trimester abortuses have chromosomal abnormalities and the majority of them are aneuploidy or polyploidy. Chromosomal abnormalities detected at 16–18 weeks of gestation during amniocentesis are higher than in newborns. Fifty percent of cases with primary amenorrhea have some sort of chromosomal abnormalities. A number of surveys suggest that 20% of mental retardation is due to chromosomal abnormalities, while chromosomal aberrations in sterile males were at 10%. For the past decade or so, a rapid method for fetal chromosome analysis has gained much popularity from chorionic villus biopsy where the procedure is performed between 8–10 weeks of gestation. However, in recent years amniocentesis performed earlier (8–10 weeks) than 16 weeks is preferred over chorionic villus sampling (Simpson and Elias, 1993).

The recent revolution in microdissecting the specific chromosome band and amplification by polymerase chain reaction (PCR) has further refined syndromes at the single gene level and are referred to as contiguous genes or segmental aneusomy syndromes (Schmickel, 1986; Kao and Yu, 1991; Meltzer et al., 1992). This approach has obviously resulted in detecting genetic defects which could not be demonstrated at the chromosomal level (Fiedler et al., 1991; Peuch et al., 1992; Yu et al., 1992).

MOLECULAR BASIS FOR DETECTION OF DISEASES

The molecular basis for detection of human diseases can be traced back to 1949 when Pauling and his associates demonstrated that an abnormal protein is due to allelic change in a single gene. Currently, the genetic and molecular basis of several hundred diseases are being explored and understood (Chehab, 1993). The contiguous gene syndrome which results is being further refined and/or defined (Verma, 1993). The availability of single gene probes have helped to characterize single gene disorders. In many cases the restriction fragment length polymorphisms have been used to establish the marker with its respective targeted genes. A variety of techniques have been established to detect mutation at a single nucleotide level. Allele-specific oligonucleotide (ASO) probes can be synthesized, which are much shorter than cloned probes. Diseases can not only be characterized at the DNA or RNA level, but at the protein level (Rossiter and Caskey, 1991). Production of unlimited quantities of proteins from cloned genes have opened new avenues toward the understanding of molecular mechanism of diseases based on gene products.

The use of the PCR, where a new mutation can be identified rapidly, has reduced diagnosis time and expenses (Vosberg, 1989). The technique is so sensitive that it can even be used with a single cell, including both diploid somatic cells and haploid germ cells (Li et al., 1988), and can be used for prenatal diagnosis at embryonic stages including preimplantation (Remick et al., 1990). Carrier screening has resulted in prevention of many disorders (Verlinsky et al., 1990). Even many diseases—whose onset after 30 years or later can be diagnosed—can in many cases, have predictive values established (Verma, 1993). It has become increasingly true that many genetic tests have infiltrated the practice of medicine in such a refined manner that these molecular diagnostic tools have become a way of life in modern medicine (Grody and Hilborne, 1992). Techniques which are routinely used are Southern blot, dot blot/reverse dot blot, Northern blot, PCR, denaturing gradient gel electrophoresis, single-strand conformation polymorphism, ribonuclease protection assay, DNA sequencing, and *in situ* hybridization. These procedures can detect point mutations, deletions, duplications, tandem repeat amplifications, rearrangements, translocations, chromosomal aberrations, and extranuclear changes. DNA-based technology has just emerged where every patient may be tested for the genetic basis of a disease. Hopefully, these techniques will be automated in the not-too-distant future and will eminently be implemented (Billings et al., 1991; Wolcott, 1992). Obviously, the unfolding of the molecular era will provide us with an accurate diagnosis of diseases which will be the ultimate goal of treating these diseases at a genetic level.

ADVANCES IN PRENATAL DIAGNOSIS

For the past quarter century, amniocentesis has been a traditional approach for prenatal diagnosis of chromosomal and biochemical disorders. Advanced maternal age or couples with balanced translocations who have an increased risk of producing a genetically imbalanced fetus are strongly recommended for prenatal diagnosis. Traditionally, genetic amniocentesis is performed around 15 to 16 weeks gestation. Nevertheless, with the availability of high resolution ultrasound equipment, many centers are offering amniocentesis earlier than 15 weeks gestation (e.g., 8–12 weeks). Prenatal diagnosis can also be performed in the first trimester using chorionic villi sampling (CVS) by transabdominal–transcervical biopsy of fetal tissues (Jackson et al., 1992). In general, the abnormal chromosome outcomes from CVS were comparable with those seen with midtrimester amniocentesis (Jackson and Wapner, 1993). Fetal blood sampling, which is usually performed after 12–18 weeks gestations, has been an additional approach in detecting genetic disorders (Orlandi et al., 1990). The fetal loss rate is approximately one percent following blood sampling (Daffos et al., 1985).

There are a number of disorders which can be studied from fetal skin biopsies which are usually performed between 17 and 20 weeks gestation (Elias et al., 1993). Fetal liver biopsies have been used to diagnose inborn errors of metabolism (Simpson and Golbus, 1992). The routine screening of amniotic fluid alpha-fetoprotein (AFP) and maternal serum alpha-fetoprotein (MSAFP) has become an essential tool as many genetic disorders can be screened rapidly and individual risks can be assessed using a combination of both tests (Palomaki et al., 1990). The level of human chorionic gonadotropin (HCG) in maternal serum is twice the level in the second trimester of pregnancies with trisomy 21 (Down syndrome), while unconjugated estriol (UE3) levels are on average 25% lower (Bogart et al., 1987; Canick et al., 1988). The most discriminatory measurement used to increase screening efficiency was established by combining HCG, UE3, and AFP tests. Levels of UE3 can be measured as early as 9 to 11 weeks gestation (Cuckle et al., 1988).

During the past 15 years, continued strides have been made to understand the molecular mechanisms of inherited metabolic disorder (Desnick and Grabowski, 1981). Very specific enzymatic defects have been localized in hundreds of biochemical disorders with inborn errors of metabolism (McKusick, 1992). Therapeutic interventions have been devised which have become a curative approach for many disorders (Desnick, 1980). New avenues have emerged to understand the inheritance of many life threatening diseases and antinatal diagnosis, which once was inconceivable, is now routinely practiced. Through proper counseling and planning, couples at risk have been identified and the risk for another pregnancy with devastating consequences can be eliminated through heterozygous detection (Milunsky, 1979). It is now possible to characterize DNA specific molecular mutations of a single gene (Antonarakis, 1989; Kazazian, 1990) where enzymatic or immunologic procedures are either supplemented or completely eliminated, allowing cells from CVS or amniotic fluids to be used directly to asses inborn errors of metabolism.

PREIMPLANTATION GENETICS

The recent technical advances in detecting human genetic diseases have been extended to the period of embryonic development prior to implantation. A number of techniques have been devised allowing the isolation of a single cell from as early as the eight-cell stage of the embryo, and to perform polar body and polar trophectoderm biopsies (Edwards and Hollands, 1988; Verlinsky et al., 1990). The genetic composition of a single cell can be evaluated for a targeted gene and an abnormal egg can be identified at the pre-embryonic stages as opposed to the current practice of CVS or amniocentesis. At least a dozen genetic disorders have been identified using preimplantation genetics. The DNA from a single cell can be amplified rapidly and with such precision that a diagnosis based on DNA alone can be achieved by PCR technology. Cells can also be cultured for chromosome analysis, and by fluorescence *in situ* hybridization genetic defects can be localized instantly (Critser, 1992; Grifo et al., 1992).

MITOCHONDRIAL DNA AND HUMAN DISEASES

Molecular technologies have led to a renewed interest in the genetics of extrachromosomal inheritance. For example, the mitochondrial genome has proven to be the most intriguing organelle with its many functions, including oxidative phosphorylation, fatty acid oxidation, amino acid metabolism and the Kreb's cycle. This magnificent machinery is associated with maternal inheritance as only small amounts of mitochondrial DNA (mtDNA) are contributed from sperm. Certain diseases are manifest beyond childhood as abnormal mitochondria accumulate over time (Verma, 1990; Bendich, 1993). It is believed that mtDNA has a higher mutation rate than nuclear DNA. mtDNA is small, consisting of circular, nonidentical strands containing 16,569 base pairs. It does not have intervening sequences (introns), while replication is bidirectional and asynchronous, having its own initiation and termination codons. Many disorders that affect protein impair mitochondrial function (see Table 1). In general, mitochondrial diseases can be categorized into three classes: point mutation, insertion and deletion, and depletion of

Table 1. Diseases Due to Mutations and Deletions in Mitochondrial DNA

Abbreviation	MIM No.	Designation
LHON	535000	Leher's hereditary optical neuropathy
MELAS	540000	Mitochondrial encephalomyopathy
	540050	Lactic acidosis with stroke-like signs
MERRF	545030	Myoclonic epilepsy and ragged red fibers
MMC	590050	Maternally inherited myopathy end cardiomyopathy
NARP	551500	Neurogenic muscular weakness with ataxia and retinitis pigmentosa
CEOP	258470	Progressive external ophthalmoplegia
KSS	530000	Kearns-Sayre syndrome (ophthalmoplegia, pigmental degeneration of the retina, and cardiomyopathy)
PEAR	557000	Pearson syndrome (bone marrow and pancreatic failure)
ADMIMY	157640	Autosomal dominant inherited mitochondrial myopathy with mitochondrial deletion in the D loop (type Zeviani)

Source: (MIM: Mendelian Inheritance in Man, McKusick's Catalog of Human Genes and Genetic Disorders. 11th edn., Johns Hopkins University Press, Baltimore, 1994.) [after Passarge, 1995]

mitochondria from the cell (Clarke, 1992; Wallace, 1992, 1993). The best known mitochondrial diseases resulting from point mutations are Leber's hereditary optic neuropathy (LHON), neurogenic muscle weakness, ataxia and retinitis pigmentosa (NARP), mitochondrial encephalopathy, lactic acidosis, MELAS, myoclonic epilepsy and ragged red fibers (MERRF), and maternally inherited myopathy and hypertrophic cardiomyopathy (MMC) (Wallace et al., 1992; Hammans and Harding, 1993). The diseases associated with deletion and insertion of mtDNA are chronic progressive external ophthalmoplegia (CPEO), Kearns-Sayre syndrome, and Pearson syndrome (Harding and Hammans, 1992). A significant decrease in the number of mitochondria per cell can lead to some lethal diseases during early infancy (Clarke, 1992; Zeviani, 1992). The role of mitochondria in neoplastic transformation has just begun to unfold.

CODON EXPANSION AND HUMAN DISEASES

Recent discoveries associated with a number of bewildering traits whose inheritance does not follow simple Mendelian rules have caused much amazement. For example, unstable expansion of trinucleotides in a number of diseases has set the basis for an unusual inheritance pattern. These findings have led to new concepts which perplexes us all (Richards and Sutherland, 1994; Kawaguchi et al., 1994; Willems, 1994; Ashley and Warren, 1995; Hummerich and Lehrach, 1995; Larsson and Clayton, 1995; McMurray, 1995; Passarge, 1995; Oostra and Halley, 1995; Oostra and Willems, 1995; Wells, 1996). During the past two years a number of diseases have been identified which fall into this category (Table 2).

Although the mechanisms of trinucleotide expansion are unknown, the diseases they cause share many unusual characteristics including accelerated age of onset,

Table 2. Genetic Diseases of Trinucleotide Reiteration

Disease	Chromosome Locus	Trinucleotide Repeat	Normal Range	Disease Range
Fragile X	Xq27.3	CGG	6–54	50–1,500
Spinal and bulbar muscular atrophy	Xq21.3	CAG	11–33	40–62
Myotonic dystrophy	19q13.3	CTG	5–37	44–3,000
Huntington's disease	4p16.3	CAG	9–37	37–121
Spinocerebellar ataxia type 1	6p22	CAG	25–36	43–81
Spinocerebellar ataxia type 3	14p23	CAG	6–36	40–100
Dentatorubral-pallidoluysian atrophy (DRPLA)	12q23-24	CAG	7–23	62–68
Machado-Joseph disease	14q32.1	CAG	13–36	68–79

severity of disease in successive generations, and unusual parental biases during transmission (Green, 1993; Martin, 1993; Schalling et al., 1993; Trottier et al., 1993; Koide et al., 1994; Miwa, 1994). The mutation causing repeat expansion of variable size represents a unique phenomenon and may be found in other diseases as well.

GENOMIC IMPRINTING

Two serendipitous events have occurred which have challenged Mendel's laws of inheritance. Codon expansion causing human diseases is one (Richard and Sutherland, 1992), while genomic imprinting is the other phenomenon where the behavior of genes do not follow the normal pattern of inheritance. The term genomic imprinting is the other enigma where the behavior of genes do not follow the normal pattern of inheritance. The term genomic imprinting implies the differential expression of genes, depending on the parental origin (Monk, 1988; Reik, 1988; Solter, 1988; Sapienza, 1989). There is compelling evidence that genomic imprinting does occur in parts in the human (Moore and Haig, 1991; Nicholls, 1991). For example, in human triploidy, the phenotypic expression depends upon genetic factors (McFadden et al., 1993) and the origin of two diploid complements (maternal versus paternal). In the case of Prader-Willi syndrome, both chromosomes are maternal without deletion (Nicholls et al., 1989), while genomic segmental deletions produce Prader-Willi and Angelman's syndromes (Sapienza, 1989; Searl et al., 1989). Similarly, in many tumors the loss of heterozygosity, depending upon parental chromosomes, has variable consequences (Feinberg, 1993). Also, the inheritance of mutated genes from different parents will have different degrees of disorders in a variety of diseases (Clarke, 1990; Hall, 1990). These findings have caused us to examine the morbid anatomy of the genome at a single gene level whose make-up may depend upon parental origin (Hoffman, 1991; Durcan and Goldman, 1993; Peterson and Sapienza, 1993).

GENOME MAPPING AND GENOME TARGETING

For the past eight years there has been an international undertaking to produce genetic, physical, and DNA sequence maps of the human genome at the highest resolution. With a variety of advanced molecular techniques, the 3 billion bases will be sequenced and the location of approximately 100,000 genes will be determined. The linkage map of genetic traits will be constructed for each chromosome. It is estimated that 1cM is equivalent to approximately 1 million based pairs (or 1 megabase; mb). On average, human genes contain 30 kb with a range of 800 nucleotides to more than 1 million bases. Close to 4,000 candidate genes have been identified through positional cloning. Our present task is to improve the existing technologies and add a few more for sequencing the entire genome quickly with great cost effectiveness. Unparalleled advances have been made in understanding many diseases whose inheritance is quite obscure. The conservation of genes which are also found in humans have added a new dimension in identifying new genes (Cotton and Malolm, 1991). In eukaryotes, the manipulation of genes has been carried out for a number of years (Capecchi, 1989). With the advent of new technology, using embryonic stem cells derived from the preimplantation blastocyst, it is possible to manipulate the gene in mammalian systems (Mansour, 1990; Robbins, 1993). A number of animal models for human diseases have been proposed through precise manipulation of the mammalian genome (Karp and Broder, 1994; Lowenstein, 1994) and have begun to yield some fruitful results helpful in understanding the genetic basis of human diseases.

GENE THERAPY

Treatment of diseases by transferring normal genes into the patient's body represents the start of a new era in medicine. Gene therapy can be conducted in two forms: Normal genes can be transferred either to germ cells (termed germ line therapy), or recombinant genes can be injected into somatic cells (called somatic gene therapy) (Miller, 1992; Weissman, 1992). Germ line therapy has not come of age yet but somatic gene therapy is underway for a variety of disorders (Morgan and Anderson, 1993). A variety of strategies have been utilized to accomplish the adoption of new genes into the system (Miller, 1990). Such experiments have been performed through the application of retroviral mediated gene therapy (Ledley, 1992). Obviously, new endeavors are not free from problems and effectiveness and safety of the gene delivery systems are currently being evaluated by a number of health agencies. Based on successful trials, this technology will be applied to treat a wide variety of human diseases (Tolstoshev, 1993). A number of animal models have been developed to understand the molecular mechanism of the human diseases (Capecchi, 1994). Gene therapy, once a scientist's fantasy, will become a routine procedure which will alter the practice of medicine in the 21st century.

ETHICS IN THE GENETICS REVOLUTION

Technical complexities coupled with many ethical issues have threatened advances in modern medicine. Clearly, some concerns may be well-founded, while others are unwarranted (Fost, 1993). Some genetic diseases are complex and differ from other disorders as they can be inherited by their progenies, hence they need special attention in diagnosis. It is highly imperative that patients should be informed and their views must be taken into consideration when counseling them about clinical alternatives. Individuals with genetic diseases must be educated about the advantages and risks associated with technology. Who are we to decide as to who should have or not have? However, technology should be applied for the welfare of mankind. Treatment protocols with genetically engineered genes should have guidelines which should be regulated carefully by health agencies. Knowledge and fame do come between patients and care providers and every attempt should be made to separate the practice of medicine from the advancement of science. Sometimes overcaution can be harmful and deserving patients may not be benefited by the advancement of technology. Therefore, ethical issues should proceed hand-in-hand with legal and practical implications. In my opinion, every patient should have the right to receive quality care. Certain genetic diseases have enormous emotional stakes and it is undoubtedly a difficult task to guide the patients. Therefore, a highly sensitive approach towards disclosure must be applied. Does a patient have the right to know or is the health provider obligated to tell the individuals who may be at risk? There is no question that DNA has become a "double-edged" helix and ethical issues associated with it can jeopardize the well-being of many geneticists and mankind at large. Of course, when it comes to forensic genetics, the double helix has occupied a unique status and molecular biologists throned it as an avenging angel, though critics have tainted its legendary might with grimmer facts. The ability to direct the expression of genes *in vivo*, as well as *in vitro*, clearly provides an important challenge for gene therapists. The implementation of technology and its usage in clinical practice from laboratory to the bedside has clearly provided a ray of hope. The concept of a customized approach is remarkably straightforward and revolutionary. The strategy will pay off faster than anyone had dared to think. The development of many genetically designed drugs and their successful testing in animals has provided a powerful challenge for combating many genetic diseases. The animal rights groups, most of whom are themselves non-vegetarians, are creating unfounded social disorder. They should not live in illusions because longevity is good for mankind. Ethicists should know better than everyone that we all want to go to heaven but no one wants to die. The most brilliant minded patients can be undermined by the draining effect of some ethical conflicts that no one presently wants to acknowledge. Finding the genes which threaten survival is the best hope for curing genetic diseases. However, it can not be denied that certain genetic tests are so powerful that we have invaded the body's most intimate secrets and, in turn, highly sensitive information has been unleashed which can be used for the wrong reasons. Every attempt should be made to safeguard the confidentiality of the genetic make-up of patients and family members. Genetic diseases whose onset is at a later date can create a new group whose fundamental rights in gaining employment, health insurance, etc. can be denied and even the pending eugenic movement can be revived, consequently scientific racism can be injected in the society. These are very fundamental questions whose answers are tainted with traditional values and one day "an equal opportunity employer" will have to prove unequivocally that they do not discriminate because of one's genetic make-up. Genetic literacy should be brought to the common level so a layman does not have to go to a geneticist for reproductive options to reveal the body's most intimate secrets. Obviously, the "privacy rule" will play an important role for a future diary.

SUMMARY

Today, genetics occupies a special place in the practice of medicine. Diseases which were once thought to be trivial have now been clearly identified in a genetic sense. Determining the expression of genes from the simplest to the most complex at the single nucleotide level, has clearly facilitated understanding the mechanisms of many disorders, thus benefiting basic scientists and clinicians alike. Cloning and sequencing specific candidate genes have become routine practice in the clinical laboratory. These efforts have clearly emerged as linking chariots between phenotypes and nucleotide sequences and solely diagnosing a patient at the syndrome level has become a dinosaurous approach. The evolutionary concept will revolutionize the detection rate of diseases causing gene mutation; an idea which was once inconceivable has now become readily accessible.

Recent discoveries in a number of bewildering traits whose inheritance do not follow simple Mendelian rules have deluged us with amazement and the universality of Mendel's laws are continuously being challenged. For example, there are a number of diseases whose triplet repeat mutation and amplification in future generations have intrigued us all. It has become much clearer than ever before, that approximately 75,000 genes composed of three billion nucleotides will sooner or later be deciphered. No question this is a Herculean task, but the human genome mapping project will yield numerous major medical breakthroughs. Gene therapy, once a scientist's fantasy, will become a routine procedure that will encompass the geneticist's bench top to the patient's bed side. In many cases, presymptomatic diagnosis of genetic diseases has become possible as this is the dawn of a new era of preventive medicine. We have gone back to the grave to search our geneology and the cause of death of our loved ones. Remarkable success in prenatal diagnosis of single gene disorders for many untreatable diseases offers new hope for future generations. The feasibility and utility of new technology is associated with many ethical issues and reproductive options should be made available to all. Obviously, there are many profound issues concerning intervention of numerous genetic disorders. For example, prior to marriage, heterozygous detection of genetic diseases in couples may become a law where "love" will be warned by geneticists and their findings will be challenged by ethicists. Undoubtedly, the double-edged helix has occupied a pristine podium in forensic genetics and the gene has become a social icon. It cannot be denied that reverberations will continue, but we should move with great caution concerning the role of genetics in criminality, seductiveness, and shyness, etc. Nevertheless, it is one technology that will play a major role in cutting health care costs, while other approaches will eventually fall short. In order to deliver credible diagnoses, medicine will be shifted to the back burner, while genetics will hold the high ground and patients will begin to utter that "Doc, my genes are hurting me."

REFERENCES

Antonarakis, S.E. (1989). Diagnosis of genetic disorders at the DNA level. N. Engl. J. Med. 320, 153–163.

Ashley, C.T., & Warren, S.T. (1995). Trinucleotide repeat expansion and human disease. Ann. Rev. Genet. 29, 703–728.

Bendich, A.J. (1993). Reaching for the ring. The study of mitochondrial genome structure. Curr. Genet. 24, 279–290.

Billings, P.R., Smith, C.L., & Cantor, C.R. (1991). New techniques for physical mapping of the human genome. FASEB J. 5, 28–34.

Bogart, M.H., Pandian, M.R., & Jone, O.W. (1987). Abnormal material serum chorionic gonadotropin levels in pregnancies with fetal chromosome abnormalities. Prenat. Diagn. 7, 623.

Canick, J.A., Knight, G.J., Palomaki, G.E., & Haddow, J.E., et al. (1988). Low second trimester maternal serum unconjugated estriol in Down syndrome pregnancy. Br. J. Obst. Gynecol. 95, 330–333.

Capecchi, M.R. (1989). Altering the genome by homologous recombination. Science 244, 1288–1292.

Capecchi, M.R. (1994). Targeted gene replacement. Sci. Am. 270, 52–61.

Chehab, F.F. (1993). Molecular diagnostics: Past, present and future. Hum. Mutation 2, 331–337.

Clarke, A. (1990). Genetic imprinting in clinical genetics. Development (Suppl.), 131–139.

Clarke, L.A. (1992). Mitochondrial disorders in pediatrics. Clinical biochemical and genetic implications. Ped. Clin. North Am. 39, 319–334.

Cotton, R.G.H., & Malcolm, A.D.B. (1991). Mutation detection. Nature 353, 582–583.

Critser, E.S. (1992). Preimplantation genetics. An overview. Arch. Path. Lab. Med. 116, 383–387.

Cuckle, H.S., Wald, N.J., & Barkai, G., et al. (1988). First trimester biochemical screening for Down syndrome. Lancet II, 851–852.

Daffos, F., Capella-Povolsky, M., & Forestier F. (1985). Fetal blood sampling during pregnancy with use of a needle guided by ultrasound: A study of 606 consecutive cases. Am. J. Obst. Gynecol. 153, 655–660.

DeGrouchy, J., & Turleau, C. (1984). In: Clinical Atlas of Human Chromosomes. pp. 2–284, John Wiley, New York.

Desnick, R.J. (1980). In: Therapy in Genetic Diseases. Alan R. Liss, New York.

Desnick, R.J., & Grabowski, G.A. (1981). Advances in the treatment of inherited metabolic diseases. Adv. Hum. Genet. 11, 281–369.

Durcan, M.J., & Goldman, D. (1993). Genomic imprinting: Implications for behavioral genetics. Behavior Genet. 23, 137–143.

Edwards, R.G., & Holland, P. (1988). New advances in human embryology: Implications of preimplantation diagnosis of genetic diseases. Hum. Reprod. 3, 549–556.

Elias, S., Simpson, J.L., & Holbrook, K.A. (1993). Fetal skin, liver and muscle sampling. In: Essentials of Prenatal Diagnosis (Simpson, J.L., & Elias, S., eds.), Churchill Livingstone, New York.

Feinberg, A.P. (1993). Genomic imprinting and gene activation in cancer. Nature Genet. 4, 110–113.

Fiedler, W., Claussen, U., Ludecke, H.J., et al. (1991). New markers for the neurofibromatosis-2 region generated by microdissection of chromosome 22. Genomics 10, 786–791.

Fost, N. (1993). Genetic diagnosis and treatment. Am. J. Dis. Child. 147, 1190–1195.

Green, H. (1993). Human genetic diseases due to codon reiteration: Relationship to an evolutionary mechanism. Cell 747, 955–956.

Grifo, J.A., Boyle, A., Tang, Y.-X., & Ward, D.C. (1992). Preimplantation genetic diagnosis. In situ hybridization as a tool for analysis. Arch. Path. Lab. Med. 116, 393–397.

Grody, W.W., & Hilborne, L.H. (1992). Diagnostic applications of recombinant nucleic acid technology: Genetic diseases. Lab. Med. 23, 166–171.

Hall, J.G. (1990). Genomic imprinting. Arch. Dis. Child. 65, 1013–1015.

Hammans, S.R., & Harding, A.E. (1993). Mitochondrial DNA and disease In: Advances in Genome Biology: Morbid Anatomy of the Genome, Vol. 2 (Verma, R.S., ed.), pp. 41–68, JAI Press, Greenwich.

Harding, A.E., & Hammans, S.R. (1992). Deletions of the mitochondrial genome. J. Inherit. Metab. Dis. 15, 480–486.

Hoffman, M. (1991). How parents make their mark on genes. Science 252, 1250–1251.

Hummerich, H., & Lehrach, H. (1995). Trinucleotide expansion human disease. Electrophoresis 16, 1698–1704.

Jackson, L.G., Zachary, J.M., Fowler S.E., et al. (1992). A randomized comparison of transcervical and transabdominal chorion villus sampling. N. Eng. J. Med. 327, 594–598.

Jackson, L., & Wapner, R.J. (1993). Chorionic villus sampling. In: Essentials of Prenatal Diagnosis (Simpson, J.L., & Elias S., eds.), pp. 45–61, Churchill Livingstone, New York.

Jones, K.L. (1988). In: Smith's Recognizable Patterns of Human Malformation 4th Ed., W.B. Saunders Co., Philadelphia.

Kao, F.T., & Yu, J.W. (1991). Chromosome microdissection and cloning in human genome and genetic disease analysis. Proc. Natl. Acad. Sci. USA 88, 1844–1888.

Karp, J.E., & Broder, S. (1994). New directions in molecular medicine. Cancer Res. 54, 653–665.

Kawaguchi, Y., Okamoto, T., Taniwaki, M., et al. (1994). CAG expansion in a novel gene for Machado-Joseph disease at chromosome 14q32.1. Nature Genet. 8, 221–228.

Kazazian, H. (1990). Current status of prenatal diagnosis by DNA analysis. Birth Defects 26, 210.

Koide, K., Ikeuchi, T., Onodera, O., Tanaka, H., et al. (1994). Unstable expansion of CAG repeat in hereditary dentatorubralpallidoluysian atrophy (DRPLA). Nature Genet. 6, 9–13.

Larsson, N-G., & Clayton, D.A. (1995). Molecular genetic aspects of human mitochondrial disorders. Ann. Rev. Genet. 29, 151–178.

Ledley, F.D. (1993). Current status of somatic gene therapy. Genet. Hormones 8, 1–5.

Lejeune, J., Gautier, M., & Trpin, R. (1959). Etude des chromosomes somatiques des neuf enfants mongoliens. D.R. Acad. Sci. 248, 1721–1722.

Li, H., Gyllensten, U.B., Cui, X., Saiki, R.K., Erlich, H.A., & Arnheim, N. (1988). Amplification and analysis of DNA sequences in single human sperm and diploid cells. Nature 335, 414–417.

Lowenstein, D.H. (1994). Basic concepts of molecular biology for the epileptologist. Epilepsia 35 (suppl 1), 817–819.

Mansour, S.L. (1990). Genome targeting in murine embryonic stem cells. Introduction of specific alterations into the mammalian genome. Genet. Anal. Tech. Appl. 7, 219–227.

Martin, J.B. (1993). Molecular genetics of neurological diseases. Science 262, 674–676.

McFadden, D.E., Kwong, L.C., Yam, I.Y.L., & Langlois, S. (1993). Parental origin of triploidy in human fetuses. Evidence for genomic imprinting. Hum. Genet. 92, 465–469.

McKusick, V. (1992). In: Mendelian Inheritance in Man. 10th Ed. John Hopkins University Press, Baltimore.

McMurray, C.T. (1995). Mechanisms of DNA expansion. Chromosoma 104, 2–13.

Meltzer, P.S., Guan X.-Y., Burgess, A., & Trent, J.M. (1992). Rapid generation of region specific probes by chromosome microdissection and their application. Nature Genet. 1, 24–28.

Miller, A.D. (1990). Progress toward human gene therapy. Blood. 76, 271–278.

Miller, A.D. (1992). Human gene therapy comes of age. Nature 357, 455–461.

Milunsky, A. (1979). In: Genetic Disorders and the Fetus. Plenum, New York.

Miwa, S. (1994). Triplet repeats strike again. Nature Genet. 6, 3–4.

Monk, M. (1988). Genomic imprinting. Genes Dev. 2, 921–925.

Moore, T., & Haig, D. (1991). Genomic imprinting in mammalian development: A parental tug-of-war. Trend. Genet. 7, 45–49.

Morgan, R.A., & Anderson, W.F. (1993). Human gene therapy. Annu. Rev. Biochem. 62, 191–217.

Nicholls, R., Knoll, J., Butler, M., Karam, S., & Lalande, M. (1989). Genetic imprinting suggested by maternal heterodisomy in non-deletion Prader-Willi syndrome. Nature 342, 281–285.

Nicholls, R. (1991). Uniparental disomy as the basis for an association of rare disorders. Am. J. Med. Genet. 41, 273–274.

Oostra, B.A., & Halley, D.J.J. (1995). Complex behavior of simple repeats: The fragile X-syndrome. Pediatric Res. 38, 629–637.

Oostra, B.A., & Willems, P.J. (1995). A fragile gene. BioEssays 17, 941–947.

Orlandi, F., Damiani, G., Jakil, C., et al. (1990). The risks of early cardocentesis (12–21 weeks): Analysis of 500 procedures. Prenat. Diag. 10, 425.

Palomaki, G.E., Knight, G.J., Holman, M.S., & Haddow, J.E. (1990). Maternal serum alpha-fetoprotein screening for fetal Down syndrome in the United States: Results of a survey. Am. J. Obstet. Gynecol. 162, 317–321.

Passarge, E. (1995). Color Atlas of Genetics. Thieme Medical Publishers, New York.

Peterson, K., & Sapienza, C. (1993). Imprinting the genome. Imprinted genes and a hypothesis for their interaction. Ann. Rev. Genet. 271, 7–31.

Peuch, A., Ahnine, L., Ludecke, H.J., et al. (1992). 11p15.5 specific libraries for identification of potential gene sequences involved in Beckwith-Wiedemann syndrome and tumorigenesis. Genomics 13, 1274–1280.

Reik, W. (1988). Genomic imprinting: A possible mechanism of the parental origin effect in Huntington's chorea. J. Med. Genet. 25, 805–808.

Remick, D.G., Kunkel, S.L., Holbrook, E.A., & Hanson, C.A. (1990). Theory and applications of the polymerase chain reaction. Am. J. Clin. Pathol. 93 (Suppl 1) 49–54.

Richards, R.I., & Sutherland, G.R. (1992). Heritable unstable DNA sequences. Nature Genet. 1, 7–9.

Richards, R.I., & Sutherland, G.R. (1994). Simple repeat DNA is not replicated simply. Nature Genet. 6, 114–116.

Robbins, J. (1993). Gene targeting: The precise manipulation of the mammalian genome. Circulation Res. 73, 3–9.

Robinson, A., & Linden M.G. (1993). In: Clinical Genetics Handbook 2nd edn., Blackwell Scientific Publication, Oxford.

Rossiter, B.J.F., & Caskey, C.T. (1991). Molecular studies of human molecular diseases. FASEB J. 5, 21–27.

Sapienza, C. (1989). Genome imprinting and dominance modification. Ann. N.Y. Acad. Sci. 564, 24–38.

Schalling, M., Hudson, T.J., Buetow, K.J., & Housman, D.E. (1993). Direct detection of novel expanded trinucleotide repeats in the human genome. Nature Genet. 4, 135–141.

Schmickel, R.D. (1986). Contiguous gene syndromes: A component of recognizable syndromes. J. Ped. 109, 231–241.

Searle, A.G., Peters, J., Lyon, M.F., et al. (1989). Chromosome maps of man and mouse. Ann. Hum. Genet. 53. 89–140.

Simpson, J.L., & Golbus, M.S. (1992). In: Genetics of Obstetrics and Gynecology 2nd edn., W.B. Saunders, Philadelphia.

Simpson, J.L., & Elias, S. (1993). In: Essentials of Prenatal Diagnosis, Churchill Livingstone, New York.

Solter, D. (1988). Differential imprinting and expression of maternal and paternal genome. Ann. Rev. Genet. 22, 127–146.

Tjio, J.H., & Levan, A. (1956). The chromosome number of man. Hereditas 42, 1–6.

Tolstoshev, P. (1993). Current status and future directions in human gene therapy. Adv. Genome Bio. 2, 389–415.

Trottier, Y., Devys, D., & Mandel, J.L. (1993). Fragile-X syndrome. An expanding story. Curr. Biol. 3, 783–786.

Verlinsky, Y., Pergament, E., & Storm, C. (1990). The preimplantation genetic diagnosis of genetic diseases. J. *In Vitro* Fert. Embryo. Transf. 7, 1–5.

Verma, R.S., & Babu, A. (1989). In: Human Chromosomes: Manual of Basic Technique, Pergamon Press, New York.

Verma, R.S. (1990). Mitochondrial genome. In: The Genome (Verma, R.S., ed.), pp. 161–182, VCH Publishers, New York.

Verma, R.S., ed. (1993). In: Advances in Genome Biology. II. Morbid Anatomy of the Genome. JAI Press, Greenwich.

Vosberg, H.-P. (1989). The polymerase chain reaction. An improved method for the analysis of nucleic acids. Hum. Genet. 83, 1–15.

Wallace, D.C. (1992). Diseases of the mitochondrial DNA. Ann. Rev. Biochem. 6, 1175–1212.

Wallace, D.C. (1993). Mitochondrial diseases. Genotype versus phenotype. Trend. Genet. 9, 128–133.

Wallace, D.C., Lott, M.T., Shoffner, J.M., & Brown, M.D. (1992). Diseases resulting from mitochondrial point mutation. J. Inherit. Metabol. Dis. 15, 472–479.

Weissman, S.M. (1992). Gene therapy. Proc. Natl. Acad. Sci. USA 89, 11111–11112.

Wells, R.D. (1996). Molecular basis of genetic instability of triplet repeats. J. Biol Chem. 271, 2875–2878.

Willems, P.J. (1994). Dynamic mutations hit double figures. Nature Genet. 8, 213–215.

Wolcott, M.J. (1992). Advances in nucleic acid-based detection methods. Clin. Microbiol. Rev. 5, 370–386.

Yu, J., Hartz, J., Xu, Y., et al. (1992). Isolation, characterization and regional mapping of microclones from a human chromosome 21 microdissection library from human chromosome 2q35-q37. Am. J. Hum. Genet. 51, 263–272.

Zeviani, M. (1992). Nucleus-driven mutations of humans of human mitochondrial DNA. J. Inherit. Metab. Dis. 15, 456–471.

RECOMMENDED READINGS

Cooper, D.N., & Krawczak, M. (1993). In: Human Gene Mutation. Bios. Scientific Publishers, Oxford.

McKusick, V.A. (1992). In: Mendelian Inheritance in Man: Catalogs of Autosomal Dominant, Autosomal Recessive and X-linked Phenotypes. The Johns Hopkins University Press, Baltimore.

Robinson, A., & Linden, M.G. (1993). In: Clinical Genetics Handbook. Blackwell Scientific Publications, Cambridge.

Scriver, C.R., Beaudet, A.L., Sly, W.S., & Valle, D. (1989). In: The Metabolic Basis of Inherited Disease. McGraw-Hill, New York.

Singer, M., & Berg, P. (1991). In: Genes & Genome: A Changing Perspective. Blackwell Scientific Publication, Oxford.

Thompson, M.W., McInnes, R.R., & Willard, H.F. (1991). In: Genetics in Medicine. W.B. Saunders, Philadelphia.

Valle, D. (1989). In: The Metabolic Basis of Inherited Disease. McGraw-Hill, New York.

Verma, R.S. (1990). In: The Genome. VCH Publisher, New York.

Verma, R.S., (Ed.). (1992). In: Advances in Genome Biology. Unfolding the Genome. JAI Press, Greenwich.

Verma, R.S., (Ed.). (1993). In: Advances in Genome Biology. Morbid Anatomy of the Genome. JAI Press, Greenwich.

Winnacker, E.-L. (1987). In: From Genes to Clones. VCH Publisher, New York.

Chapter 17

The Genome Project—A Commentary

DARRYL R.J. MACER

INTRODUCTION

The genome project is an international project aimed at obtaining a detailed map and a complete DNA sequence of the genome of a variety organisms, including

Principles of Medical Biology, Volume 5
Molecular and Cellular Genetics, pages 345–376.
Copyright © 1996 by JAI Press Inc.
All rights of reproduction in any form reserved.
ISBN: 1-55938-809-9

mycoplasma, the bacteria *Escherichia coli*, the yeast *Saccharomyces cerevisiae*, the roundworm *Caenorhabditis elegans,* the fruitfly *Drosophila melanogaster,* wheat, rice, the mouse, and that of *Homo sapiens.* It is the Human Genome Project which has captured the public attention, and has been the image responsible for increased funding of genetics research since 1988–1990. It will have many scientific, medical, economic, ethical, legal, and social implications. This commentary seeks to trace the historical background, progress, and foreseeable impact of the genome project in both science and in society.

The number of human genes that have been sequenced is exponentially growing, and it is impossible to keep a count on the progress. The expected completion date of the project is around the year 2000 and most gene sequences were known by 1995. The countries that have Human Genome Projects at the time of this writing include Australia, Brazil, Canada, China, Denmark, The European Community (especially France, Italy, and the U.K.), Japan, Mexico, the Netherlands, Russia, South Korea, Sweden and the U.S. Genetic analysis of a number of other organisms is also well underway as part of the Genome Project, and the common language of genetics and similarity of many genes means that molecular and cellular genetics is being advanced by studies on all organisms.

ORIGINS OF THE GENOME PROJECT

There have been various accounts written of the origin of the Human Genome Project. The other genome projects arose in terminology at the same time or after the phrase "Human Genome Project" had been coined. However, all of these projects involve genetic mapping and sequencing which was underway long before the trendy phrase of "genome project" was used, and specific long-term projects to entirely sequence some viruses, eukaryote organelle, and bacterial genomes have been underway since the late 1970s.

Mapping of the human genome has been progressing for decades (Culliton, 1990). The beginnings of the genome project can be traced back at least to Mendel's genetic studies on peas; the mapping of the trait for color blindness to the X-chromosome of *Drosophila* by T.H. Morgan and workers; to Avery and colleagues who found DNA was the physical substance of genes; to Crick, Franklin, Watson, and Wilkins who determined the structure of DNA; to those who discovered the genetic code; to Sanger and others who developed DNA sequencing; and to many others who contributed to our knowledge of genetics and molecular biology. Therefore, no single person can claim to have initiated the goals of the genome project; and the date of the origin is obscure.

The progress in genetics since the 1960s has been rapid, and by the 1980s the goal of mapping and sequencing the complete genome of particular organisms became distant but within reach. The first organisms to have their DNA sequenced were viruses, then the complete mitochondrial DNA of some organisms was sequenced. Projects in Europe, Japan, and North America began which were aimed at automating DNA sequencers in preparation for the project. In scientific conferences in the 1980s people discussed the idea of a human genome project. In 1988 the polymerase chain reaction (PCR) was devised, which increased the rapidity of DNA manipulation enormously (see chapter 12). By 1992 the project had resulted in physical genetic maps of human chromosomes 21 and Y, a refined complete genetic map of human DNA; and the complete sequence of a yeast chromosome.

The Genome Project has political dimensions as could be expected for any project requiring much funding, with the huge potential rewards for the biotechnology industry from the sequencing and future application of genes to medicine and agriculture. The country putting forward the most money for the project is the U.S., and some would say that the genome project as a specific image arose in the U.S. There were several people in the U.S. who saw the goals of the genome project as ideal for initiating the first large scale biological research project with a definite endpoint (Cantor, 1990). The Human Genome Project, or Human Genome Initiative, is the collective name for several projects begun in the late 1980s in several countries, following the U.S. Department of Energy (DOE) decision to create an ordered set of DNA segments from known chromosomal locations, to develop new computational methods for analyzing genetic maps and DNA sequence data, and to develop new techniques and instruments for detecting and analyzing DNA (OTA, 1988). The DOE has had a long-term interest in assessing the effects of radiation on human health, and sequencing the genome may allow detection of mutations in the DNA. Whether the motive was to fill vacant DOE laboratories, to provide renewed emphasis for science, or to put U.S. biotechnology companies in a better international position (Lewin, 1990), the idea itself was sure to catch the imagination of politicians. Some biologists commented that they do not think physicists can do good biology, so the project should not be left to the DOE, and because the National Institutes of Health (NIH) is the major funder of U.S. biomedical research, the NIH joined the project. The activities of the NIH and DOE have been coordinated for genome research since they signed a Memorandum of Understanding on October 1, 1988.

The genome project is often compared to the Apollo space project. The analogy shows the glamor of the project. Not only will the genome project lead to the development of useful new technology but, unlike the Apollo project, the goal itself is also of immense direct practical use. The importance of reaching the projects goal is also different—people would not have gone to the moon if a positive decision had not been made—but the human genome map will be obtained, with or without a positive effort, though over a longer period of time if undirected (Macer, 1991). The original time frame was 15 years, which has since been considerably shortened.

There have been numerous scientists who have contributed to our knowledge, and it will also be fitting that to complete this project will require the joint collaborative work of innumerous international scientists. In the perspective of who initiated the project, who does the work, and whose knowledge is needed, the answer is clearly that many people are, and will be, directly responsible for the mapping and sequencing, and the later interpretation of the data.

GENETIC MAPPING

The DNA that is being sequenced is a composite of different human tissue cell lines. About 0.3–0.5% of the nucleotides in our DNA vary between different people. These differences vary from person to person, therefore it does not matter whose genome is actually sequenced. Different laboratories often use different human tissue culture cell lines. However, by the characterization of standardized marker regions, the DNA between different individuals will be able to be compared, and a single general map and sequence produced.

A detailed genetic map of the human genome is being refined prior to full scale sequencing. A map is essential to efficient sequencing so that a physical library of DNA fragments can be systematically sequenced. The distance unit used in gene mapping is called a centi-Morgan (cM), and one cM is equivalent to two markers being separated from each other in chromosome crossing over in normal reproduction 1% of the time. The actual physical length of one cM varies, being approximately 1 million base pairs (Mb) in humans.

The construction of any map requires markers. The markers of sequence are specific DNA sequences, and the longer the sequence the less copies there will be in any genome. A sequence of 20 nucleotides, for example, is generally unique in a genome. There are two types of maps: maps that are drawn in picture, and physical maps which include a physical collection of sequences of known order and location. Genome analysis involves both types. For short sequences, such as a few genes, convenient markers are the specific nucleotide sequences that are cut by restriction endonuclease enzymes. A sequence of DNA can be cut into smaller specific pieces by digestion with restriction enzymes, and the pattern from digestion with several different enzymes can be used to make a map of that sequence. The use of restriction fragment linkage patterns (RFLPs) in combination with genetic linkage analysis allowed the construction of linkage maps for each human chromosome with an average spacing of 10–15 cM in 1987 (Doris-Keller et al., 1987).

The current mapping paradigm is based on a proposal in 1989 to use physical sequence-tagged sites (STSs) as the map labels (Olsen et al., 1989). These sequences are longer than those utilized by restriction enzymes, and may be 100–1,000 nucleotides long, so that a single tagged site in a genome can be defined by this short nucleotide sequence. Different researchers use different cloning vectors for gene analysis, so the exchange of DNA pieces in these different vectors, or DNA clones, is not possible. What is possible using the DNA PCR is to generate DNA sequences for any DNA if short sequences are known for each STS, from which primers for the PCR can be made. Therefore, the ends of large DNA fragments should be sequenced, and the data combined, to make a STS map of each chromosome. This approach means that researchers can continue to use different methods, and develop better procedures, while the information obtained can be integrated to produce the actual physical map. This will avoid the need to exchange different clones of DNA between laboratories, because each laboratory can use the marker sequence as a starting point. This approach was used to make a map of the human genome with an average spacing of 5 cM in 1992 (Weissenbach et al., 1992; NIH/CEPH, 1992; Vaysseix and Lathrop, 1992). This means that a gene associated with a disease may be linked to one marker, and the amount of DNA which needs to be examined is only about 5 Mb, rather than 15 Mb, as in 1987. By 1994 the average density had fallen to 0.7 cM (Cooperative Human Linkage Center, 1994). As the map progresses, the average spacing will be decreased much more so that the length of sequence between markers will only be 0.1 Mb.

The initial approach was for different research groups to concentrate on different human chromosomes in order that they could all have the complete map in a shorter time. There are actually 24 chromosomes to be sequenced, 22 autosomes and the X and Y chromosomes. An initial goal is a map with STS markers spaced at about 100,000 base pairs, and the assembly of overlapping contiguous cloned sequences (called contigs) of about 2 million base pairs length. From this physical and informational library system, the sequencing can be started. Data management technology must also improve, such as programs to search the DNA sequence libraries, using advanced computing technology (Watts, 1990).

Physical chromosome maps (overlapping cloned pieces of the complete chromosome) of two chromosomes, 21 (Chumakov et al., 1992), and Y (Foote et al., 1992) were published in 1992. X-chromosome analysis is also advancing, and 40% was in physical cloned maps in 1992 (Mandel et al., 1992). Overlapping pieces of DNA were cloned, and the tags on each piece allowed the pieces to be put into order. This means that once a disease-causing gene is localized to a particular region of the DNA, the appropriate clone can be taken from a freezer and sequenced.

Meanwhile, in 1992, researchers in France described improved techniques which meant that a whole human genome approach, not an individual chromosome approach, could be used to construct a physical library of contigs (Bellanne-Chantelot et al., 1992). A French project, Généthon, took the lead in genome mapping in 1992–1993 because it was the world's largest gene mapping lab (Anderson, 1992a). They completed the first physical human genetic map in 1993, using automatic robotic analysis systems. The library of overlapping clones is basically completed using yeast artificial chromosomes (YACs). There are minor gaps where the DNA sequences are difficult to clone. Following its rapid progress the U.S. began setting up two large genome mapping centers, as the approach began to be directed at the whole genome rather than just individual chromosomes. The current map is on-line on Internet from the Whitehead Research Institute (<http://www-genome.wi.mit.edu>).

Using the STS approach allows small teams of researchers to contribute results. There are worldwide efforts, although the major international effort is centered around the U.S./Canada, Europe, and Japan. There are many people from Australia and Latin America who are also involved, via access to international DNA databanks. There will be contributions from many other countries for particular genome projects that have regional economic importance; for example, China has a rice genome project.

Model Organisms

Smaller genomes are useful models for the human genome, and several organisms are models for the sequencing project. Although it is interesting to give the proportion of the genomes that have been sequenced, such figures are constantly increasing, and readers are referred to the expanding body of literature to read about the completion of sequencing efforts of model organisms. Many viral genomes have been completely sequenced, including the 200 kb cytomegalovirus genome.

By mid-1992 the EMBL databank contained 76% of the 5 Mb *E. coli* genome sequence, compared to 27% of the 15 Mb *S. cerevisiae* (yeast) and 0.6% of the 3,000 Mb human sequence. In 1995, the first two complete bacterial genomes to be sequenced were reported by the U.S. company Human Genome Sciences Inc., those of *Haemophilus influenzae* (Fleischmann et al., 1995) and *Mycoplasma genitalium* (Fraser et al., 1995). In 1996 the 2.8 Mb *Staphylococcus aureus* genome was completed by Human Genome Sciences Inc. The complete sequence of chromosome 3 of yeast (which has 16 chromosomes) was the first to be completely sequenced (Oliver et al., 1992). It was sequenced by the combined efforts of 35 European laboratories, and was 315 kb of contiguous sequence. It is expected to be the first eukaryote organism to have a fully sequenced genome, and may be completed in 1996.

The first multicellular organism that is expected to be fully sequenced is the roundworm *C. elegans*. This genome project is being undertaken at the MRC Laboratory of Molecular Biology in Cambridge, U.K., and at Washington State University, with other international collaborators. It is being used as a model project as a prelude to the sequencing of the human genome. The total sequence is about 100 Mb. This worm is one of the best characterized animals in the world, with the complete pattern of cell divisions for the whole organism (959 nuclei) and the full connections of the 302 nerve cells known. The life cycle is 3.5 days, which makes it easy to study mutations. The work on the physical map began in the early 1980s, and the genome is arranged in less than 50 contigs (pieces of contiguous cloned DNA sequence) (Coulson et al., 1991). In 1992, about 1 Mb of sequence was produced, and the complete sequence is expected within several years. The cost of sequencing per nucleotide has fallen to about 5 cents, a hundredth of the cost in 1988 when the genome project began.

The construction of a YAC contig library for the model plant organism, *Arabidopsis thaliana*, has been made. By 1994 about 2 Mb of sequence was obtained and all the cDNAs will be sequenced by 1997. Japan and China have rice genome projects, which are still at a mapping stage. Other crops such as wheat, soybean, and corn are also being mapped, but their genomes are very large. Similarities in genome organization aid the progress in all species.

Another model organism that has much historical significance to genetics is the fruitfly, *Drosophila*. The size of the *D. melanogaster* genome is estimated to be 165 Mb, with about 5–15,000 genes (Merriam et al., 1991). A physical cloned map has been completed, and the sequence may also be completed before the human sequence. *Drosophila* has been used in mutation studies of gene function for many decades, and because of a short generation time will continue to be a major research organism for genetics. The genetic sequencing will allow detailed genetic mutation studies at a nucleotide level, which allows studies of the effects of chemical mutagens and ionizing radiation.

The mouse genome is the same size as that of humans, yet there is also a mouse genome project because of the use of mice in genetics research. Most of the mouse project is underway in the U.S., and a 1 cM resolution genetic map is expected by 1995. The maps and sequence markers of the mouse and humans share many similarities. Cattle breeders have also begun a cow genome project, pig breeders a pig genome project, and behavioral geneticists have begun a dog genome mapping project to examine genes that are linked to the selected behavioral traits of inbred dog strains.

GENETIC SEQUENCING

For many purposes there does not need to be any repository of DNA information; what is required is a computer data bank of the sequence. The project requires the establishment and constant improvement of databases containing the sequences of genes and their location. There are several international databases, and the information should be openly shared among them to make the best and most up-to-date database possible. However, the physical cloned genes may be more efficiently obtained from cloned libraries, and such cloned libraries are becoming available from both the public and private sector. For example, the YAC contig library made at Généthon has been copied for several laboratories and is intended to be openly available to all people.

The total sequencing project for a genome of 3,000 Mb will take a lot of time, and is currently expected to be completed by 2002 A.D. However, by sequencing cDNA, which is complementary DNA to mRNA, it is possible to quickly and cheaply sequence a large number of genes. cDNA represents the expressed genes of a tissue or organism that the mRNA was obtained from. Several different cDNA libraries are being robotically sequenced around the world, and the number of genes sequenced from these is so large that by the end of 1994 most of the expressed genes

in the human genome were sequenced. Short sequences of a few hundred nucleotides from a cDNA library are automatically sequenced, and these are called
expressed sequence tags (ESTs). The sequences are tags because they can be used
to identify a gene sequence, and as a probe from which the full gene can be cloned.
Groups in several countries are pursuing this research approach, including a group
in the NIH. A list of worldwide web (Internet) sites for EST maps is at
<http://www.ncbi.nlm.nih.gov/dbEST/index.html>. Until 1992 a group at the NIH
sequencing a human brain cDNA library was led by Dr. C. Venter, and patent
applications on many of these sequences has led to international controversy.

This work is also being pursued by companies. In 1992 the Genomic Research
Institute was founded in the U.S., which is directed by Dr. C. Venter (Anderson,
1992b). He estimates that during 1990–1992 the NIH identified 8,000 human genes.
In April 1993, they reported that they sequenced about 500 to 1,000 new gene
sequences a day, and he estimates that the total in the genome is only 75,000 genes.
The facility involves 50 robotic DNA sequencers, and compares to the French
genetic mapping approach in its automation. The cost has fallen tremendously
because of the large scale. The representatives of the new institute say that they will
not apply for broad patents, but intend only to apply for patents on a few key genes
with suggested utility, and will publish results within six months of discovery. The
company name is Human Genome Sciences, Inc., which has since sold license
rights for human cDNA maps to SmithKline-Beecham. In 1995, most of the data
was published as the "Genome Directory" by *Nature* journal.

The number of genes sequenced is growing exponentially, and estimates of the
total number of human genes range from 75,000 to 150,000. This comprises only
about 5% of the total DNA in the human genome. The rest of the DNA is of
unknown function, and much is thought to be nonfunctional. The total sequence is
about 2.8 billion linear bases on 23 chromosomes.

FUNDING

Biomedical research is performed and funded in many countries. For many years
science has not only been the individual pursuit of people with unusual ideas, but
rather it involves the funding of research by public or private money, consisting of
taxes, charities, and business investments. We should not be surprised, therefore,
to hear the justifications for the funding of the project in terms of the business
opportunities. The U.S. Congress was partly convinced of the usefulness of funding
the project by the opportunity to boost U.S. biotechnology (OTA, 1995). However,
in France, much of the money for the genome project comes from charity, which
has different implications for the way that the data will be shared, as will be
discussed later.

The U.S. portion of the project (possibly 50% of the total) was initially estimated
to cost $3 billion over the next 15 years, to the intended completion date in 2005.
However, new technology suggests the date will be the year 2000 and the cost

of sequencing will be US $300–400 million. The total cost is unknown because the project is being broadened to include other organisms as models, private companies are also spending money, and many countries are contributing additional money to it. Data handling will be another important portion of the total. In 1992, U.S. government funding specifically for the human genome project (not including other indirect research on genetic mapping) was $164 million (NIH $105 million/DOE $59 million), and this figure is increasing. When one compares this with the cost of the development of a single drug, at $50–100 million, or the annual U.S. health care expenditure of over $600 billion, it is a small price to pay for such a large amount of information. Biotechnology is very big business, and the projected average $200 million annually for the human genome project in the U.S. (National Research Council, 1988) is minor. The scientific methodology for sequencing DNA is routine, and the costs of $1–5 for each nucleotide were reduced before the major sequencing effort began.

The initial fear by scientists was that the money would come from other biological research, which traditionally involves many small projects encouraging many individual scientists. Critics would prefer the project to stop after generating a general genetic map of the chromosomes, and after the cDNA had been sequenced. Identifying a particular disease-causing gene can take several years of intense investigation, as seen in the tracing of the cystic fibrosis gene. It has been estimated that it has cost about $30 million to trace this gene, yet if a map had existed the cost may have been only $200,000. A more detailed map decreases the amount of DNA that must be searched through for each gene. Various leading scientists have called for an evaluation of the project priorities after the map is completed, to use the money in the best way to encourage research (Leder, 1990). This also means that the first regions of the human genome to be sequenced will be ones that are known to have disease-associated genes present. In the initial phase some of the funds for the DOE and NIH projects came from existing research funding. The NIH is spending the equivalent to 2–3% of its total budget on the genome project which, in view of the importance of a coordinated mapping project, is worth the cost (Cook-Deegan, 1990).

Given the potential direct medical benefits of the project, the research money that is being spent on this project, and possibly more, can be ethically justified from the principle of beneficence. From the ethical principle of justice, other countries who can afford to pay and will benefit should share the cost of the project. The 1990 funding in Japan was $10 million, and it is expanding. France spent about $30 million in 1992, more than half from the French muscular dystrophy association. The U.K. government provided £11 million over 1991–1993, and has stated that it hopes this is enough to buy its stake in the use of the information. The Canadian genome research program may be about C$60 million over 5 years. The reasons for other countries to join in the project is not entirely because of their recognition of the principle of justice, but includes more the elements of potential economic benefit, and fear of being denied access to the U.S.-based databases.

The databases are being internationally funded. Japan gave $600,000 to the Genome Data Base in Baltimore in 1992. This was still less than half of the European contribution, but a sign of the international cooperation. The site of a new European animal and human genome sequence database, the European Bioinformatics Institute, is in Cambridge, U.K. Access to these databases, and the U.S. GenBank and EMBL Heidelberg gene sequence databases is possible from other countries. Japan is also host to an international genome database, in communication with these other databases, providing further access for scientists in many countries of the world.

In 1993 the U.S. funding for the human genome project from government agencies was about $170 million, with about $80 million from industry. In 1992 eight genome-related companies were established, as the project became more commercial (Anderson, 1993). There are private companies embarking on the project (Kanigel, 1987). The company Genome Cooperation was created by Walter Gilbert, with the intention of selling databases that contain sequences of key segments of the genome. Private companies should perform such research and be able to recover costs if they are still more economical than government research, providing the results of mapping and sequencing are openly accessible. In practice much of the project will be publicly funded, but contracts to do the work may be awarded to whoever is the most competitive.

In France two private nonprofit groups (Centre d'Etude des Polymorphismes Humain and Généthon) are spending more money than the French Government on genome mapping, in pursuit of genes for humanitarian reasons, rather than economic ones. The role of nonprofit private organizations is also very important in biomedical research in other countries, leaving less room for economically motivated private companies. Time magazine (February 8, 1993) has said that the French genome project is on higher moral ground than the U.S. project, because it is donating the results to the United Nations and is not seeking gene patents. In 1994 the U.S. company Merck Inc. announced it would create a public EST database for human gene sequences, following the release of the access conditions to the Human Genome Sciences EST data via SmithKline-Beecham. However, at the time of writing much of the information is available on the Internet.

DATA-SHARING AND PATENTING

The international mapping and sequencing is being coordinated by the Human Genome Organization (HUGO). Coordination of the international effort is needed to avoid duplicity of effort. The European Commission is also coordinating research in Western Europe, and UNESCO has interests in coordination of worldwide research. Although people recognize the importance of mapping, many researchers remain more interested in pursuing specific disease-causing genes. Therefore more attention has been paid to chromosomes 21, 7, and X, which have unknown disease-causing genes.

There are special needs for the information to be freely shared, though the former director of the U.S. NIH genome project, Dr. James Watson, earlier threatened that countries that do not contribute funds may not get the information immediately. This remark was aimed at encouraging the Japanese government to provide funds to HUGO. However, this has been widely criticized by those who believe the information resource belongs to no country, but to the world for its use in medicine. The first round of threats resulted in wider international funding.

A much stronger threat to data-sharing came following applications by the NIH for patents on cDNA sequences of several thousand genes, before their function was known. The U.S. Patent Office rejected the first cDNA patent applications on all three grounds needed for patents: novelty, nonobviousness, and utility. They said that they could find some partial sequences in existing databases so the application lacked nonobviousness; and they also said it lacked novelty since they used a publicly available cDNA library (Roberts, 1992). However, lawyers suggest it may be possible to overcome the Patent Office objections. Several U.S. companies have also filed patent applications on cDNA fragments (Anderson, 1993), and one Japanese company has applied for patents on 60 partial sequences of full-length cDNA clones which have had their gene function inferred (Swinbanks, 1993).

A secretary of the British MRC has suggested that copyright law might provide researchers with a better way to protect commercial applications while avoiding any obstruction in data flow. But the MRC was not planning to pursue it. However, the U.K. MRC applied for patents in the U.K. in 1992 on 1,100 cDNA gene fragments. The ethical arguments against copyright would be similar to those against patenting that are described in the next section; basically the information is naturally occurring and common property. Following this, the NIH attempted to apply for patents on a further 4,000 genes, but a U.S. lawyer in the Department of Health and Human Sciences blocked the application. Varying points of view have been expressed by patent lawyers (Adler, 1992; Eisenberg, 1992; Kiley, 1992), and although broad patent claims have been withdrawn by NIH and MRC, the legal issue remains unresolved.

There is the already existing problem of data-sharing from the viewpoint of individual competing scientists. The new technology, for example, automated DNA synthesizers and the PCR, makes replication of results very rapid, which could encourage researchers to delay publication while they get more of a head start on the next stage of the research. In a system where academic jobs depend on the number of publications, everyone wants to get papers published. The self-interest of the researchers must be considered for the sake of their autonomy and future work. Researchers may not reply to letters requesting data, or just reply within a selected peer group. The U.S. NIH and DOE guidelines stipulate that data and materials must be made publicly available within 6 months of generation, and earlier if possible. The guidelines are rules for those receiving grants from the NIH or DOE genome project, but encourage researchers to share information, to avoid bureaucracy. The results of discovering a disease-causing gene, or a detailed map which advances the discovery of such genes by many years must outweigh the short-term interests of individual scientists.

One example of the collaboration is between the 30 research teams working on chromosome 21 (it contains the Down's syndrome and Alzheimer genes), who produced a physical contig map (Chumakov et al., 1992). In this respect the genome map may be the ultimate collaborative research project. The most rapid progress will come from immediate data-sharing, and it is a chance for "scientific altruism" on a global scale. There is an ethical obligation by researchers, especially those using public money, to share data as soon as it is available. However, the financial size of the biotechnology industry that stems from molecular genetics is large enough to have already affected the way that data is shared. The net effect has probably been positive, because of the large share of genetics research that is being performed in biotechnology and pharmaceutical companies.

ETHICS AND LAW OF PATENTING DNA

The question of who legally owns the data is very topical because some of the work is funded by businesses. The question of patenting of genetic material remains a contentious issue with some European researchers delaying data submission to U.K. or U.S. databases following the U.K. and U.S. cDNA patent applications. Behind this is a pressure in many countries towards privatization of research funding. The U.S. Congress wants publicly funded science to be commercialized, and during the 1980s intellectual property rights were decentralized from government to research institutions to create commercial incentives (Cook-Deegan, 1990).

There are two basic approaches to applying patent law to biotechnology inventions. In the U.S., Australia, and many other countries, the normal patentability criteria shall apply, that is, the invention has the attributes of novelty, nonobviousness, and utility, and the invention should be deposited in a recognized depository (OTA, 1989). While a country may accept the first type of criteria, some countries have specifically excluded certain types of invention. What is ethical is not the same as what is legal, though we can attempt to reduce the difference (Macer, 1991). Concerns over ethics have affected patent laws; for example, European countries who joined the European Patent Convention have barred the patenting of plants or animals. Denmark has an even stronger worded exclusion in its national law. There is public rejection of the idea of patenting animals in some countries, and the patenting of human genetic material is potentially more contentious.

To qualify for a patent an invention must be novel, nonobvious and useful. If the claimed invention is the next, most logical step which is clear to workers in that field, then it cannot be inventive in the patent sense. If a protein sequence is known, then the DNA sequences that code for it will not in general be patentable, unless there is a sequence which is particularly advantageous, and there is no obvious reason to have selected this sequence from the other sequences that code for the protein (Carey and Crawley, 1990). In the case of natural products there are often difficulties because many groups may have published progressive details of a molecule or sequence, so it may have lost its novelty and nonobviousness. These are essentially short pieces of the human genome. There are also patents on protein molecules which have medical uses. In this case the protein structure is patentable if it, or the useful activity, was novel when the patent was applied for. The invention must also be commercially useful. There are patents on short oligonucleotide probes used in genetic screening. If someone can demonstrate a use for a larger piece of DNA then they can theoretically obtain a patent on it. An example of a larger patentable section of genetic material would be a series of genetic markers spread at convenient locations along a chromosome. Another set of genetic markers on the same chromosome can be separately patented if they also meet those criteria. The direct use of products, such as therapeutic proteins, is well established. The genetic information can also be used to cure a disease, for example using the technique of gene therapy with a specific gene vector.

With the completion of the genome sequence of many organisms, including humans, most genetic material will no longer be novel as it will be available in a database. The completion of the genome maps and sequences of many organisms will have many implications for the future of biotechnology patents.

The public attitudes to the patenting of human genetic material is rather negative (Macer, 1991, 1992a,b, 1994a). The negative reaction reflects the general feeling that genetic material is special, and should be different from other types of information. Patenting is said to reward innovation, which is a basis of the successful modern democratic and Asian economic systems. They do recognize property rights in inventions. However, there is an existing difference in the protection of property rights compared with other rights in international law and declarations of human rights. Property rights are not absolutely protected in any society because of the principle of justice; for the sake of "public interest," "social need," and "public utility," societies can confiscate property (Sieghart, 1985). Therefore there is an existing precedent for exemption from property ownership—which is the point of the exclusiveness of patents—when some property is of great benefit to the public.

Using a more positive argument, the knowledge gained should be considered as the common property of humanity. There is an existing legal concept that things which are of international interest of such a scale should become the cultural property of all humanity. It can be argued that the genome, being common to all people, has shared ownership, is a shared asset, and therefore the maps and sequence should be open to all. All people can say that the sequence is 99% similar to their own. In the United Nations Declaration of Human Rights, Article 27, there are two basic commitments that many countries in the world have agreed to observe (in their regional versions of this declaration). These are (italics added for emphasis): 1.) everyone has the right freely to participate in the cultural life of the community, to enjoy the arts and *to share in scientific advancement and its benefits*, and 2.) everyone has the right to the *protection of the moral and material interests resulting from any scientific,* literary or artistic *production of which he is the author* (Sieghart, 1985). An important question arising from section (2) is whether all people are the author of information that is shared by all people? The writers of this declaration may not have considered DNA, but it would certainly be in the spirit of the Declaration to interpret the DNA sequence as shared ownership. In section (1), everyone has the right to freely share in scientific advancement. This article expresses two important and relevant guiding statements of law that reflect the ethical principles of justice and beneficence, and the idea of legal property rights.

The common claims for authorship of the genome should be considered in all aspects of the genome project, especially in the questions of who should make the decisions in the project and the use of data (Macer, 1991). Some of the common factors that derive from the shared ownership are that the utilization must be peaceful, access should be equally open to all while respecting the rights of others, and the common welfare should be promoted (Knoppers and Laberge, 1990).

The authorship of the genome can be answered in two ways. The DNA could be viewed as a random sequence of bases, and the author is the sequencer, but this is not what we would normally talk of as an author or inventor; rather the sequencers are discoverers. The sequencers of DNA are not sequencing unowned land but rather they are sequencing uncharacterized land; the name of mappers is rather suitable for this analogy. The DNA is not random, it is merely unknown. Only the method for sequencing, or mapping, can be invented and patented, and whether that side of the project can be ethically patented lies more with the question of benefit and utility as discussed above.

If these arguments are insufficient to dissuade the private ownership of genome data, in addition to the precedents for exclusion of patents, public opinion could force a policy change over the patenting of such genetic material. In the U.S. the commercialization of human cells and tissues is generally permissible *unless it represents a strong offense to public sensitivity.* The sale of the human genome map and sequence data may be a strong offense to many and incite adverse public reaction forcing legislators to exclude it from patenting (Macer, 1991). The debate will continue, as companies will naturally desire to obtain some information protection for their investment, but they will have to be sensitive to strong public feelings that could easily be aroused, which as argued, has an ethical backdrop.

The idea that the human genome sequence should be in the public trust and therefore not subjected to copyright is the position of the French government, and was also the conclusion of the U.S. National Research Council (1988), and the American Society of Human Genetics (Short, 1988). The European Parliament "Human Genome Analysis" program limits contracting parties' commercial gains with the phrase "there shall be no right to exploit on an exclusive basis any property rights in respect of human DNA" (European Parliament, 1989). This idea would also include the option that the donor of genetic information, in terms of a cell line, should be able to make that information publicly available, which is usually a reasonable interpretation of the motives for patients to provide material for medical research, the motive to aid humanity in general rather than a commercial interest. The European Parliament attempted to introduce a directive on biotechnology patents in 1995, which failed, but would have allowed "farmer's privilege," so that farmers pay royalties on a patented gene in a seed or livestock only once, not every time the gene is passed to a descendant.

Some human genes and proteins that have medical use are patented, and this type of patent has brought about huge profits for some proteins. The annual market for the protein erythropoietin in the world is over $500 million, and some similarly profitable proteins will be found in the future. Some awards are required for promotion of the biotechnology industry and medicine. However, many people object when a routine medical technique is withheld from people who cannot afford to pay for it. Disease-related genes will be sequenced by the project, and some patents may be issued on screening methods and genes. Patent fees are being requested on the cystic fibrosis gene, even though the patent is not yet awarded. For some researchers, such patent fees may discourage studies. European researchers are challenging these royalties, and we can expect this issue to be as controversial as the cDNA patents discussed above.

The scientist behind the data of the NIH cDNA patents, Dr. C. Venter, has since signed a joint letter calling for an end to "patents of naturally occurring gene sequences" in favor of gene patents only on the uses of those patents, which was also approved by participants in the first South-North human genome conference held in Brazil in 1992. Venter left the NIH to begin a nonprofit genome research institute that is funded by Human Genome Sciences Inc. (Anderson, 1992b).

The most rapid progress will be obtained if data are shared between all researchers. The full value of one part of the sequence is only known when compared to the rest. Even if one government declines to support such a project, the information still belongs to all people of that country and it is ethical for other countries to share it with them.

BENEFITS

There are major applications and implications of the genome project for much more than molecular and cellular genetics. It will be a huge resource of information for medicine in the next century. There will be much useful information arising prior to the completion of the project, as growing numbers of disease-causing and susceptibility genes are sequenced and the mutations characterized. Most of the major single gene disorders and some of the genes involved in complex diseases should be known within the decade. It will be possible to develop DNA probes to diagnose any known genetic disorder, and it will also be easier to characterize new disorders. The list of diseases that have probes is growing (McKusick and Amberger, 1993). It will expand the number of human proteins that can be made by genetically-modified organisms, which would allow conventional symptomatic therapy for many more diseases, which could be supplemented by somatic cell gene therapy when appropriate. It would also expand our basic knowledge of human biology, which allows medical treatments to be developed. We may not be able to predict when therapies will emerge after the genes are discovered, because there can still be a long delay in clinical applications following biochemical understanding. The amount of new knowledge is hard for us to comprehend—it will take decades to process it all—but it offers the potential understanding of all genetic diseases sometime during the next century.

The rest of this chapter outlines some examples of the benefits, together with some of the impact of the genetic knowledge. There are many ethical, legal, and social impact (ELSI) issues that need to be addressed, and some of these are also discussed. The genome project is also proving to be important for research on these ELSI issues because of the large level of funding being spent on bioethics research compared to before the project began. A combined total of over $9 million a year was spent annually for ELSI research on agricultural biotechnology and the Human Genome Project in the U.S. alone in 1992 and 1993. Other governments are also spending some proportion of their genome project research funding on ELSI research and education (Macer, 1992d).

GENETIC SCREENING

Every person has a different genetic sequence, except for identical twins. Genetic screening involves the use of the binding between the nucleotides A-T and G-C. A sample of DNA is taken from a cell, and then the DNA is split into single chains. The bases in this single-stranded DNA will bind to the complementary bases of a nucleotide probe. The presence or absence of particular DNA sequences that represent different genes or mutations can be screened for.

There are many types of genetic disease, some minor and others untreatable and fatal. One of the most common genetic diseases is inherited forms of cancer. Some genes that increase susceptibility to cancer have been shown to be very common in the population, for example, one inherited breast cancer gene will result in breast cancer in about one in 170 women (Roberts, 1993).

Prenatal genetic screening involves screening of the fetus, inside the mother's womb. There are several safe and commonly used methods to allow samples to be taken from the fetus (Macer, 1990). In addition to noninvasive ultrasound imaging, and the invasive sampling methods of amniocentesis and chorion villi sampling, new methods are being developed. One of the most promising is the sampling of fetal cells from the maternal bloodstream, which has already been practically performed. There are always some fetal cells that enter the mother's blood, and these cells can be separated. When the method is developed to a simpler procedure, this will allow convenient and safe genetic examination of the fetus by a simple blood sample from the mother (Roberts, 1991). We can also screen for genetic disease in embryos before they are transferred after *in vitro* fertilization, but less than 100 babies have been born after preimplantation diagnosis. This technique may only be applicable to a few people because it is expensive and has a low pregnancy rate.

Prenatal genetic screening of the fetus can be used to detect characteristics of the fetus, and should only be performed for serious diseases. These include diseases that would result in serious mental or physical health damage to the fetus. In most cases the fetus is found not to be afflicted, thus removing much anxiety from the parent's mind. Without the use of such tests to confirm the absence of disease, some mothers would have an abortion. In the case that the fetus is found to be afflicted from a disease there are two different possible courses of action. If there is a therapy available and treatment before birth is necessary to avoid health damage, the fetus can be treated for the detected disease. There have been increasing numbers of operations performed on fetuses in the womb (*in utero*), which have avoided permanent health damage to the fetus. Sometimes the fetus is removed from the womb, operated on, and then replaced to complete normal gestation. If the detected disease is untreatable, and is serious, then the option of selective abortion is available.

There is considerable variability in the abortion laws of different countries around the world, but in general selective abortion of fetuses suffering from a genetic disease is a respected moral choice of the mother. In many European countries routine prenatal diagnosis is offered under national health schemes to all women at high risk of having a fetus with genetic disease or chromosome abnormality. This includes all women older than about 35 years of age, and there is strong support from those women for the provision of these services to screen for serious diseases to all pregnant women.

There is clear public support for prenatal genetic screening in most countries that have been surveyed, and that it also should be available under government funding (Macer, 1992a; 1994a). One of the most striking results of opinion surveys conducted in Japan is the similarity in the responses of different groups of people; the public, high school teachers, and scientists have very similar mixtures of opinions. This suggests that the results obtained may be close to the real opinion of the Japanese population, and that education does not play the dominant role in determining people's attitudes. In 1993 I conducted an International Bioethics Survey with collaborators, including Australia, Hong Kong, India, Israel, Japan, New Zealand, The Philippines, Russia, Singapore, and Thailand (Macer, 1994a). In most countries about two thirds of the population supports prenatal genetic testing and one third does not, or is unsure of their attitude. Despite the support, we

We need to elevate the importance of individual autonomy, especially in reproduction. Technology will allow the production of cheap and very simple, for example colorimetric, genetic screening testkits. Should these be available to the public, such as do-it-yourself pregnancy tests are today in many countries? There will need to be more serious consideration given to personal reproductive decisions in the future, making life more complicated while hopefully improving its quality (Macer, 1990). While the people who can make decisions regarding the availability of such kits cannot claim to understand the social consequences of such a move, the general public also cannot understand the broader consequences of their combined individual actions. In this situation there is a case for control of public property in order to avoid doing harm. While it may be possible to regulate the use of such kits via the intermediatory control by health care workers, there could be particular testkits which may not even be made because of the fear of misuse.

There does need to be control over the use of cosmetic screening and therapy (that has no compelling medical reason), especially when it affects children. However, an important question in ethics and public policy is how will this control be effected? In the 1990 German Embryo Protection Law there is specific mention of Duchenne muscular dystrophy as a serious genetic disease that genetic screening (for preimplantation diagnosis) can be performed for. There has been criticism of this approach because of possible increased discrimination towards the handicapped who suffer from the legally designated "serious" diseases, for which embryos suffering from such diseases can be discarded. Life will be further complicated by combinations of various diseases which may be "judged" permissible for parental selection after genetic screening or for treatment using gene therapy. The extremes of a free market approach or a total ban on genetic testing are both strongly undesirable. The question of who decides the application of technology in individual cases must be addressed, whether individual genetic counselors, codes of practice, legally established regulatory committees, parents, and whether it is freely available to all or only to those who can pay, or only to those judged to be at "significant" risk.

The human genome project raises similar ethical and legal issues to those in current genetic screening, such as confidentiality of the results. However, it will lead to screening on a huge scale, for many disease traits and susceptibility to disease (Holtzman, 1989; Muller-Hill, 1993; Knoppers and Chadwick, 1994). It is important that we deal satisfactorily with the test cases, before we are faced with all this new information. The technology may change the way we think. The amount of information obtained will overwhelm existing genetics services and geneticists. More training of genetics (as well as ethics) will be required for health care workers, scientists, and the general public (Macer, 1990).

Presymptomatic genetic testing is important for some diseases such as Huntington's disease and polycystic kidney disease, but it should only be performed with extensive counseling. Huntington's disease generally afflicts people in their

40s, and is very disturbing. The disease-causing allele of the gene is dominant, and the children have a 50% chance of inheriting it. The children may be close to marrying age when the parent first shows symptoms, and they may desire testing before marriage, or testing of their fetus. It is interesting to see that the level of acceptance is lower than for genetic testing in general, suggesting that people do recognize the complexity of knowing whether they will get sick in the future, or die at a premature age. Such services should be available, but only where adequate psychological counseling can be ensured. These diseases are particularly difficult because the people may experience 40 years of unaffected life (though often they experience the disease in one of their parents). Should they alter life plans if they have the gene? On the benefit side, there is considerable personal relief for people who were at risk if they learn they do not have the disease-causing allele. In practice, only about half the people at risk have used the 95% accurate Huntington's disease genetic test.

PRIVACY AND DISCRIMINATION

The information about whether an individual has a particular DNA sequence and gene can be very powerful, especially in the diagnosis of genetic disease. It is very important that privacy is respected, because the information in a person's genes identifies some of our risk to disease that medical insurance companies and employers could use to discriminate against (Holtzman, 1989; Macer, 1990; Muller-Hill, 1993). There are already cases of discrimination of individuals after genetic testing in North America (Billings et al., 1992; Geller et al., 1996).

There is less experience with presymptomatic genetic testing of disease risk, such as for Huntington's disease. This is another area where the data need to accumulate before we will be able to make reasonable predictions about the more widespread use of such testing. The other questions arising from the screening of children or adults for disease susceptibility is somewhat easier to address. As mentioned above, from the principle of justice we should work against genetic discrimination, and establish national health schemes, and equal access to employment (except when there is an actual, current, risk of third party harm). There should be a right to privacy of genetic information. Some employment performance-based testing can be used, when there is a reasonable and potential risk of harm, but not mandatory genetic tests. There needs to be guidance over the storage of genetic information for legal purposes, in immigration and in crime.

There are important questions about who has the right to know our individual genetic makeup (Zimmerli, 1990). Of course, our general genetic makeup will be common knowledge. It could be argued that because we all share in the information to be made public, we all have a say in the discovery and presentation of it. The sequence must be protected from abuse; it will be the most detailed common knowledge about every individual, and will provide many opportunities for abuse. Though it is not unique in this, for example, psychologists have understood common complicated features of the human mind for many years, and such knowledge is of similar risk to that being unravelled by geneticists.

DNA sequences can be used to identify individuals, and have been used for forensic cases. When the DNA is specifically cut up by restriction enzymes and the pieces are separated, unique patterns are made, called DNA fingerprints. They have been used for hundreds of court cases around the world. In several countries a genetic register of criminals has been established. DNA fingerprints are also used for immigration cases to prove genetic parenthood. There have been disputes about the probability of two individual's DNA fingerprints matching, and calculations have recently been revised, but there is still a low probability of such fingerprints matching (when the size of the fragments of DNA made after digestion with the restriction enzymes are the same in two individuals).

The question of fairness in the use of genetic information with respect to insurance, employment, criminal law, adoptions, the educational system, and other areas must be addressed. In those countries with private medical insurance, some people may be put into high risk or uninsurable groups because of genetic factors (for example, high blood cholesterol, or family history of diseases). Some insurance companies and employers perform screening in the U.S. (OTA, 1990). Some legislation has been passed in the U.S., such as the Americans with Disabilities Act in 1990, or the proposed Human Genome Privacy Act. However, there is only one solution, that is a national health insurance system with equal access to all and no discrimination based on disease risk factors that people have no choice over. Privacy is the key ethical issue to protect individuals, and until we can guarantee privacy and no abuse of the information, we should not establish genetic registers.

EUGENICS

Our genetic information is very important in determining much of our physical character and intellectual capacity. This is especially clear in cases of people suffering from genetic disease. The effect that these people have on a family can be good or bad, some people can cope with it and some can not. Some of the suffering that such people have is the suffering that healthy people have in their own minds, as the handicapped people may not know life to be any different to what they have; that is, others can impose our life goals into their lives. However, some suffering can be real, especially when they suffer much pain all their life. The suffering of the family can also be very real, and preoccupying.

The word eugenics was coined by Sir Francis Galton, and is derived from the Greek word "eugenes" which means "well born" or "hereditarily endowed with noble qualities." Eugenics differs from other human activities in that it is an activity in which we are trying to change ourselves, not the environment or other creatures, and therefore is particularly challenging (Macer, 1990). The idea of some groups of humans being inferior to others was often based on intelligence, or a method prescribed to define this. The rational was called superior to the animal, thus

Aristotle claimed women and slaves were inferior by nature because of diminished reason and being closer to an animal state. The 18th century biologists claimed to prove that Negroes' skulls and physiognomy most clearly resembled those of apes thus justifying slavery. Superiority is often judged by how close people approach to someone's "ideal" of intelligence and rationality.

Galton defined eugenics as the science of improving the "stock" and the study of agencies under social control that may improve or impair the racial qualities of future generations either physically or mentally. He intended eugenics to extend to any technique that might serve to increase the representation of those with "good genes," in this way accelerating evolution. A major motivation underneath many eugenicists was an idea of human progress, that we must be progressing genetically as well as in our knowledge. This was boosted by the theory of evolution, the survival of the fittest was equated with the survival of the best. The best were the best people to cope with modern life. Galton was a cousin of Charles Darwin. Social Darwinists' tended to equate a person's genetic fitness with their social position. Social Darwinist ideology provided a good climate for eugenic thought, and many qualities such as intelligence, temperament, and behavior were believed to be inherited (Kelves, 1985; Macer, 1990). The eugenicist's concept of the best human was their idea of the "perfect man," which tended to be an intelligent white male of northern European stock, who had been said to have a larger brain. Interestingly, the primary genome sequence used in the genome project will also be of European origin. Some 18th century philosophers had believed in the possibility of human perfectibility. There was also a fear that the "stock" was deteriorating. Some 20th century philosophers and scientists may also share in the dream of perfectibility, but most would argue that the knowledge of genetic variation between individuals already shows that there is no such possibility of the perfect person, in all 100,000 genes and regulatory sequences!

A principal concern of the eugenicists was the lower fecundity of family "stocks" from wealthy, and more educated, families. As these people were from this type of family they had fears of their progeny being swamped by large numbers of progeny from uneducated, and thus genetically unfit, classes. These ideas were around before 1900, but people had been ignorant of the process of heredity. Chromosomes were known to be carriers of the genes only around 1900. This gave a rule for the transmission of traits, so instead of relying on ideas from animal breeders, they now had a biological theory. There were eugenic policies in about 40 countries, many including sterilization programs to stop the "unfit" from breeding, and immigration policies to select desirable breeding stock into new countries such as the U.S. They were particularly prevalent in Northern European countries, such as Scandinavia and Germany, and were rejected in Britain, Holland, and some central European countries. We can imagine what may result from the knowledge at the year 2000, of all the genes, if there is a genetically deterministic model of life.

The individual's right to reproduce has been prevented in the past by compulsory sterilization measures. Because of opposition there has been a growing move to voluntary measures, though these measures may be more enforced by peer group pressure to conform and by financial costs if the state does not provide medical care. If health care becomes centered on private medical insurance companies, there could be more pressure not to bring disabled children into the world, as the insurance companies could insist on prenatal screening.

The major use of eugenic selection occurred together with the move to a more scientific world view. Recently developed scientific techniques offer more than sterilization operations or artificial insemination by donor sperm; there is genetic screening to gene therapy in the immediate future. Eugenic measures often end up with racial or social group overtones, more than breeding from the "best genes." The model chosen depends on the society, for instance Spartans wanted good soldiers; geneticists from the middle-class; want well-behaved middle-class; the Nazis wanted blonde, blue-eyed, Aryans; and the 1980s Singapore government wanted academically minded parents (Macer, 1990). We must have a clear view of human dignity founded in individuals possessing equal value not dependent on their ability or performance of some task.

For all of history people have preferred to have their children born free of genetic disease. With modern medicine many handicapped people are living much longer; to avoid the need to have these people born, genetic screening is developing. The actual number of people born with genetic disease has decreased in Europe, because of more intercultural marriages which have diluted harmful recessive alleles, and from prenatal screening programs and better nutrition for pregnant women. There will be less severely handicapped newborns if selective abortion is increasingly used, and eugenicists will have new methods. The name has also changed, from eugenics to genetic counseling; we must ensure the focus has also changed, from not just the interests of societies, but to protecting individuals.

The concerns of society are often placed above the rights of the individual when eugenics is developed, and this is the fear held by many today. Eugenic measures have been used in societies under different circumstances, and eugenicists have included both sides of political opinion. We must be aware of our modern medical practice in the light of eugenics and the associated attitudes. The opposition to eugenics may come from the concern for the rights of individuals, both those born, and fetuses; belief that we should not interfere very much with nature or God's purposes or chance; or that it conflicts with some political view, such as the earlier Russian Marxist view that all humans are given equal ability. The Christian opposition is based more on the view that all humans are given equal status or rights, which is not the same as ability. This is not just to say that the new eugenics is all bad, in fact most people support some genetic counseling. The lesson from history may be that we must be very careful where we draw the line, and that it remains voluntary. There are some important areas of reproductive choice which should be left up to the individual or couples.

The definition of what is a healthy or meaningful life varies widely, and depends on the circumstances people have experienced. Population-wide screening among Caucasian people for cystic fibrosis, with the aim of lowering the incidence of children born with this disease, is being considered in North America and Europe. However, in the U.S., families whose members suffer from cystic fibrosis appear unwilling to personally use selective abortion (Wertz et al., 1991), though they support the right of others to use prenatal diagnosis and selective abortion. Nowadays the life expectancy of persons with cystic fibrosis is about 40 years of relatively normal life, bringing into question the "need" for selective abortion of fetuses with this disease. There is a fear, especially by handicapped people groups, that the use of prenatal diagnosis and selective abortion will worsen the discrimination against handicapped adults; therefore, they oppose prenatal diagnosis being used at all (Wikler and Palmer, 1992).

However, we should remember that many people become handicapped by accidents or diseases after birth, and these will continue even if some people use selective abortion. The outlook for a handicapped person born in Japan is, like many countries, not always good. Some families are unable to emotionally or financially support a child suffering from serious disease, and these may be the major reasons given for selective abortion. In some opinion surveys some respondents have suggested that they would not support selective abortion if the social services in the country for care and education of afflicted children and families were better. Other families may find the opposite situation, that if they keep the fetus, the child that is born may become a blessing to the family in other ways.

We need to avoid abuses of such screening for nondisease conditions, such as sex selection, and for eugenic purposes, by some regulations. An extra insurance against abuse is to improve social support services for handicapped people, and to continue to educate people away from discrimination that is based on any apparent difference. Such education measures can change social attitudes towards the "handicapped" (Macer, 1992a).

GENE THERAPY

Many genetic diseases may be treated by correcting the defective genes through gene therapy. Gene therapy is a therapeutic technique in which a functioning gene is inserted into the somatic cells of a patient to correct an inborn genetic error or to provide a new function to the cell. There are over one hundred human gene therapy trials currently underway for several different diseases, including several cancers. There have been many human gene transfer or therapy trials approved by the Recombinant DNA Advisory Committee (RAC) in the U.S., and also trials approved in China, France, Germany, Italy, Japan, The Netherlands, New Zealand, and the U.K. It is still an experimental therapy, but if it is safe and effective, it may and the U.K. It is still an experimental therapy, but if it is safe and effective, it may prove to be a better approach to therapy than many current therapies, because gene therapy cures the cause of the disease rather than merely treating the symptoms of a disease. Also, many diseases are still incurable by other means.

Currently, such gene therapy is not inheritable; we need to have much wider discussion about the ethics and social impact before we start inheritable gene therapy (Macer, 1990, 1994b). However, noninheritable gene therapy to treat patients involves similar issues to any other therapy, and if it is safer and more effective, it should be available to patients who consent to it. There are many approaches being developed and we can expect rapid introduction of these tech- niques, because of more than 20 years of preparatory experiments on animals, and the success of some of the current clinical trials.

We can distinguish gene therapy from genetic engineering for ethical debate, by the therapeutic motive. It is better to leave the term human genetic engineering for genetic intervention which is not aimed at direct therapy. However, we may still think of cases where indirect medical benefit can be given, such as improvement of the efficiency of the immune system by genetic means or genetic vaccination. Cosmetic uses of medicine are commonly available from the private medical sector, and it may be difficult to limit the cosmetic use of gene therapy in the future when cheap vectors and delivery systems have been developed. Public opinion supports the use of gene therapy for treating disease (Macer, 1992a,c, 1994a; Macer et al., 1995).

GENETIC DETERMINISM

There is increasing reference made to genetics in popular culture and the media, and a growing sense of genetic determinism and reductionism. Medical problems, both physical and mental, are increasingly attributed to genetic causes. The envi- ronment also plays a necessary part in shaping the phenotype of an organism, yet genetic explanations are used even when only genetic susceptibilities are involved. We all will be shown to have some increased susceptibility compared to average for particular multigene conditions, such as cancer, or particular mental states. Therefore the difference between a 20% increased risk and the power of a dominant disease-causing gene needs to be emphasized. Additionally, there should also be education to show that despite all the information, we should not expect disease to be cured within 20 years, and it will not be a panacea for the world's woes. A critical paper by Lippman (1992) says that the emphasis placed on the genome project may have negative eugenic effects on society and health care. Genetic explanations are increasingly given as the major causes of disease, and the genome project reinforces this. This restricts the concepts of health and illness.

Most religious approaches support the rationale for obtaining better genetic information, which can be used to alleviate human suffering. The question is how to use it properly. There are dangers in any large scientific project, that they take control of the people in becoming the sole ideal for progress. The possibility of mastery and control over the human DNA raises the issue of genetic selection. Ideas of eugenics could be explored. We need to maintain a distinction between diagnosis and treatment of disease, and selection for desirability, while at the same time we need to discuss where molecular biology will take the human species.

All people should benefit from the results of the genome project, but will only do so if we avoid stigmatization or ostracism. Some people hope that the knowledge that we are all equal in our genetic differences might end discrimination. However, this will require much education and laws to ensure such an equality is respected. There will be a change in attitudes to ourselves also, and genetic determinism might become popular. A danger with simple-minded adherence to genetic hypotheses for behavior is that it oversimplifies the complex interaction of genetics and environment. In the extreme, determinism eliminates the idea of genuine choice, leaving no room for the belief that we can create, or modify ourselves, or that we can make moral choices. The question whether higher human attributes are reducible to molecular sequences is a controversy in the philosophy of biology. The knowledge of human genetics will make scientific understanding of human life much more sophisticated. There may be alteration in social customs, especially if the genetic information is misunderstood by the public as occurred at the beginning of this century.

In existing genetic services we recognize a right not to know our genes. Could this right be extended to general genome sequencing? We may have a right not to know that we will develop Huntington's disease, but what about the right not to know we are at high risk for schizophrenia, alcoholism, or a life in academia? Do we have a right not to know that the common human DNA sequence "programs" us to die at 85 years of age? The difference seems to be in features that distinguish us from the norm in our society, common knowledge such as general life expectancy are not hidden, though death may be a taboo topic in most countries. Are people still afraid of being different, even in the supposedly individualistic Western societies? This is another question that needs answering before we can work out appropriate means of regulation.

Genetic mapping of humans is being extended to all races of the world in efforts aimed to complement anthropological studies. This project is called the Human Genome Diversity Project, and is headed by Luca Cavalli Sforza (Cavalli Sforza and Cavalli Sforza, 1995). It involves taking samples from at least 25 unrelated individuals in 400 populations around the world, and analyzing their DNA for about 100 loci. It will also mean the generation of numerous cell lines which could be used in future research. Genetic lineages can be drawn to trace the migration of people around the world. Such studies have a potential for racial discrimination if misused, but will answer questions that many scientists want to know. However, people of those tribes may not desire such an answer, and consent may not be asked, raising further ethical questions. The tracing of individual family lineages raises the question of privacy for those individuals, but the privacy issue has yet to be applied to such global studies.

The ability to analyze small quantities of DNA means that ancient DNA can be analyzed to trace changes in the DNA that may give useful data about evolution of various species. The oldest DNA to be analyzed to date is about 30 million years old, from insects preserved in amber. The technology is expected to be extended, and even within a time frame of a million years would provide valuable insights for molecular genetics. Although this will shed great light on our genetic origins, in recent history political and religious ideals have had a stronger effect on human behavior and history than genes.

The message for health care workers regarding genetic determinism is more clinical. We must be clear that a pursuit for our lives to be free of physical suffering is not going to make the ideal world. Genetic defects have a smaller effect on people than the moral, spiritual defects and lack of love. The goal of healing is to benefit the patient—the actual patients are the "ends" rather than the "means." In this case individuals are more important than hypothetical individuals (ones as of yet unconceived), or the human species in general. For example, using gene therapy to improve the health for sick children is not going to trade away a part of our humanity that is worth preserving. Love can be shown in many other ways than through sympathy to people suffering from genetic disease, and may be shown more by the cured patients. But the highest type of love is unconditional love, which means accepting all the people who are different from us in the world. The act of love may involve genes, but more generally it will involve social support, diet, and a clean and peaceful environment.

EDUCATION

As already discussed, there are numerous ELSI issues, and the resolution of these issues will take longer than the physical process of sequencing the human genome. Additionally, the relatively low cost of ethical and legal studies of the implications of the project compared to the biological research has encouraged some funding bodies to provide funding to ensure society is more prepared for the data (Annas and Elias, 1992). In the U.S. the NIH and DOE have awarded research funds for study of these issues, and have established a joint working group on ELSI issues. In the U.S., 5% of the 1992 and 1993 NIH Human Genome Project budget has been allocated to research on ELSI issues and education. In Europe a similar proportion is being spent, and in Canada 7.5% is allocated to this. In Japan less than 1% of the human genome research grants are being spent on these issues (Macer, 1992d; Fujiki and Macer, 1992; Fujiki and Macer, 1994; Okamoto et al., 1996). In Europe, a funding program for Human Genome Project scientific research was developed only after establishing a system to allow research funding of the ELSI issues (MacKenzie, 1989). There are also ELSI studies in Australia.

Even with these ELSI studies, it is likely that the implementation of major genetic screening programs will be delayed for ethical reasons until people, both providers and consumers, are educated. Discussion regarding how we use the information has been occurring among the public in several countries, including North America, Europe, and Japan. It is essential for widespread education to be available in a way that the public can understand it, so that they can be involved in decisions about their project. An adequately prepared lay community is the best way to ensure that misuse of genetics does not reoccur. Most people have a poor knowledge of genetics, which must be improved before they will be able to understand the new knowledge. Incomplete knowledge can be very dangerous when combined with existing discrimination, as seen with eugenic programs earlier this century. We should all realize that we are genetically different, and normality is very culturally defined, perhaps as those who can live comfortably, or anonymously, in a given society. Education of social attitude together with science is required.We need to ask what is ethical biotechnology? (Macer, 1995).

Universal laws, for example Article 23 of the International Covenant on Civil and Political Rights, states that "the right of men and women of marriageable age to marry and found a family shall be recognized." That has been signed by over 75 countries, and should guarantee that compulsory eugenics is not introduced. It is a very strong statement based on the ethical principle of respect for human autonomy. However, social pressures are very difficult to control. Such a law needs to be supported by equal access to social and health services in order to make it effective. In the same covenant there is also supposed recognition of equal access to healthcare, but what is required is wording such as "equal access to *equal* health care." This is one avenue that action could be taken from well-accepted ethical principles, but education about technology and how to make decisions in an ethical way is also necessary.

GENETICS: THE FOUNDATION OF MEDICINE?

Will the next generation benefit from being genetically selected? We can ask questions such as is a life suffering from serious disease better than no life? As is the case today, these questions need to be answered by individual parents, but they will become much more apparent with the increasing number of conditions that can be diagnosed and ease of screening, though the answers may not. It may be easier to try to answer the question of whether the parents who use such screening on their gametes, embryos, and fetuses will benefit from it? There are benefits to avoiding problems of the extra time that they may need to spend with children, and the extra costs, but they also change themselves by using such screening, by making presumed health a condition of acceptance of children. It is also quite debatable whether the extra time parents spend with children and the potentially greater opportunity to give love, is time better spent than pursuing previous life goals such as time with other children or careers. While we should not be afraid for society to change, we should be wary of change when adverse social attitude changes may occur. We need data to measure the effects on personal, family, and social attitudes.

Will society allow individuals to have free choice over the use of genetic manipulation and screening when there is no medical reason for it? Screening can be used both to reduce individual differences to others, and to highlight differences. The arguments used against genetic intervention which has no therapeutic value include: it would be a waste of resources, may present risks to offspring, it will promote a bad family attitude, will be harmful support of society's prejudices, and may reduce social variability. It will probably not have any significant effect on genetic variability as there will be plenty of alternative healthy alleles. There could also be the idea of a natural genetic autonomy, that we should let the genes come together naturally, and let the individuals develop their genetic potential without unnecessary interference by parents or society (Macer, 1990). A criteria for trans-generational ethics is that not only must a gene alteration be safe, but it must be good therapeutic sense over many generations. There must be unquestionable objectives and benefits for many generations (Macer, 1994b).

A common feature of many issues raised by the human genome project data is that we need to consider the effects of knowledge and technology on future generations. The beneficiaries and those at risk may not yet be existing. In the sense of benefits and risks, it is their genome project more than ours. We have an obligation to the future from the principle of justice (Rawls, 1988). Our traditional view of morality only involves short-term consequences. The ethics of long-range responsibility are needed. It means that researchers may be held accountable for secondary consequences of their research. Of course it may be very difficult to predict what will happen in the future; the social pressures and thinking are already very distinct between different countries. If social ideas change, then so may the pressures, such as the desire to use genetic enhancement. We need to ensure future generations retain the same power over their destiny as we do, while benefiting from the culture and technology we have developed (Macer, 1991).

Decisions on the use of genetic manipulation in one country will affect other countries, because people move, changing their countries. It is therefore imperative that the decisions about any future germline genetic manipulation, especially of humans, take into account people's opinions worldwide. This may be best handled by an international forum, with which national committees should interact. The coordination needed for the genome project may aid this process, but the developing countries need to be adequately represented, especially because they represent such a large proportion of the world's population. The international nature of the project and its universally applicable results make it a project of all humanity. Many countries are unable to significantly contribute material resources to the scientific project, but they share in the material that is being sequenced—their genes—and must be involved in the project's benefits and decisions. Society's interests not only should transcend proprietary rights, but the special nature of the genome project and the claims that we can all make upon the genome should make the shared authorship and ownership legally compelling (Macer, 1991).

Increasingly health care workers will be forced to use genetic solutions to problems which involve genetic factors. For many conditions such genetic therapy represents a better alternative than other treatments, or early death. However, the simplest and most economical therapy for many conditions may be behavioral change in diet and exercise. Changes in lifestyle and diet have increased the incidence of some genetic diseases, including some types of diabetes, cancers, and heart disease. The genome project offers much to molecular genetics and medicine, but good medicine must be much more than this.

SUMMARY

The genome project is a collective name for genetic mapping and sequencing efforts being pursued in various organisms by scientists around the world. These projects, such as the Human Genome Project, arose out of past decades of genetics research, and will result in knowledge of the complete genetic sequence of some bacteria, yeast, plants, animals, and human beings. The data generated is being shared by researchers around the world, but the quantity involved requires advanced genetic and computing technology.

The benefits of this grand project of molecular genetics include the rapid and economical identification of disease-related genes, and the expansion of genetic diagnosis and therapy. There will be numerous benefits for agriculture and industry. However, there are numerous ethical, legal and social impact (ELSI) issues magnified by the enormity of the project. These include privacy and genetic testing, genetic discrimination, eugenics, and genetic determinism issues. The potential economic rewards of utilization of the gene sequences has led to questions about patenting of cDNA and other genetic data, which are still to be resolved. There has already been a shift in society during the course of this project to focus on the genetic causes of phenotype, and this trend could result in ethical abuses of the information, as well as a different view of humanity. We need to examine the role of molecular and cellular genetics in medicine and society, and educate people how to utilize genetic knowledge for the physical, mental, and spiritual health of all people.

REFERENCES

Adler, R.G. (1992). Genome research. Fulfilling the public's expectations for knowledge and commercialization. Science 257, 908–914.

Anderson, C. (1992a). New French genome centre aims to prove that bigger really is better. Nature 357, 526–527.

Anderson, C. (1992b). Wall Street remains bearish on value of genome project. Nature 358, 180.

Anderson, C. (1993). Genome project goes commercial. Science 259, 300–302.

Annas, G.J., & Elias, S. (eds.) (1992). Gene Mapping. Using Law and Ethics as Guides. Oxford University Press, New York.

Bellanne-Chantelot, C., et al. (1992). Mapping the whole human genome by fingerprinting yeast artificial chromosomes. Cell 70, 1059–1068.

Billings, P.R., Kohn, M.A., Cuevas, M., Beckwith, J., Alper, J.S., & Natowicz, M.R. (1992). Discrimination as a consequence of genetic testing. Am. J. Hum. Genet. 50, 476–482.

Cantor, C.R. (1990). Orchestrating the Human Genome Project. Science 248, 49–51.

Carey, N.H., & Crawley, P.E. (1990). In: Human Genetic Information: Science, Law and Ethics (Ciba Foundation Symposium 149), pp. 133–147, Elsevier North Holland, Amsterdam.

Cavalli-Sforza, L.L., & Cavalli-Sforza, F. (1995). The Great Human Diasporas: The History of Diversity and Evolution. Addison-Wesley, Reading, MA.

Chumakov, I., et al. (1992). Continuum of overlapping clones spanning the entire human chromosome 21q. Nature 359, 380–387.

Cook-Deegan, R.M. (1990). In: Genetics, Ethics and Human Values (Bankowski, Z., & Capron, A.M., eds.), pp. 56–71, CIOMS, Geneva.

Coulson, A., Kozono, Y., Lutterbach, B., Showkeen, R., Sulston, J., & Waterston, R. (1991). YACs and the C. elegans genome. BioEssays 13, 413–417.

Culliton, B.J. (1990). Mapping terra incognita (humani corporis). Science 250, 210–212.

Doris-Keller, H., et al. (1987). A genetic linkage map of the human genome. Cell 51, 319–337.

Eisenberg, R.S. (1992). Genes, patents, and product development. Science 257, 903–908.

European Parliament (1989). "Human Genome Analysis program," COM, section 3.2 (13th Nov. 1989), 532.

Fleischmann, R.D., et al. (1995). Whole-genome random sequencing and assembly of Haemophilus influenzae Rd. Science 269, 496–512.

Fraser, C.M., et al. (1995). The minimal gene complement of Mycoplasma genitalium. Science 270, 397–400.

Foote, S., Vollrath, D., Hilton, A., & Page, D.C. (1992). The human Y chromosome: Overlapping DNA clones spanning the euchromatic region. Science 258, 60–66.

Fujiki, N., & Macer, D.R.J. (eds.) (1992). Human Genome Research and Society. Eubios Ethics Institute, Christchurch, N.Z.

Fujiki, N., & Macer, D.R.J. (eds.) (1994). Intractable Neurological Disorders, Human Genome Research and Society. Eubios Ethics Institute, Christchurch, N.Z.

Geller, L.N., Alper, J.S., Billings, P.R., Barash, C.I., Beckwith, J., & Natowicz, M.R. (1996). Individual, family, and societal dimensions of genetic discrimination: A case study analysis. Science & Engineering Ethics 2, 71–88.

Holtzman, N.A. (1989). Proceed With Caution: Predicting Genetic Risk in the Recombinant DNA Era. John Hopkins University Press, Baltimore.

Kanigel, R. (1987). The genome project. New York Times Magazine (13th Dec.), 43–44, 98–101, 106.

Kevles, D.J. (1985). In the Name of Eugenics. Knopf, New York.

Kiley, T.D. (1992). Patents on random complementary DNA fragments. Science 257, 915–918.

Knoppers, B.N., & Laberge, C.M. (1990). In: Genetics, Ethics and Human Values (Bankowski, Z., & Capron, A.M., eds.), pp. 39–55, CIOMS, Geneva.

Knoppers, B.M., & Chadwick, R. (1994). The human genome project: Under an international ethical microscope. Science 265 (1994), 2035–2036.

Leder, P. (1990). Can the human genome project be saved from its critics . . . and itself? Cell 63, 1–3.

Lewin, R. (1990). In the beginning was the genome. New Scientist (21st July), 34–38.

Lippman, A. (1992). Led (astray) by genetic maps: The cartography of the human genome and health care. Social Sci. Med. 35, 1469–1476.

Macer, D.R.J. (1990). Shaping Genes: Ethics, Law and Science of Using Genetic Technology in Medicine and Agriculture. Eubios Ethics Institute, Christchurch, N.Z.

Macer, D. (1991). Whose genome project? Bioethics 5, 183–211.

Macer, D.R.J. (1992a). Attitudes to Genetic Engineering: Japanese and International Comparisons. Eubios Ethics Institute, Christchurch, N.Z.

Macer, D. (1992b). Public opinion on gene patents. Nature 358, 272.

Macer, D.R.J. (1992c). Public acceptance of human gene therapy and perceptions of human genetic manipulation. Human Gene Therapy 3, 511–518.

Macer, D. (1992d). The "far east" of biological ethics. Nature 359, 770.

Macer, D.R.J. (1994a). Bioethics for the People by the People. Eubios Ethics Institute, Christchurch, N.Z.

Macer, D.R.J. (1994b). Universal bioethics and the human germ-line. Politics & Life Sciences14, 27–29.

Macer, D.R.J. (1995). Bioethics and biotechnology: What is ethical biotechnology?, In: Modern Biotechnology; Legal, Dynamic and Social Dimensions, Biotechnology, Volume 12, pp. 115–154 (Brauer, D., ed.). VCH, Weinheim, Germany.

Macer, D.R.J., Akiyama, S., Alora, A.T., Asada, Y., Azariah, J., Azariah, H., Boost, M.V., Chatwachi-rawong, P., Kato, Y., Kaushik, V., Leavitt, F.J., Macer, N.Y., Ong, C.C., Srinives, P., & Tsuzuki, M. (1995). International perceptions and approval of gene therapy. Human Gene Therapy 6, 791–803.

MacKenzie, D. (1989). European Commission tables new proposals on genome research. New Scientist (25th Nov.), 6.

Mandel, J.L., Monaco, A.P., Nelson, D.L., Schlessinger, D., & Willard, H. (1992). Genome analysis and the human X chromosome. Science 258, 103–109.

McKusick, V.A., & Amberger, J.S. (1993). The morbid anatomy of the human genome: Chromosomal location of mutations causing disease. J. Med. Genet. 30, 1–26.

Merriam, J., Ashburner, M., Hartl, D.L., & Kafatos, F.C. (1991). Toward cloning and mapping the genome of *Drosophila*. Science 254, 221–225.

Muller-Hill, B. (1993). The shadow of genetic injustice. Nature 362, 491–492.

National Research Council (1988). Mapping and Sequencing the Human Genome. National Academy Press, Washington D.C.

NIH/CEPH Collaborative Mapping Group (1992). A comprehensive genetic linkage map of the human genome. Science 258, 67–86, 148–162.

Okamoto, M., Fujiki, N., Macer, D.R.J. (eds.). (1996). Protection of the Human Genome and Scientific Responsibility. Eubios Ethics Institute, Christchurch, N.Z.

Oliver, S.G., et al. (1992). The complete sequence of yeast chromosome III. Nature 357, 40–47.

Olsen, M., Hood, L., Cantor, C., & Botstein, D. (1989). A common language for physical mapping of the human genome. Science 245, 1434–1435.

OTA-U.S. Congress, Office of Technology Assessment (1988). Mapping Our Genes- The Genome Projects: How Big, How Fast? U.S.G.P.O., Washington D.C.

OTA-U.S. Congress, Office of Technology Assessment (1989). New Developments in Biotechnology, 4: Patenting Life. U.S.G.P.O., Washington D.C.

OTA-U.S. Congress, Office of Technology Assessment (1990). Genetic Monitoring and Screening in the Workplace. U.S.G.P.O., Washington D.C.

OTA-U.S. Congress, Office of Technology Assessment (1995). Federal Technology Transfer and the Human Genome Project. Background Paper. U.S.G.P.O., Washington, D.C.

Rawls, J. (1988). A Theory of Justice. Oxford University Press, 8th impression, Oxford.

Roberts, L. (1991). FISHing cuts the angst in amniocentesis. Science 254, 378–379.

Roberts, L. (1992). Top HHS lawyer seeks to block NIH. Science 258, 209–210.

Roberts, L. (1993). Zeroing in on a breast cancer susceptibility gene. Science 259, 622–625.

Sieghart, P. (1985). The Lawful Rights of Mankind. Oxford University Press, Oxford.

Short, E. (1988). Proposed American Society of Human Genetics position on mapping/ sequencing the human genome. Am. J. Hum. Genet. 43, 101–102.

Swinbanks, D. (1993). Institute files for patents on first Japanese sequences. Nature 361, 576.

Vaysseix, G., & Lathrop, M. (1992). A second-generation linkage map of the human genome. Nature 359, 794–801.

Watts, S. (1990). Making sense of the genome's secrets. New Scientist (4th August), 37–41.

Weissenbach, J., Gyapay, G., Dib, C., Vignal, A., Morissette, J., Millasseau, P., Wertz, D.C., Rosenfield, J.M., Janes, S.R., & Erbe, R.W. (1991). Attitudes toward abortion among parents of children with cystic fibrosis. Am. J. Public Health 81, 992–996.

Wikler, D., & Palmer, E. (1992). In: Human Genome Research and Society: Proceedings of the Second International Bioethics Seminar in Fukui, 20–21 March 1992 (Fujiki, N., & Macer, D.R.J., eds.), pp. 105–113, Eubios Ethics Institute, Christchurch, N.Z.

Zimmerli, W.C. (1990). In: Human Genetic Information: Science, Law and Ethics (Ciba Foundation Symposium 149), pp. 93–110, Elsevier North Holland, Amsterdam.

RECOMMENDED READING

Collins F., & Galas, D. (1993). A new five-year plan for the U.S. Human Genome Project. Science 262, 43–46.

Cook-Deegan, R.M. (1994). The Gene Wars. Science, Politics, and the Human Genome. Norton, New York.

Freidmann, T. (1990). The Human Genome Project—Some implications of extensive "reverse genetic" medicine. Am. J. Hum. Genet. 46, 407–414.

The Genome Directory (Nature 377, 6547S Supplement, 28 Sept., 1995).

Gibbs, R.A. (1995). Pressing ahead with human genome sequencing. Nature Genetics 11, 121–125.

Hoffman, E.P. (1994). The evolving genome project. Current and future impact. Amer. J. Human Genetics 54, 129–136.

Jordan, E. (1993). The Human Genome Project: Where did it come from, where is it going? Am. J. Hum. Genet. 51, 1–6.

Juengst, E.T. (1994). Human genome research and the public interest: Progress notes from an American science policy experiment. Amer. J. Human Genetics 54, 121–128.

McKusick, V.A. (1989). Mapping and sequencing the human genome. New Engl. J. Med. 320, 910–915.

Watson J.D. (1990). The human genome project: Past, present and future. Science 248, 44–49.

Chapter 18

Transgenic Regulation in Laboratory Animals

SANDRO RUSCONI

Principles of Medical Biology, Volume 5
Molecular and Cellular Genetics, pages 377–401.
Copyright © 1996 by JAI Press Inc.
All rights of reproduction in any form reserved.
ISBN: 1-55938-809-9

INTRODUCTION

Mammalian development can be compared to a complex computer program of which we have identified and characterized (cloned) only a handful of subroutines (the genes). Expression of cloned genes in wild type or mutated form in tissue culture cells (subroutine testing) has led to several conclusions about their role in the cell metabolism and in the developing organism. However, tissue culture cell systems allow only a limited recapitulation of all the complex feedback that occurs within and between tissues. Therefore, a conclusive assessment of the hypotheses involving tissue culture cells still depends on the experimentation in the whole animal (main program run in real time). Transgenic techniques are particularly suited since they allow the addition/subtraction of defined genes to/from the mammalian genome. The approaches that are currently available can be subdivided into three broad categories: the gain-of-function (overexpression or ectopic expression of a particular gene), the change-of-function (expression of dominant mutant alleles), and the loss-of-function (site directed genomic alteration or expression of molecular antagonists aimed toward the reduction of specific gene activities). This review indicates how some results have undoubtedly confirmed earlier hypotheses and opened interesting perspectives, while others have inexorably produced less positive conclusions. The current decade will be certainly remembered as a crucial period of verification (or rejection) for many established theories dealing with the functional role of particular genes in the organism.

WHY TRANSGENICS AND NOT SIMPLY CELL CULTURES?

Somatic and Germline

The life cycle of every multicellular living being is dependent on the equilibrium between two major classes of cells: somatic and germ cells. The somatic cells

---

Figure 1. Germ-line and somatic line. (A) The life cycle of a mammal. Development starts with a diploid zygote (2n, lozenge) that rapidly divides. At the blastocyst stage, several commitments may already have been made, such that some cells remain totipotent and other lose their potency (squares, circles). In the adult, the only true totipotent cells are represented by the gametes (n) that can undergo fusion and restart the life cycle. The organism is mortal while the germ line continues with the next generation. (B) Scheme of cell division after fertilization. An external factor inducing a genetic alteration (black lightning bolt) into a pluripotent cell will lead to a mosaic somatic line composed of intact and mutated (filled symbols) cells. The germ line remains intact. (C) A mutation affects a totipotent cell whose descendants partly develop into the germ line. The mutation is passed to the next generation with the gametes. If some descendants of the original mutant cell develop into somatic cells, the somatic line will also be mosaic.

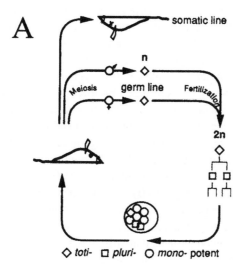

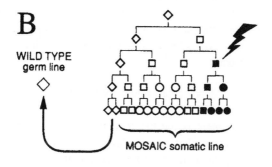

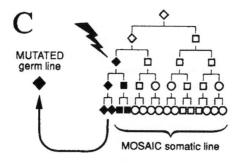

constitute the bulk of the body that sustain all the vital functions, including mating. The germ cells are devoted to the perpetuation of the species, since they are the repository of the transmissible genetic material. Therefore, while the somatic cell line ends after each cycle, the germ line carries the genetic information and is the starting material for the subsequent generation (see Figure 1A). Every multicellular eukaryote starts as a totipotent single cell (the zygote) that results from the fusion of the gametes. Totipotency (represented by lozenges in Figures 1B and 1C) is progressively lost in the cells that will participate in the somatic line, since they shift to pluripotent (squares), monopotent (circles) and finally differentiated cells. Totipotency is of course strictly maintained in germ cells. Every genetic alteration (filled symbols in 1B) introduced into the cell committed to the somatic line will have an effect on the clonal descendants of the affected cell, but will leave the germ line in its original state (see Figure 1B). On the other hand, if the genetic content of a totipotent cell is changed, then it is likely that this alteration will be transmitted to the next generation (see Figure 1C). These concepts are essential for understanding some basic methodologies that will be described later. Cell cultures can be derived from many different tissues, but are condemned to die after a certain number of passages unless they become immortalized. Immortalization can occur spontaneously or can be obtained upon infection with transforming viruses. However, the process of immortalization usually alters the properties of the cells in such a way that they cannot always be considered to be a convincing model for their physiological counterparts.

Concerted Gene Regulation

Ontogenetic development is the result of the concerted temporal and spatial regulation of many functions that are encoded by gene products. In the last decade several hundred eukaryotic genes have been characterized at the molecular level.

Figure 2. The various levels of regulation of gene expression. Thick arrows pointing to the top indicate the flow of genetic information; the figure should be read the same way. In order to be expressed, the chromosomal segment containing the gene must be organized in a somewhat loose chromatin structure (stretched line, euchromatin), whereas nonexpressed regions are more densely packed (heavy waved line, heterochromatin). For reviews on chromatin organization see Svaren and Chalkley, 1990. The *cis*-elements that comprise a regulated cistron are: the coding segment (Gene), a short regulatory region overlapping with the transcription initiation (Pr, promoter) and more distantly located regulator segments that can act positively (Enh, enhancers; Serfling et al., 1985; Dynan, 1989) or negatively (Sil, silencers; Baniahmad et al., 1990). The conversion from heterochromatic to euchromatic state ("off/on" state) and the maintenance of the latter are thought to be mediated by elements called locus control regions (LCR; Orkin, 1990; Weissman, 1992, see also text). (Continued)

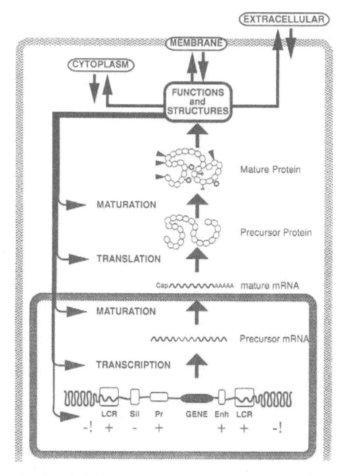

Figure 2. (Continued) The first step is the transcription of a specific DNA strand into its complementary RNA (short wavy line). Transcription regulation is the most common quantitative regulatory event and certainly also the most studied (Serfling et al., 1985; Mitchell and Tjian, 1989; He and Rosenfeld, 1991). The "+" and "–" signs under the regulatory signals indicate their general positive/negative influence over transcription. The mRNA is subject to several modifications (splicing, capping, polyadenylation, and editing), and other functions (degradation, transport). Thus, in the course of RNA maturation several additional regulatory steps are also possible. The same applies to translation (conversion of nucleotide code into aminoacid code) and the posttranslational events (Protein maturation). The final product defines the functions and structures that constitute cell metabolism and its interactions with the environment (secretion, import, signaling). Many gene products are devoted to the regulation of gene expression at either level (see arrows feeding back to upper levels). The nuclear membrane is symbolized by the heavily shaded border, the plasma membrane is symbolized by a wavy-shaded border. This figure is modified from Rusconi, 1993, with permission from J.R. Prous, S.A.

Extensive analysis of the cloned segments has led to the widely accepted notion that the differential regulation of most genes occurs at the level of transcription, although several other levels of fine tuning also apply.

As shown in Figure 2, a typical eukaryotic gene consists of various operationally distinguishable portions (*cis*-acting elements). Thus one identifies: (a) a coding portion (gene) which is partly preserved in the mature transcripts (the exons, comprising the exons and the introns which are spliced out during the RNA maturation process); (b) proximal *cis*-control elements, the promoter (Pr, that lies close to the transcription start site, and specifies quantity, accuracy of initiation and polarity of transcription); (c) some more distal *cis*-elements that act positively or negatively and are called enhancer (Enh) or silencer (Sil) elements, which, in cooperation with the promoter, specify the rate of transcription; and (d) additional remote control elements (locus- or dominant-control regions, LCR or DCR) which may specify the overall on/off state of the chromatin encompassing the transcribed unit. Further explanation of the terms are given in the legend of Figure 2.

The *cis*-acting elements are clusters of binding sites for *trans*-acting factors that are specific DNA binding proteins playing an essential role in the regulative process. The scheme is rendered very complex by the notion that some gene products are themselves *trans*-acting factors and are able, therefore, to feed back on their own expression and the one of other genes (see arrows feeding back in Figure 2). It is important to remember that the position and the relative importance of each *cis*-element varies enormously from gene to gene. Therefore, the final level of expression will depend also on posttranscriptional (maturation, transport, translation), as well as posttranslational events (maturation, storage, turnover, secretion), each of them often depending on extracellular signals.

We have seen that the level of expression of any particular gene product is dependent on many different metabolic conditions and molecular signals. Many of these signals are still operating in tissue culture systems and thereby allow to some extent cell-specific studies. Typically in these experiments, the presumed regulatory region is cloned in front of a reporter function that encodes a product which is readily distinguishable, like an antigen, a unique mRNA, or an enzyme. Such reporter constructs can be transfected in cells by a variety of methods and have shown that both promoters and enhancers can work in a cell-specific manner. For instance, an immunoglobulin enhancer will be rather meaningless in fibroblasts while conferring a high level of expression in B-cells (Banerji et al., 1983; reviewed by Serfling et al., 1985), or a glucocorticoid-dependent reporter gene will be active only when tested in cells containing the receptor and only in the presence of the ligand (reviewed by Beato, 1989). Furthermore, it can be shown that certain gene products, such as most oncogenes, can specifically alter the properties of a cell line in a predictable fashion. Thus, in several cases it has been possible to obtain either regulation or an effect that is consistent with the supposed role of these genes in the intact organism.

In spite of many convincing examples, cell culture systems cannot always recapitulate all the situations that occur in a developing organism. In many cases the signals depend on simultaneous actions of neighboring cells (paracrine or synaptic systems) or on far remote organs (endocrine systems). For instance, it is not possible to measure the impact of the overexpression of a homeotic gene that changes vertebrae identity or limb formation because the phenotype will be only visible in the organism. The same applies to many genes involved in erythropoiesis, osteogenesis, liver development, neurogenesis, autoimmunity, let alone genes that are involved in fertility or behavior control. Therefore, the study of specifically altered genes in the animal represents the ultimate possibility of explaining their role in development. We must also add that, due to their limitations, cell culture systems are bound to give a simplified picture of the role of the gene of interest. As discussed later, many such inconsistencies have become apparent in the last two years. We will also see that, besides the mere academic or medical interest, the transgenic techniques represent a very attractive technological application.

TRANSGENICS: WHICH EXPERIMENT AND WHERE TO PUBLISH?

In the "B.C." literature ("before cloning"), we already find several attempts to generate transgenic organisms like fish, barley, or *Drosophila* that have been performed with undefined DNA preparations (reviewed by Rusconi, 1991). However, the first reports in which vertebrates have been transformed with cloned DNA made their appearance only at the beginning of the 1980s. After the ice had been broken we witnessed a spectacular development of the field (see Figure 3, solid line). From the half-dozen papers in 1983 we now count more than 700 publications per year (excluding work with transgenic plants). We can also notice that the focus of attention has moved from the initially most relevant studies on tissue-specificity of reporter genes expression to the effect of the newly inserted genes on the development of the host organism (heavily shaded line in Figure 3).

In fact, in recent years (see heavily shaded line in the years 1991–1992), only about 10% of the publications report the use of transgenics to directly study gene expression patterns in mice. Among the most important discoveries regarding gene expression per se is the identification of the locus control regions (LCR, see patterned line in Figure 3 at bottom) whose importance could not be detected with cell culture systems. In any case, the bulk of today's literature describes approaches aimed at changing some metabolic or developmental feature. This latter point has been greatly facilitated by the emerging techniques of site-directed gene knock out (see faintly shaded line in Figure 3 at top) in which chromosomal genes can be specifically altered and the effect of the mutation can be studied in the context of the entire organism, as discussed below. Of course, the profile of publications not only reflects the trends of the scientific community but also the naturally changing interests of the journal editors. For instance, a paper describing a "no phenotype"

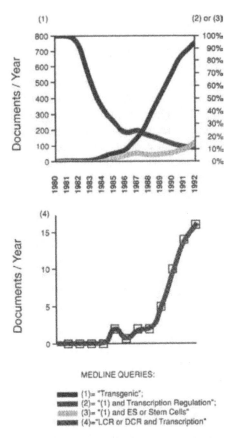

Figure 3. Profile of publications dealing with transgenic mouse techniques in the years 1980–1992. The publications were found by searching the database MEDLINE with the queries indicated at the bottom. Top, documents dealing with different aspects of transgenic techniques: global (black line); focusing on transcriptional regulation of the transgene (heavily shaded); dealing with ES-cell mediated transgenesis (lightly shaded). Scale at left, documents per year (= rate of publication, referring to black line); scale at right, percent of documents dealing with sub-topics (shaded lines). Bottom, publications per year dealing with the characterization of locus control regions (LCR, also initially called DCR).

after targeted disruption of the PrP gene (Büeler et al., 1992) made it to a cover story in an international journal in 1992. A few months later an equally important negative result regarding the knock out of the retinoic acid receptor alpha had been more anonymously published (Li et al., 1993). We must deduce that in the future, gene knock outs that do not produce a visible phenotype will not be publishable in internationally known journals. Similar anecdotes are particularly abundant in the field of transgenic research and have contributed to the swelling of the GUH (generation of unsung heroes). Perhaps it is time to think of a specialized journal devoted to the publication of all those experiments that are potentially very informative but came slightly too late for the scoop of the first negative result.

HOW TO MAKE TRANSGENICS?

The basic aim of transgenics is to introduce a defined change in the host genome and to follow the behavior of the changed element or the consequences thereof during the development of the transformed organism. In my opinion the word "transgenic" is etymologically difficult to defend and remains a rather unfortunate description (probably the result of trying to distinguish it from the more general "transformed"). The term "exogenic" would have been more appropriate, since it indicates the acquisition of an exogenous gene. In any case, transgenic was introduced by a leading team in 1983 and has now become part of the common vocabulary. Today's available techniques are the result of outstanding achievements in both mammalian embryo manipulation and gene cloning/splicing strategies. We distinguish two kinds of approach: (a) the older and most straightforward microinjection of defined DNA into mouse zygotes (which I shall refer to as "conventional") and (b) the recently developed techniques based on manipulation of pluripotent embryo stem cell lines. The essential features of either method are outlined in Figure 4.

Conventional Transgenesis

Successful transgenesis (Figure 4 at left) was first reported at the beginning of the 1980s. It was based on the possibility of microinjecting a few hundred molecules (1–2 ng/μl of DNA for less than 200 femtoliter volume) into the pronucleus of a freshly collected zygote (Figure 4A,B). The foreign DNA is integrated with high frequency, albeit at random positions into the host chromosomes. The manipulated embryos are transferred into the oviduct of a female where they are allowed to develop to term. The efficiency of transgenesis among the new-born still largely depends on the skill of the experimenter and it is around 15%. Many additional losses have to be taken into account in this type of work and finally, only a few dozen embryos out of 1,000 manipulated zygotes will be born with the desired genomic acquisition (see Table 1).

The integration into the host chromosome is thought to occur during random break-repair. As a direct consequence, many DNA constructs will not be capable of autonomous regulation, since their expression will be strongly influenced by the flanking regions of the host chromosome. These position effects (reviewed by Wilson et al., 1990) have been noticed in many experiments and seem to be different for each gene segment tested. Moreover, it has been suggested that the presence of prokaryotic vector sequences in the transferred DNA may exacerbate these position

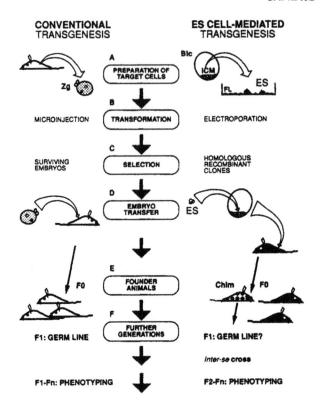

Figure 4. The two fundamental strategies for transgenesis. At left, from top to bottom, essentials of conventional transgenesis; at right, the steps necessary for ES cell-mediated transgenesis; central column, categorization of the different steps. Explanation of symbols: thickening arrows indicate magnification of represented item and vice versa; Zg, zygote(s); Blc, blastocyst(s); ICM, inner cell mass; ES, embryo stem cells; FL, feeding layer cells; Chim, chimera(s). Other explanations of displayed items are in the text. This figure is reproduced in modified form from Rusconi, 1993, with permission from J.R. Prous, S.A.

effects (Palmiter and Brinster, 1986). Position effects are unpredictable and can sometimes vary during passage on the next generation, a phenomenon that is reminiscent of the position-variegation described in *Drosophila*. Although the position effects pose a serious drawback when aiming toward a consistent behavior of the newly acquired gene, they have been successfully exploited to disclose specific regulatory regions. In this type of experiment, an "inert" gene marker (preferably an enzymatic function such as the bacterial LacZ that can be monitored histochemically) is inserted at random position into the host genome. The constructs bear either a minimal promoter (enhancer trap vectors) or they are depleted of a

Table 1. Features of Conventional Transgenesis

Feature	% Average Success	Remaining (from 1,000 Initial Embryos)
Properly staged	80	800
Survive injection	70	560
Successful transfer	80	450
Successful foster mother	80	360
Developing to term	60	210
Carrying the transgene	15	40
Expressing the transgene	variable	5–30?

start codon and possess splice acceptor sites that can give rise to functional fusion polypeptides when inserted in the correct orientation and spliced in the correct reading frame (gene trap constructs reviewed by Skarnes, 1990). In either situation the expression of the LacZ marker is followed histochemically in the transgenic line. The DNA of the mice giving an interesting spatial-temporal expression pattern is isolated and the sequences flanking the newly inserted DNA are subsequently cloned. This is a very straightforward protocol to isolate genes whose expression is developmentally regulated. In the most fortunate cases (that is, when the insertion inactivates the host gene) the phenotype of the homozygously bred transgenics can be used to infer the role of the altered gene in development.

Several attempts to render foreign genes more independent of the influence of flanking sequences have been made. In many cases an attempt has been made to maintain the foreign gene in an episomal (extrachromosomal) state. These protocols have been only partially successful and several claims still await confirmation by independent laboratories. A system based on a modified polyoma replication vector that efficiently replicates in embryonic stem (ES) cells (M. Gassmann, personal communication) seems promising in this regard. Another possibility of rendering the transfected gene more independent from the local environment is to increase the size of its natural flanking sequences. Although some gene constructs behave very consistently even if they include only few hundred base pairs of the natural promoter, some other genes appear to require larger amounts of natural flanking sequences to operate position independently (reviewed by Rusconi, 1991). One can facilitate the introduction of large genomic segments in the host chromosomes in at least three ways: (a) the use of cosmid vectors that permit the culture and manipulation of 30 to 40 kb of DNA (Braun et al., 1990; Rusconi, 1991; Tripodi et al., 1991); (b) the locus reconstitution approach, in which overlapping segments are coinjected and do reconstitute into a large segment via recombination of the homologous overlapping sequences (Pieper et al., 1992); and (c) the use of yeast artificial chromosomes (YACs) (reviewed by Capecchi, 1993) that allow the culturing and manipulation of up to a few hundred kilobases of genomic sequences.

This latter technique has been pictured as "coming to the rescue" (Capecchi, 1993) and is bound to become extremely popular in the future.

In some paradigmatic cases, position independence could be tracked down to DNA segments lying several kilobases upstream or downstream of the promoters of interest. These segments have been termed dominant control regions (DCRs, reviewed by Orkin, 1990) or more recently locus control regions (LCRs, see Figure 2 (Dillon and Grosveld, 1993)). It has been shown that these segments are able to convey a consistent tissue-specific expression level to the linked construct. Initially one had thought that DCR/LCR segments could have worked in a generalized manner, that is, in being able to confer a generalized position independence. It was shown later that these elements are themselves highly tissue-specific (Orkin, 1990; Skalnik et al., 1991). In any case, DCR elements have aroused an increasing interest. Publications dealing with this topic have increased from two in 1988 to a total of more than 60 in 1992 (see Figure 3 at bottom). The correct expression of many genes probably depends on the presence of these important flanking elements. This has a profound impact on the strategies to be applied not only in transgenic organisms but also in long-term somatic gene therapy approaches, where the newly inserted corrective genes also have the tendency to be progressively silenced (Weissman, 1992). As discussed above, there are several ways to ensure that even remote LCRs are included in the transgene.

ES Cell-Mediated Transgenesis

The alternative approach for transformation of mice relies on the possibility of maintaining the pluripotent state of ES cells during the time necessary for specific transformation *in vitro* (Figure 4 at right). ES cells are derived from the inner cell mass (ICM) of blastocysts and can be cultured for a relatively long time on feeder layer (FL) cells (Evans and Kaufman, 1981). The ES cells can be transformed (Figure 4B) by a variety of routine methods and the transformants are re-introduced into a host-blastocyst (Figure 4D), where they are expected to contribute to normal development. The resulting organism is a mosaic (chimera) of different cell types and the distinction between transformed and host cells is simply obtained by the use of genetic pigmentation markers (see Figure 4E). If the totipotency of the ES cells is maintained along the entire manipulation, then in a certain proportion of the chimeric organisms the transformed cells will contribute to the gametogenesis. Obviously, the maintenance of totipotency is the key issue that can make the ES cells an efficient vehicle for gene transfer.

The most interesting possibility by far involves the introduction of specific genomic alterations via homologous recombination between the newly transferred DNA molecule and the resident locus in the host genome. Several methods exist (reviewed by Capecchi, 1989). One of the most widely used is illustrated in Figure 5. The protocol has been mainly developed in the laboratory of Capecchi and

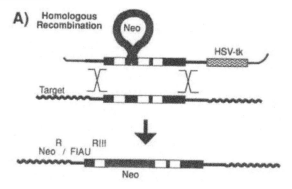

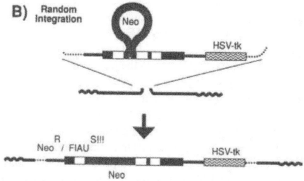

Figure 5. The double drug selection, a widely used approach for efficient targeted gene disruption. The DNA construct used for gene targeting includes a relatively large (10–15 kb) portion of the targeted gene (filled boxes, exons; empty boxes, introns; thick line, 5' and 3' untranslated region) in which (i) a selectable marker (Neo, bacterial neomycin resistance gene) is inserted or substituted for an exon (dashed thick circling line), and (ii) an additional selectable marker is engineered at a certain distance (HSV-tk, wavy lined box, herpes virus thymidine kinase). The HSV-tk enzyme possesses high affinity for some potentially toxic drugs such as gancyclovir or FIAU (Borrelli et al., 1988; Capecchi, 1989). The constructs are electroporated into pluripotent ES cell lines and drug resistant clones are selected. **(A)** If precise homologous recombination occurs at both sites flanking the Neo marker, then the HSV-tk marker is lost and the cell clone displays the double resistance (symbolized with R) and will be Neo^R/FIAU^R. The exactness of recombination must of course be ascertained by DNA blotting or polymerase chain reaction techniques. If conveniently placed, the Neo marker leads to a recessive loss-of-function of the targeted gene. That means the phenotype will be visible only when the chromosome bearing the alteration is bred homozygously. **(B)** Random integration of DNA by chromosomal break-repair mechanisms usually result in insertion of one or several intact copies of the cloned DNA. Therefore, most of the cell clones generated by such events will retain the HSV-tk segment and will be susceptible to treatment with FIAU (Neo^R/FIAU^S). This figure is reproduced from Rusconi, 1993, with permission from J.R. Prous, S.A.

colleagues (Mansour et al., 1988) and exploits the possibility of a double selection (positive/negative selection). A relatively large cloned portion of the gene to be targeted is modified by conventional gene-splicing techniques such that a Neomycin (Neo) marker is inserted into an essential exon. An additional, negatively selectable marker is added to the construct at a certain distance from the Neo gene. Routinely, the HSV thymidine kinase gene is used, since this enzyme generates toxic metabolites in the presence of the drugs gancyclovir or FIAU (Borrelli et al., 1988). Other laboratories have also used some toxin-coding genes instead of HSV-tk. The idea is that this second marker should be lost if homologous recombination (= double mitotic cross-over or gene conversion) occurs in the regions that are flanking the Neo gene (see Figure 5A). The modified construct is electroporated into cultured ES cells and subsequently Neo-resistant/FIAU-resistant clones are selected (Figure 5A). This approach eliminates the majority of random insertion events in which the entire construct is introduced via chromosomal breaks, thus producing a $Neo^R/FIAU^S$ phenotype (Figure 5B).

The frequency of homologous versus random integration varies widely among different genomic segments, probably due to the presence/absence of recombinational hotspots. In fact, noncoding genomic regions evolve relatively rapidly and their polymorphism can reduce the efficiency of homologous recombination by several orders of magnitude (see Table 2) (te Riele et al., 1992). Therefore, it is emphasized to include in the construct rather large flanking genomic segments (10 to 15 kilo base pairs) that have been isolated from the same cell line that is used for the targeting. It is not yet clear whether chromosomal loci that are not expressed in ES cells are significantly more difficult to target than loci which are expressed. The correctly transformed cell clones are identified by polymerase chain reaction or blotting techniques and micro-injected into the recipient blastocysts as soon as possible to avoid unnecessary passages that can disturb totipotency. The newly introduced ES cells mingle with the resident cells of the inner cell mass and the manipulated blastocysts are allowed to develop into chimeras (see Figure 4). As pointed out, maintenance of pluripotency is essential, because the aim is to obtain mosaic offspring (F0 mice) producing gametes that bear the genomic alterations generated in the ES cells. We mentioned earlier that mutation of the germ line is essential for the perpetuation of the genotype and for a consistent study of its phenotype (Figure 1). When this is achieved, mice that are heterozygous for the targeted mutation (F1) can be subsequently generated by conventional mating (see Figure 4 at bottom). The F1 mice are crossed inter se such that they will generate 25% homozygous wild type, 50% heterozygous, and 25% homozygous mutant. Since in most cases the gene disruption produces a recessive mutation, the phenotype is visible only in the corresponding 25% of the second generation. When the mutation is lethal at the homozygous state, the corresponding fraction of the F2 will be missing. In these cases, analysis of F2 embryos will reveal at which developmental stage the lethality occurs. To conclude, the process of obtaining mice that

Table 2. Features of ES-Mediated Homologous
Recombination/Transgenesis

Appearance	Frequency
Drug resistance (per cell)	$\sim 10^{-4}$
Homologous integration (nonisogenic DNA, per clone)	$\sim 10^{-2}$
Homologous integration (isogenic DNA, per clone)	$> 10^{-1}$
Successful transfer/development (per embryo)	$\sim 25\%$
Chimeric embryos (per newborn)	10–60%
Germ line transmission (per chimera)	5–50%

are homozygous for specific alterations is still long and riddled with several technical obstacles that make it accessible so far only in relatively few, specialized laboratories (see Table 2). Nevertheless, the information that one can gather is so important, that this kind of experiment has been started by many research groups.

The Three Types of Approaches

The potential of transgene effect studies has been outlined in very early experiments by Palmiter and colleagues who were able to show that transgenic mice that overexpress a growth hormone gene are of a significantly larger size than their nontransgenic littermates (Palmiter et al., 1982). This example illustrates very appropriately the concept behind the so-called gain-of-function assay. In this approach, the expression of a gene is forced in the tissues of interest or in the entire animal and the metabolic or ontogenetic consequences are observed.

Analogously, instead of expressing an intact version of a gene, one can express a mutant or any other molecule that interferes with the proper function of the gene product. In order to be effective, the mutant allele should exhibit *trans*-dominant properties. These properties are relatively easy to achieve when the gene product is part of an oligomeric functional complex (reviewed by Herskowitz, 1987). Other types of dominant antagonists can be obtained by expressing a high level of antisense RNA that matches with the gene of interest (reviewed by Munir et al., 1990) or specific RNA molecules (Cech, 1988) that, besides being able to hybridize with a specific target mRNA, are capable of cleaving it. Although very elegant on paper, the latter approach has never been shown to be successful in a real *in vivo* situation. Regardless of the molecule used for these interference experiments, we shall refer to this approach as the change-of-function assay.

Finally, it is noteworthy that it is also possible to completely suppress gene function by introducing specific alterations via homologous integration into the genome (see preceding section, Figure 5) or after fortuitous integration in the vicinity of a developmentally important gene. Since the mutation obtained in these cases is usually recessive, one has to breed the transgenic strain to obtain progeny that are homozygous for the mutant alleles. The phenotype arising from these crosses is the result of a general loss-of-function. Examples of each type of assay will be given in the next section.

THE GAIN-OF-FUNCTION ASSAY

Cell culture studies have provided the basis by which the expression of a reporter gene can be targeted to particular tissues. It must be acknowledged that some promoter enhancer combinations like the lens crystallin, the elastase, or the immunoglobulin constructs (reviewed by Rusconi, 1991) have been particularly suited for the straightforward kind of approach, in that they showed a very consistent autonomous behavior (predicted by work with cell cultures) even when inserted at random positions in the host genome. Other constructs were either naturally contained or supplemented with LCRs (see Figure 2, reviewed by Dillon and Grosveld, 1993) in order to become position independent.

As for the effect of the expressed genes, several examples of targeted expression of dominant oncogenes like the early SV40 region or other oncogenes (reviewed by Cory and Adams, 1988; Muller, 1991) have been used to study the formation of specific tumors in model systems. Also, the gain-of-function approach has been useful in studies of the phenomena of allelic exclusion of immunoglobulin gene rearrangements in B cells and in obtaining many clues about how tolerance and autoimmunity are operating (reviewed by Parham, 1988; Hanahan, 1990; Bluethmann, 1991). Finally, the effects of overexpression or ectopic expression of several natural products have been studied; these have revealed some new features of the genes of interest. Some examples include the bleeding phenotype obtained by deregulated expression of tissue plasminogen activator genes, the creation of a mouse model for Alzheimer's disease upon overexpression of the gene for the amyloid precursor beta-APP751, the hyperplasias obtained by overexpression of TGF-alpha, the homeotic transformation observed by ectopic expression of a homeobox gene, or the hypertension generated by overexpression of renin genes in transgenic rats (reviewed by Kurtz, 1993; Rusconi, 1993). In all these cases, the consequences of deregulated expression of wild-type versions of the gene of interest were quite evident, and in several instances it has been possible to use the transgenic organisms to obtain model systems for known diseases.

A further variant in the gain-of-function approach is the possibility of obliterating or depopulating particular cell lineages by targeting toxic molecules to them. Such molecules can be bacterial toxins (e.g., the diphtheria toxin; Palmiter et al., 1987; Breitman et al., 1990) or enzymes that are capable of converting low-toxic compounds to lethal metabolites (like the already mentioned HSV-tk that can kill cells by metabolizing the drugs Gancyclovir or FIAU). The result is actually a loss of function of the entire cell lineage.

THE CHANGE-OF-FUNCTION ASSAY

I shall define change-of-function as the topical inactivation of gene products. In fact, it is possible to interfere with gene expression in selected tissues. One relatively simple way to inhibit a certain gene is by expressing antisense RNA (reviewed by Munir et al., 1990; Weintraub, 1990). This is done as follows: a portion of the gene to be inhibited is linked in inverted orientation to a strong transcriptional regulatory region, such that the nascent RNA is complementary to the natural mRNA. Strong expression of the antisense mRNA can effectively reduce the expression of the cognate gene, although the precise mechanism (splicing inhibition, transport, translation inhibition) is still under debate (Weintraub, 1990). There is still no standard protocol that guarantees a successful reduction of gene expression with antisense RNA and the approach has to be optimized case-by-case. For unknown reasons, this approach has been very successful in plants (reviewed by van der Krol et al., 1988), where a number of extremely convincing and even patentable applications have been reported. In mice, however, there are still only relatively few reports, even though the same technique is being used. Further information about the use of antisense RNA is available in Chapter 8.

Besides antisense RNA, results obtained with *trans*-dominant negative (TDN) mutants are also relatively encouraging. In this case the molecular antagonist is not an RNA but a defective protein that can reduce the action of the wild type in various ways (reviewed by Herskowitz, 1987). This has been worked out in several cases in cell culture systems, for example, with the fos/jun transactivator complex (Ransone et al., 1990; Lloyd et al., 1991) or has been documented with many natural mutants or variants such as the thyroid-receptor TDN viral variant v-erbA (Desbois et al., 1991) or mutants of the p53 anti-oncogene (Milner and Medcalf, 1991). In our laboratory, we have been able to generate glucocorticoid receptor mutants possessing a strong TDN property (Lanz et al., 1995) and we are testing them in transgenic animals (Verca, unpublished observations). Since the expression of the TDN molecule can be directed to particular tissues, it works as a sort of tissue-specific gene knock-out. Examples of the use of TDN mutants in mice are still few, although some work has been already done to study chondrogenesis and osteogenesis (Metsaranta et al., 1992).

THE LOSS-OF-FUNCTION ASSAY

Generalized loss of gene activity can be obtained by fortuitous integration or by targeted integration into the host chromosomes. After successful gene targeting (see Figures 4 and 5), the mutation can be bred to homozygosity to assess the effect of the cistron elimination. We can distinguish at least four types of result (see Table 3): (a) the alteration affects the tissues in which the gene is normally expressed; (b) the effect of gene disruption is stronger or milder than expected or is pleiotropic; (c) the alteration affects other, unexpected tissues; and (d) the mutation has no apparent effect on the mice raised under laboratory conditions or even after experimental challenging (e.g., behavioral or immunological tests).

As examples for the "expected" phenotypes we shall cite the effects obtained by disrupting the antioncogene p53 (increase of tumor incidence; Lavigueur et al., 1989), the glucocerebrosidase (impaired lysosomal storage; Tybulewicz et al., 1992), the apolipoprotein-E (hypercholesterolemia; Zhang et al., 1992) or the c-fos gene (defective bones and cartilage formations; Wang et al., 1992). In some of these cases the phenotypes can serve as animal models to study inheritable human disorders (discussed below).

Caution must be exercised when the knock-out produces phenotypes which are much milder than expected (See Table 3 for examples). This is the case with interleukin-2 (Schorle et al., 1991) where only a minor fraction of the immune system seemed to be affected. This was an unexpectedly mild phenotype, considering the plethora of reports claiming a specific role of this cytokine in the most disparate pathways (reviewed by Smith, 1988). Analogously, the disruption of methyl-transferase (Li et al., 1992) still allowed normal formation of most tissues, thus casting doubt about the presumptive role of DNA methylation in the control of differential gene expression during development (reviewed by Bird, 1992). Interpretation, of course, becomes more difficult when the genomic alteration produces a pleiotropic phenotype, such as transforming growth factor β (Shull et al., 1992).

In other cases the phenotype, though evident, is not the same as that expected (see examples under "Not anticipated" in Table 3). For example, disruption of the gene encoding the supposed myogenic factor Myf-5 (Braun et al., 1992) resulted in normal muscular tissue but a missing rib cage. Thus, the exact role of the Myf-5 gene needs to be redefined. A completely out-of-the-blue effect has also been observed by the disruption of the c-src gene (Soriano et al., 1991). The loss of c-src unexpectedly affected bone formation, while leaving unaltered the hematopoietic and nervous systems where this gene is normally expressed at high levels. Also in this case the supposed role of the gene in hematopoiesis must certainly be reassessed.

Table 3. Possible Phenotypes After Gene Disruption

Phenotype	Examples
Anticipated	p53; glucocerebrosidase; int-1; apolipoprotein-E; c-fos; myelin; interferon γ
Partly anticipated; Milder/Stronger; or Pleiotropic	IL-2; β2 -microglobulin; TGF; cystic fibrosis; methyl-transferase; pp59fyn; c-abl; TCR-beta
Not anticipated	Myf-5; Rb; c-src
Not apparent	PrP; tenascin; HPRT; MyoD; c-yes; TCR-alpha

Note: Modified from Rusconi, 1993, with permission from J.R. Prous, S.A.

It is most frustrating when the gene knock-out does not generate any measurable change in the physical appearance and behavior of the transgenic mouse (see the many examples under "Not apparent" in Table 3). This situation is particularly perplexing when the gene of interest is evolutionarily conserved, its spatial-temporal regulation strictly controlled, and several *ex vivo* experiments suggest that it is involved in a particular developmental process. To cite some examples, this situation is encountered after disruption of the neuro-specific gene PrP (Büeler et al., 1992), the extracellular matrix protein tenascin (Saga et al., 1992), the retinoic acid receptor alpha (Li et al., 1993), or the myogenic transcription factor MyoD (Rudnicki et al., 1992). When the phenotypic alteration remains invisible even after systematic challenging, we have to conclude that the gene product is functionally redundant, or that the conservation of the gene in evolution is due to phylogenetic rather than to ontogenetic requirements. We can easily conceive that mutations which reduce the reproductive or survival fitness by a minimal but consistent percentage are rapidly lost in evolving populations. Since evolutionary trends are not currently amenable to controlled experimentation in the laboratory, we will have to wait for future techniques to assess the precise role of some conserved genes.

CONCLUSIONS AND PERSPECTIVES

The immediate future in the field of transgenic techniques promises to be very exciting, in particular since the pioneering era seems to be definitely over. Many laboratories are currently engaged in this kind of experimentation. Transgenesis in plants has already led to patented applications and several improved diagnostic or therapeutic tools are being developed, thanks to transgenic techniques.

Future of Conventional Transgenesis

Many technical points have been improved, in particular, the manipulation of large genomic segments (see above). This permits more reproducible expression patterns of the transgene. Moreover, the tissue targeted expression of *trans*-dominant molecules allows specific alterations of metabolic or developmental pathways. Future directions will include sophisticated strategies in which target transgene and regulatory transgene are first inserted in separate mouse lines, creating the so-called "binary systems" (Ornitz et al., 1991; Lakso et al., 1992). In these examples an inactive oncogene construct is first introduced in one chromosome. Its inactive state having been determined in the first model (Ornitz et al., 1991) by its particular promoter that contains binding sites for a nonmammalian transcription factor, and in the second model by an intervening sequence that is flanked by bacterial recombinase target sites (Lakso et al., 1992). The activation is due to the concomitant expression of the specific transactivator or the specific bacterial recombinase.

Unfortunately, these systems, while conceptually elegant, do not allow reversible induction of the target transgene. This has been only partly achieved by employing metal-responsive promoter elements (Palmiter et al., 1982), a model system that suffers from a relatively high basal expression level. A more likely solution may be offered perhaps by the application of inducible systems of invertebrates. Currently, there is an interest in steroid receptors of insects that could be used to generate ligand-dependent *trans*-activating chimeras. It should also be remembered that efforts are still being made to test the possibility of maintaining the transgenic construct in an episomal state. This would be the most straightforward way to render the transgene totally independent of the surrounding chromosomal sequences.

Future of Gene Targeting Techniques

The refined techniques of ES-mediated transgenesis make it possible to alter genomic sequences by homologous recombination. These elegant methods are not yet available in every laboratory, since the maintenance of totipotency of ES cells still rests on partly empirical methods. Nevertheless, one can expect that in the immediate future the consequences of many more gene disruptions will be known. Current studies involve almost all the aspects of a living being, including behavioral features (Grant et al., 1992). In particular, new strategies will have to be developed for challenging apparently "missing phenotypes." On the other hand, undesired all-too-strong phenotypes arising from some disruptions will have to be circumvented by techniques that allow the tissue-selective gene knock-out (for example with *trans*-dominant-negative molecules) or by making more subtle genomic alterations with the emerging "hit-and-run" protocols (Hasty et al., 1991; Valancius and Smithies, 1991; Smithies, 1993). It is not yet clear, however, to what extent these complicated approaches allow maintenance of the totipotency of the altered ES cell lines.

Medical and Pharmacological Perspectives

It has already been pointed out that many models for inheritable diseases can be obtained by specific alterations. Reconstitution of pathological conditions was first achieved with classical transgenesis by overexpressing or aberrantly expressing particular constructs. Some examples of disease models obtained by gain-of-function or change-of-function are given in Table 4. Situations which recapitulate human diseases have also been obtained by either fortuitous or site-directed gene disruption (see general loss-of-function examples in Table 4). Generation of a pathological state by a specific genomic alteration allows closer analysis of the disease and it lays the experimental ground for systematic trials of somatic gene therapy. The major obstacle in somatic gene therapy is the low efficiency and inconstancy of the

Table 4. Human Disease Models Obtained With Transgenic Animals[*]

Category Mechanism	Examples
Gain-of-function	
Overexpression	Alzheimer; Menetrier's; hypertension; arthritis; Parkinsonism
Deregulated expression	Diabetes; liver fibrosis; autoimmunity
Change-of-function	
trans-dominant	Chondrodysplasia; osteogenesis imperfecta; sickle cell anemia; anemia; spondylo-arthropathy; metastasis; leukencephalopaty
Loss-of-function	
Fortuitous disruption	Renal disorders; Dwarfism
Targeted disruption	Cholesterolemia; Gaucher's disease; cystic fibrosis

Note: [*]Modified from Rusconi, 1993, with permission from J.R. Prous, S.A.

currently available gene/cell delivery protocols (reviewed by Weissman, 1992). Other difficulties include the possible host immune response that can kill the transformed cells expressing newly acquired genes. These hurdles can only be overcome by systematic studies conducted in model systems that permit reproducible conditions. A very recent report in which a transgenic mouse model for cystic fibrosis has been employed to test the correction of the disorder by aerosol liposome delivery could be cited as pioneering work (Hyde et al., 1993).

Finally, it is worth noting that transgenic organisms can be engineered to produce pharmacologically interesting proteins, be they vaccines, antibodies, cytokines, etc. Particularly attractive is the possibility of generating large farm animals that secrete pharmacologically or diagnostically relevant molecules in their milk, a strategy that merits the name "gene pharming" (reviewed by Clark et al., 1989; Bialy, 1991) and one that may become a most outstanding technological achievement of this decade.

Ethical/Political Aspects

The carrying out of transgenic research means animal experimentation besides gene manipulation. Hence, transgenic research today is under cross-fire from both anti-gene technology groups and antivivisectionists. A point that has never been made sufficiently clear to anti-animal experimentation circles is that cell culture systems are indeed seriously limited. I believe that the recent challenging results in which a disrupted gene has produced in the animal a phenotype that is totally different from that obtained in cell culture systems could become a valid argument to demonstrate in rational terms the limits of *ex vivo* experiments. Unfortunately, rational arguments sound miserably awkward in counteracting emotional disputes. In any case, the scientific community should take a stronger stand in favor of serious and responsible research. The issue is probably more important than we may currently be aware. Genetic information can be compared to an immense fossil energy reservoir that has not yet been exploited. Our generation bears the responsibility to demonstrate to future generations that one can make reasonable and safe use of this energy.

NOTE ADDED IN POOF

Since this article was written there have been numerous publications concerning the knockout of specific genes. A rich source of information is Brandon et al. (1995).

ACKNOWLEDGMENTS

I want to thank our director Walter Schaffner together with several colleagues of this institute and my research group for help in keeping track with the most recent literature and for many inspiring discussions. I am grateful to my wife, Augusta, for precious help in editing the literature list and to all my family for moral support during the writing of this chapter. I am indebted to Prous Science Publisher Inc. for permission to reproduce modified versions of copyright material presented in Figures 2, 4 and 5 and Tables 3 and 4 that have already been published (Rusconi, 1993). The material costs of this work have been met by the Schweizerischer Nationalfonds and the Kanton Zürich.

REFERENCES

Banerji, J., Olson, L., & Schaffner, W. (1983). A lymphocyte-specific cellular enhancer is located downstream of the joining region in immunoglobulin heavy chain genes. Cell 33, 729–740.

Baniahmad, A., Steiner, C., Köhne, A.C., & Renkawitz, R. (1990). Modular structure of a chicken lysozyme silencer: Involvement of an unusual thyroid-hormone receptor binding site. Cell 61, 505–514.

Beato, M. (1989). Gene regulation by steroid receptors. Cell 56, 340–344.

Bialy, H. (1991). Transgenic pharming comes of age. Biotechnology 9, 786–788.

Bird, A. (1992). The essentials of DNA methylation. Cell 70, 5–8.

Bluethmann, H. (1991). Analysis of the immune system with transgenic mice: T cell development. Experientia 47, 884–890.

Borrelli, E., Heyman, R., Hsi, M., & Evans, R.M. (1988). Targeting of an inducible toxic phenotype in animal cells. Proc. Natl. Acad. Sci. USA 85, 7572–7576.

Brandon E.P., Idzerda, R.L., & McKnight, G.S. (1995). Knockouts. Targeting the mouse genome: a compendium of knockouts. Curr. Biol. 5, 625–634 (Part I), 758–765.

Braun, R.E., Lo, D., Pinkert, C.A., Widera, G., Flavell, R.A., Palmiter, R.D., & Brinster, R.L. (1990). Infertility in male transgenic mice: Disruption of sperm development by HSV-tk expression in postmeiotic germ cells. Biol. Reprod. 43, 684–693.

Braun, T., Rudnicki, M.A., Arnold, H.-H., & Jaenisch, R. (1992). Targeted inactivation of the muscle regulatory gene Myf-5 results in abnormal rib development and perinatal death. Cell 71, 369–382.

Breitman, M.L., Rombola, H., Maxwell, I.H., Klintworth, G.K., & Bernstein, A. (1990). Genetic ablation in transgenic mice with an attenuated diphtheria toxin A gene. Mol. Cell. Biol. 10, 474–479.

Büeler, H., Fischer, M., Lang, Y., Bluethmann, H., Lipp, H.-P., DeArmond, S.J., Prusiner, S.B., Aguet, M., & Weissmann, C. (1992). Normal development and behaviour of mice lacking the neuronal cell-surface prP protein. Nature 356, 577–582.

Capecchi, M. (1993). YACS to the rescue. Nature 362, 205–206.

Capecchi, M.R. (1989). The new mouse genetics, altering the genome by gene targeting. Trends Genet. 5, 70–76.

Cech, T.R. (1988). Ribozymes and their medical implications. JAMA 260, 3030–3034.

Clark, A.J., Ali, S., Archibald, A.L., Bessos, H., Brown, P., Harris, S., McClenaghan, M., Prowse, C., Simons, J.P., & Whitelaw, C.B. (1989). The molecular manipulation of milk composition. Genome 31, 950–955.

Cory, S., & Adams, J.M. (1988). Transgenic mice and oncogenesis. Ann. Rev. Immunol. 6, 25–48.

Desbois, C., Aubert, D., Legrand, C., Pain, B., & Samarut, J. (1991). A novel mechanism of action for v-ErbA: Abrogation of the inactivation of transcription factor AP-1 by retinoic acid and thyroid hormone receptors. Cell 67, 731–740.

Dillon, N., & Grosveld, F. (1993). Transcriptional regulation of multigene loci: Multilevel control. Trends Genet. 9, 134–137.

Dynan, W.S. (1989). Modularity in promoters and enhancers. Cell 58, 705–711.

Evans, M.J., & Kaufman, M.H. (1981). Establishment in culture of pluripotential cells from murine embryos. Nature 292, 154–156.

Grant, G.N., O'Dell, T.J., Karl, D.A., Stein, P.L., Soriano, P., & Kandel, E.C. (1992). Impaired long-term potentiation spatial learning, and hyppocampal development in *fyn* mutant mice. Science 258, 1903–1910.

Hanahan, D. (1990). Transgenic mouse models of self-tolerance and autoreactivity by the immune system. Ann. Rev. Cell Biol. 6

Hasty, P., Ramirez-Solis, R., Krumlauf, R., & Bradley, A. (1991). Introduction of a subtle mutation into the hox-2.6 locus in embryonic cells. Nature 350, 243–246.

He, X., & Rosenfeld, M.G. (1991). Mechanisms of complex transcriptional regulation: Implications for brain development. Neuron 7, 183–196.

Herskowitz, I. (1987). Functional inactivation of genes by dominant negative mutations. Nature 329, 219–222.

Hyde, S.C., Gill, D.R., Higgins, C.F., Trezise A.E.O., MacVinish, L.J., Cuthberg, A.W., Ratcliff, R., Evans, M.J., & Colledge, W.H. (1993). Correction of the ion transport defect in cystic fibrosis transgenic mice by gene therapy. Nature 362, 250–255.

Kurtz, T.W. (1993). Transgenic models of hypertension: Useful tools or unusual toys? Clin. Invest. 91, 742–747.

Lakso, M., Sauer, B., Mosinger, B., Lee, E.J., Manning, R.W., Yu, S.-H., Mulder, K.L., & Westphal, H. (1992). Targeted oncogene activation by site specific recombination in transgenic mice. Proc. Natl. Acad. Sci. USA 89, 6232–6236.

Lanz, R.B., Wieland, S., Hug, M., & Rusconi, S. (1995). A transcriptional repressor obtained by alternative translation of a trinucleotide repeat. Nucl. Acids. Res. 23, 138–145.

Lavigueur, A., Maltby, V., Mock, D., Rossant, J., Pawson, T., & Bernstein, A. (1989). High incidence of lung, bone, and lymphoid tumors in transgenic mice overexpressing mutant alleles of the p53 oncogene. Mol. Cell. Biol. 9, 3982–3991.

Li, E., Bestor, T.H., & Jaenisch, R. (1992). Targeted mutation of the DNA methyltransferase gene results in embryonic lethality. Cell 69, 915–926.

Li, E., Sucov, H.M., Lee, K.-F., Evans, R.M., & Jaenisch, R. (1993). Normal development and growth of mice carrying a targeted disruption of the alpha1 retinoic acid receptor gene. Proc. Natl. Acad. Sci. USA 90, 1590–1594.

Lloyd, A., Yancheva, N., & Wasylyk, B. (1991). Transformation suppressor activity of a Jun transcription factor lacking its activation domain. Nature 352, 635–638.

Mansour, S.L., Thomas, K.R., & Capecchi, M.R. (1988). Disruption of the protooncogene int-2 in mouse embryo-derived stem cells: A general strategy for targeting mutations to non-selectable genes. Nature 336, 348–352.

Metsaranta, M., Garofalo, S., Decker, G., Rintala, M., de Crombrugghe, B., & Vuorio, E. (1992). Chondrodysplasia in transgenic mice harboring a 15-amino acid deletion in the triple helical domain of pro alpha 1(II) collagen chain. J. Cell Biol. 118, 203–212.

Milner, J., & Medcalf, E.A. (1991). Cotranslation of activated mutant p53 with wild type drives the wild-type p53 protein into the mutant conformation. Cell 65, 765–774.

Mitchell, P.J., & Tjian, R. (1989). Transcriptional regulation in mammalian cells by sequence-specific DNA binding proteins. Science 245, 371–378.

Muller, W.J. (1991). Expression of activated oncogenes in the murine mammary gland: Transgenic models for human breast cancer. Cancer Metastasis Rev. 10, 217–227.

Munir, M.I., Rossiter, B.J., & Caskey, C.T. (1990). Antisense RNA production in transgenic mice. Somat. Cell Mol. Genet. 16, 383–394.

Orkin, S.H. (1990). Globin gene regulation and switching: Circa 1990. Cell 63, 665–672.

Ornitz, D.M., Moreadith, R.W., & Leder, P. (1991). Binary system for regulating transgene expression in mice: Targeting int-2 gene expression with yeast GAL4/UAS control elements. Proc. Natl. Acad. Sci. USA 88, 698–702.

Palmiter, R.D., Behringer, R.R., Quaife, C.J., Maxwell, F., Maxwell, I.H., & Brinster, R.L. (1987). Cell lineage ablation in transgenic mice by cell-specific expression of a toxin gene. Cell 50, 435–443.

Palmiter, R.D., & Brinster, R.L. (1986). Germ line transformation of mice. Ann. Rev. Genet. 20, 465–499.

Palmiter, R.D., Brinster, R.L., Hammer, R.E., Trumbauer, M.E., Rosenfeld, M.G., Birnberg, N.C., & Evans, R.M. (1982). Dramatic growth of mice that develop from eggs microinjected with metallothionein-growth hormone fusion genes. Nature 300, 611–615.

Parham, P. (1988). Intolerable secretion in tolerant transgenic mice (news). Nature 333, 500–503.

Pieper, F.R., de Wit, I.C., Pronk, A.C., Kooiman, P.M., Strijker, R., Krimpenfort, P.J., Nuyens, J.H., & de Boer, H.A. (1992). Efficient generation of functional transgenes by homologous recombination in murine zygotes. Nucl. Acids Res. 20, 1259–1264.

Ransone, L.J., Visvader, J., Wamsley, P., & Verma, I.M. (1990). Trans-dominant negative mutants of Fos & Jun. Proc. Natl. Acad. Sci. USA 87, 3806–3810.

Rudnicki, M.A., Braun, T., Hinuma, S., & Jaenisch, R. (1992). Inactivation of MyoD in mice leads to up-regulation of the myogenic HLH gene myf-5 and results in apparently normal muscle development. Cell 71, 383–390.

Rusconi, S. (1991). Transgenic regulation in laboratory animals. Experientia 47, 866–877.

Rusconi, S. (1993). Gene expression in transgenic animals. Drug News Perspectives 6, in press.

Saga, Y., Yagi, T., Ikawa, Y., Sakakura, T., & Aizawa, S. (1992). Mice develop normally without tenascin. Genes Dev. 6, 1821–1831.

Schorle, H., Holtschke, T., Hunig, T., Schimpl, A., & Horak, I. (1991). Development and function of T cells in mice rendered interleukin-2 deficient by gene targeting. Nature 352, 621–624.

Serfling, E., Jasin, M., & Schaffner, W. (1985). Enhancers and eucaryotic transcription. Trends Genet. 1, 224–230.

Shull, M.M., Ormsby, I., Kier, A.B., Pawlowski, S., Diebold, R.J., Yin, M., Allen, R., Sidman, C., Proetzel, G., Calvin, D., Annunziata, N., & Doetschmann, T. (1992). Targeted disruption of the mouse transforming growth factor-β1 gene results in multifocal inflammatory disease. Nature 359, 693–699.

Skalnik, D.G., Dorfman, D.M., Perkins, A.S., Jenkins, N.A., Copeland, N.G., & Orkin, S.H. (1991). Targeting of transgene expression to monocyte/macrophages by the gp91-phox promoter and consequent histiocytic malignancies. Proc. Natl. Acad. Sci. USA 88, 8505–8509.

Skarnes, W. (1990). Entrapment vectors: A new tool for mammalian genetics. Biotechnology 8, 827–831.

Smith, K.A. (1988). Interleukin-2: Inception, impact and implications. Science 240, 1169–1176.

Smithies, O. (1993). Animal models of human genetic diseases. Trends Genet. 9, 112–116.

Soriano, P., Montgomery, C., Geske, R., & Bradley, A. (1991). Targeted disruption of the c-src proto-oncogene leads to osteopetrosis in mice. Cell 64, 693–702.

Svaren, J., & Chalkley, R. (1990). The structure and assembly of active chromatin. Trends Genet. 6, 52–56.

te Riele, H., Maandag, E.R., & Berns, A. (1992). Highly efficient gene targeting in embryonic stem cells through homologous recombination with isogenic DNA constructs. Proc. Natl. Acad. Sci. USA 89, 5128–5132.

Tripodi, M., Abbott, C., Vivian, N., Cortese, R., & Lovell, B.R. (1991). Disruption of the LF-A1 and LF-B1 binding sites in the human alpha-1-antitrypsin gene has a differential effect during development in transgenic mice. EMBO J. 10, 3177–3182.

Tybulewicz, V.L., Tremblay, M.L., LaMarca, M.E., Willemsen, R., Stubblefield, B.K., Winfield, S., Zablocka, B., Sidransky, E., Martin, B.M., & Huang, S.P. (1992). Animal model of Gaucher's disease from targeted disruption of the mouse glucocerebrosidase gene. Nature 357, 407–410.

Valancius, V., & Smithies, O. (1991). Testing an in-out targeting procedure for making subtle genomic modifications in mouse embryonic stem cells. Mol. Cell Biol. 11, 1402–1408.

van der Krol, A.R., Mol, J.N., & Stuitje, A.R. (1988). Antisense genes in plants: An overview. Gene 72, 45–50.

Wang, Z.Q., Ovitt, C., Grigoriadis, A.E., Möhle-Steinlein, U., Rüther, U., & Wagner, E.F. (1992). Bone and hematopoietic defects in mice lacking c-fos. Nature 360, 741–745.

Weintraub, H.M. (1990). Antisense RNA and DNA. Sci. Am. 262, 40–46.

Weissman, S.M. (1992). Gene therapy. Proc. Natl. Acad. Sci. USA 89, 11111–11112.

Wilson, C., Bellen, H.J., & Gehring, W.J. (1990). Position effects on eukaryotic gene expression. Ann. Rev. Cell Biol. 6, 671–714.

Zhang, S.H., Reddick, R.L., Piedrahita, J.A., & Maeda, N. (1992). Spontaneous hypercholesterolemia and arterial lesions in mice lacking apolipoprotein E. Science 258, 468–471.

INDEX

Printed and bound by CPI Group (UK) Ltd, Croydon, CR0 4YY

13/10/2024

01773500-0001